Photosynthetic Carbon Assimilation

BASIC LIFE SCIENCES
Alexander Hollaender, General Editor
Associated Universities, Inc.
Washington, D.C.

A Continuation Order Plan is available for this series. A continuation order
will bring delivery of each new volume immediately upon publication. Volumes
are billed only upon actual shipment. For further information please contact
the publisher.

Photosynthetic Carbon Assimilation

Edited by

Harold W. Siegelman

and

Geoffrey Hind

Biology Department
Brookhaven National Laboratory
Upton, New York

PLENUM PRESS · NEW YORK AND LONDON

Library of Congress Cataloging in Publication Data

Main entry under title:

Photosynthetic carbon assimilation.

(Basic life sciences; v. 11)
"Proceedings of a symposium held at Brookhaven National Laboratory, Upton, New York, May 31—June 2, 1978."
Includes indexes.
1. Photosynthesis — Congresses. 2. Carbon — Metabolism — Congresses. 3. Plants — Assimilation — Congresses. I. Siegelman, Harold W. II. Hind, Geoffrey.
QK882.P558   581.1'3342   78-11545
ISBN 0-306-40064-2

Proceedings of a Symposium held at Brookhaven National Laboratory
Upton, New York, May 31—June 2, 1978

© 1978 Plenum Press, New York
A Division of Plenum Publishing Corporation
227 West 17th Street, New York, N.Y. 10011

Printed in the United States of America

Preface

The photosynthetic fixation of carbon dioxide into organic compounds is mediated by the enzyme ribulose 1,5-bisphosphate (RuBP) carboxylase. The diversity of current research on this protein attests to its central role in biomass productivity, and suggests the importance of a timely and broadly based review. This Symposium was the first devoted exclusively to RuBP carboxylase and was attended by agronomists, plant physiologists, biochemists, molecular biologists, and crystallographers. Special efforts were made to involve young scientists in addition to established investigators.

It is a pleasure to acknowledge financial support provided by the Department of Energy, the United States Department of Agriculture, and the National Science Foundation, and the valued assistance of agency representatives, Drs. Joe Key, Robert Rabson, Elijah Romanoff, and Donald Senich. Thanks are due to Mrs. Margaret Dienes, without whose editorial skills this volume could not have been produced, and to Mrs. Helen Kondratuk as Symposium Coordinator. Finally, we wish to record our indebtedness to Dr. Alexander Hollaender for his tireless efforts in support of all aspects of this Symposium.

Symposium Committee
H. W. Siegelman, Chairman
B. Burr
G. Hind
H. H. Smith

Contents

Section V

Molecular Biology of RuBP Carboxylase
Chairman: H. H. Smith

FRACTION I PROTEIN AND OTHER PRODUCTS FROM TOBACCO FOR FOOD

S. G. Wildman and P. Kwanyuen

Department of Biology, University of California
Los Angeles, California   90024

INTRODUCTION

Depending on the point of view, the tobacco plant is either
extolled for the solace it brings to those who smoke and/or for
the secure economic rewards from its cultivation and manufacture,
or branded a weed of unmitigated evil for its effect on health.
The latter view now seems to be the more popular. However, there
is another possibility. Tobacco plants can be used as a source
of high-grade protein for human consumption. The exploitation of
this possibility could turn tobacco into an agricultural commodity
of undeniable value. Since the idea of using leaf protein is not
new, the purpose of this paper is to present reasons for thinking
that tobacco plants could be a superior source of supplemental
protein in the human diet compared with leaf proteins from other
plants.

GROSS PROTEIN COMPOSITION OF TOBACCO LEAVES

Like other succulent leaves such as those of alfalfa, spinach,
sugar beets, etc., tobacco leaves contain 80 to 90% water and 10 to
20% solid matter. In a single leaf (Figure 1) that has just reached
its maximum growth in area, the accumulated weight of photosynthate
(disregarding the polymers in cell walls) consists of about 3/4
protein and 1/4 small molecules (molecular weight ≤ 5000). The
leaf proteins represent as direct a conversion of solar energy into
protein as is possible in the biological world. About 50% of the
leaf proteins are insoluble in water or buffers less alkaline than
pH 10. These proteins for the most part comprise thylakoid membranes of chloroplasts and to a lesser degree insoluble proteins

1

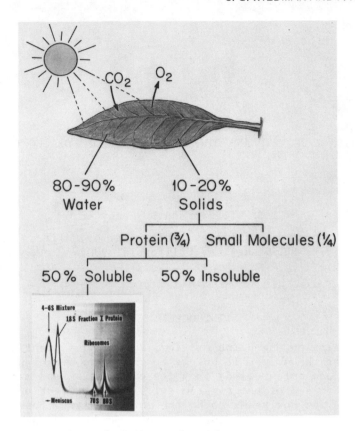

Figure 1. Diagram to show the gross composition of constituents of leaves and the ratio
of insoluble to soluble proteins.

constituting the structure of nuclei, mitochondria, and endoplasmic reticulum. Upon isolation, the insoluble proteins are
green because chlorophyll and other photosynthetic pigments remain attached to them.

As shown by the analytical centrifuge pattern in Figure 1, the
soluble proteins of leaves consist of three classes: (a) the ribosome nucleoprotein class, accounting for about 10%; (b) a mixture
of proteins with molecular weights less than 100,000; (c) a single
kind of protein with a molecular weight of about 550,000 and having
a rather inept name, Fraction I protein (F-I-p), which has been in
continuous use since 1947. F-I-p has been shown to be carboxydis-
mutase or ribulose 1,5-bisphosphate carboxylase, the enzyme that
catalyzes the first step of carbon assimilation during photosynthesis. Succulent leaves such as tobacco generally contain about 10 mg
F-I-p per gram fresh weight of leaves. F-I-p is the most abundant

single protein on earth because it is found in every organism
containing chlorophyll a.

F-I-p has been purified to a very high degree from several
kinds of plants including spinach, wheat, rice, Chinese cabbage,
clover, and tobacco. The extensive knowledge about its structure
and function is the product of fundamental research by several
groups in different institutions in the United States, Great
Britain, Japan, Australia, and New Zealand. Purification of F-I-p
was tedious and time-consuming and required expensive reagents and
equipment until 1971, when Dr. Nobumaro Kawashima of the Japan
Monopoly Corporation, working in our laboratory, succeeded for the
first time in crystallizing F-I-p from tobacco leaves (1). This made
it possible to study F-I-p in the pure state and thus to uncover
unexpected properties that would allow isolation on a scale of
several hundred pounds of crystalline F-I-p from an acre of
tobacco plants.

PROPERTIES OF CRYSTALLINE F-I-p

Figure 2 shows some properties of F-I-p crystallized from
Nicotiana tabacum leaves. Most of these results were obtained with
F-I-p from the Turkish Samsun cultivar of N. tabacum, but the
results would apply to the more than 30 other cultivars of commercial tobacco we have investigated. Large F-I-p crystals are depos-
ited as dodecahedrons (Figure 2A) and contain 80% water, which
results in a relatively huge unit cell (Figure 2B). When dissolved
in dilute NaCl solutions, the protein molecules can be seen by
electron microscopy (Figure 2C) and have dimensions of about 110 Å.
The individual molecules look granular, and all of them have the
same appearance. Constant specific enzymatic activity is achieved
by the second recrystallization of F-I-p, and no further change
occurs with additional recrystallizations (Figure 2D). After the
first recrystallization, no heterogeneity is detected when the redissolved protein is subjected to analytical centrifugation (Figure 2E). The molecular weight of F-I-p calculated from its sedimentation constant corresponds closely to the size of the macromolecules seen by electron microscopy. Pure F-I-p consists entirely
of amino acids and is water clear in solution when observed by
visible light (Figure 2F). When F-I-p crystals are subjected to
dialysis against distilled water to remove salt and buffer, they
are tasteless and odorless and appear pure white and partially
jellified. The pure crystals contain no cations other than one
molecule of magnesium per molecule of F-I-p (7).

Two other properties were observed that provided the clue to
producing crystalline F-I-p by a simple procedure (2). One is
stability to heat (Figure 3). With specific enzymatic activity
used as the most sensitive indicator of change in quaternary
structure, the data show that F-I-p loses activity at temperatures
below 25°C. The diminution is slow, requiring 20 hr to complete

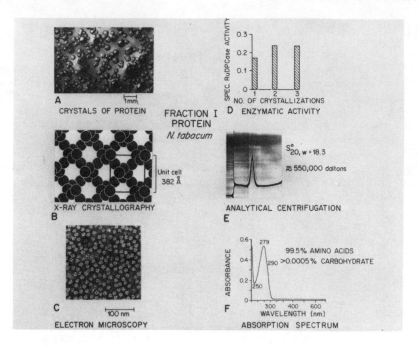

Figure 2. Properties of crystalline Fraction I protein of tobacco leaves. A, after Chan et al. (2); B, C, after Baker et al. (3); D, after Singh and Wildman (4); E, after Sakano and Wildman (5); F, after Sakano et al. (6).

when F-I-p at $25^{\circ}C$ is placed in an ice bath at $0^{\circ}C$ (Figure 3A). The loss in activity can be reversed by heating the protein to temperatures above $25^{\circ}C$. Heating at $50^{\circ}C$ for 20 min caused all of the activity to be regained (Figure 3B) without producing any turbidity in the F-I-p solution. The specific enzymatic activity could be repeatedly diminished by cold treatment and restored by 50° heat.

The second property is solubility. F-I-p catalyzes the combination of carbon dioxide with ribulose 1,5-bisphosphate (RuBP) to form the first product of CO_2 fixation, 3-phosphoglyceric acid (PGA). When crystals of F-I-p were exposed to RuBP, they immediately dissolved, and the protein became enormously soluble (>150 mg protein per ml) in low ionic strength buffer (9). As little as about 8 molecules of RuBP bound to each molecule of protein allows almost complete removal of buffer and ions by dialysis and still keeps the protein soluble. If $NaHCO_3$ is added, F-I-p converts the RuBP to PGA, and the protein becomes highly insoluble, <1 mg protein per ml buffer remaining in low ionic strength solution. However, the protein becomes highly soluble again with addition of cations. At 20 mM

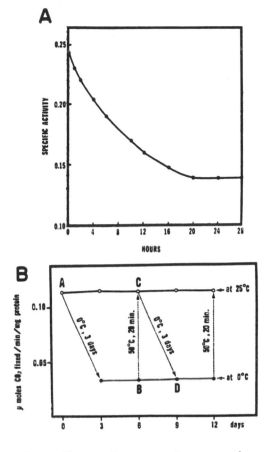

Figure 3. Effect of temperature on the specific
enzymatic activity of crystalline Fraction I protein from tobacco leaves.   A, after Singh and
Wildman (4); B, after Kawashima et al. (8).

or less of the salts shown in Figure 4, F-I-p is virtually insoluble
(<0.025 mg/ml), but it dissolves as the concentrations of the salts
are increased.   It dissolves to a much greater extent in the presence of Na^+ than K^+, to some extent in the presence of NH_4^+, and to
a very limited extent in the presence of Li^+. As seen in Figure 4,
80 mM NaCl was sufficient to dissolve nearly 1 mg of F-I-p per ml.
Since tobacco leaves contain about 10 mg F-I-p per gram leaf or
roughly 12 mg F-I-p per ml of H_2O in the leaf sap, it became apparent that the cation content of the leaf itself was probably sufficient to maintain all of the F-I-p in solution during extraction.
This finding, coupled with the observation that cold conditions were
not required to maintain integrity of the F-I-p molecules, led to a

simple procedure for preparing crystalline F-I-p not only from N. tabacum but also from more than 60 other species of Nicotiana.

ESSENTIAL AMINO ACID CONTENT OF TOBACCO F-I-p

Tobacco F-I-p is composed of 18 amino acids accounting for 99.8% of its mass. In Table 1, the amounts of the 9 amino acids essential for human nutrition found in F-I-p are listed and compared with the amounts in other proteins of animal and plant origin and in the provisional pattern of essential amino acids deemed optimal for human nutrition by the UN Food and Agricultural Organization. Except for methionine, the amounts of essential amino acids in tobacco F-I-p either meet or exceed the standards of the provi-

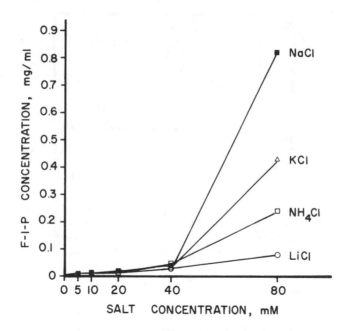

Figure 4. Solubility of Fraction I protein as a function of kind and concentration of cations. Experiment was done by suspending crystals of F-I-p in a fixed volume of salt solution, allowing 16 hr for complete solvation, and then removing undissolved protein by low speed centrifugation before determining amount of F-I-p in solution.

Table 1

Essential Amino Acids in Fraction I protein From Tobacco
Compared With Other Plant Proteins, Casein, and FAO Provisional
Pattern (Grams per 100 grams protein)

Amino acids	Prov. pat.(10)	Ca- sein(11)	Soy prot.(12)	Wheat prot.(11)	Corn prot.(11)	Rice prot.(11)	F-I-p*
Ile	4.2	7.5	5.8	4.0	6.4	5.2	4.2
Leu	4.8	10.0	7.6	7.0	15.0	8.2	8.8
Lys	4.2	8.5	6.6	2.7	2.3	3.2	5.8
Phe	2.8	6.3	4.8	5.1	5.0	5.0	4.4
Tyr	2.8	6.4	3.2	4.0	6.0	5.7	4.9
Met	2.2	3.5	1.1	2.5	3.1	3.0	1.6
Thr	2.8	4.5	3.9	3.3	3.7	3.8	5.2
Try	1.4	1.3	1.2	1.2	0.6	1.3	1.5
Val	4.2	7.7	5.2	4.3	5.3	6.2	7.2

*Calculated from amino acid analyses (13) for F-I-p from tobacco.

sional pattern. Also, tobacco F-I-p is superior in its essential
amino acid content to soy and grain proteins.

When treated with the enzymes of the digestive gut, F-I-p breaks
down into 80 to 100 small peptides. The absence of disulfide bonds
aids in its digestion. These properites of tobacco F-I-p together
with its amino acid composition and absence of taste and odor per-
suaded what was then the RANN division of the National Science
Foundation to support a project to prepare 0.6 kg of crystalline
F-I-p for biological evaluation.

CRYSTALLINE F-I-p PRODUCTION ON A 6 TO 12 g/day SCALE

Production of 0.6 kg of F-I-p crystals required for an animal
feeding test was greatly facilitated by Dr. Richie Lowe of the
University of Kentucky informing us prior to publication (14) of
his discovery that F-I-p would crystallize when clarified tobacco
leaf extracts were subjected to Sephadex chromatography.

Turkish tobacco plants were grown in a greenhouse at a spacing
of one plant per 0.25 ft^2 until they attained a height of 18 to 21
in. The leaves were detached from the stems. The blades of a gallon
size Waring blender were covered with 400 ml of water and 10 ml of
β-mercaptoethanol. Then 2 kg of fresh leaves were homogenized in
this solution. The homogenate was filtered through two layers of
fine cheesecloth supported on a 32 mesh screen. The cheesecloth
was twisted by hand to press out the green juice, which had a volume

of about 1900 ml and a pH ranging from 5.7 to 6.2 for different ex-
tractions. By comparing the amount of chlorophyll retained in the
filter cake with the amount in the green juice, it was estimated
that about 70% of the leaf cells had been ruptured by blending and
had released their contents as cell-free green juice.

In 400- to 500-ml batches, the green juice was heated in a 50°
water bath for 10 min. After it was removed from the bath, 10 ml of
50 times concentrated Tris-HCl-EDTA buffer was added to produce a
pH of 7.6 to 7.8 and a final concentration of 0.025 M Tris and
0.0005 M EDTA. When the juice was left standing at room temperature,
partial precipitation of a green sediment from a brown juice could
be seen. The neutralized juice was centrifuged at 5000 g for 15 min.
The brown juice (total volume, ca. 1700 ml) was collected and passed
through a G-25 Sephadex column equilibrated with Tris-HCl-EDTA buffer.
As the buffer passed through, the total soluble proteins of the tobacco leaves eluted in the void volume of the column. Frequently
crystals of F-I-p appeared during collection of the eluate or shortly
thereafter. The eluate was allowed to stand for 2 to 3 days at 8°
to allow more F-I-p crystals to form and settle into a thin layer
at the bottom of the flask. Then the proteins that remained dissolved in the mother liquor were decanted so that the F-I-p crystals
could be collected and washed with distilled water. The crystals
were redissolved in 0.1 M NaCl solution, which was then centrifuged
to remove slight amounts of undissolved material. The water-clear
supernatant solution was dialyzed against distilled water to cause
recrystallization of F-I-p. The recrystallized protein was transferred to flasks, and the salt-free protein was lyophilized to produce a white nonhygroscopic powder which was tasteless. The yields
of recrystallized F-I-p were consistently between 3 and 4 g dry protein per kg fresh tobacco leaves during the fall, winter, and spring
of 1976-77.

PROTEIN EFFICIENCY RATIO OF CRYSTALLINE F-I-p COMPARED WITH CASEIN

As reported by Ershoff et al. (15), when rats were fed on a
diet in which crystalline tobacco F-I-p was the sole source of protein, they grew faster during a 28-day period than did control rats
whose sole source of protein was casein. This shows that pure F-I-p
is an outstanding protein from the standpoint of nutritional value
even though it had appeared to be deficient because it contains less
methionine than does casein.

Because crystalline F-I-p from tobacco leaves is composed entirely of amino acids and contains no carbohydrates, purines, py-
rimidines, pigments, or minerals (except for the sulfur in some amino
acids) and because of its high nutritional value and absence of odor
and taste, it has the potential of being developed into important
therapeutic products for the treatment of a number of medical conditions. It should be particularly valuable in feeding patients with
various types of renal disease whose sodium and potassium intake

must be rigorously controlled, and it might well be incorporated in special diets that could reduce the required frequency of hemodialysis for patients suffering from renal failure. A number of other medical uses have been envisaged (15).

ECONOMIC POTENTIAL OF TOBACCO GROWN PRIMARILY AS A SOURCE OF PROTEIN FOR HUMAN NUTRITION

Tobacco plants are remarkably efficient converters of solar energy into plant biomass solids. The seeds of the plants are very small (ca. 14,000 seeds/g) and retain a germination capacity close to 100% for several years.

Starting from a dry seed weighing ca. 7×10^{-5} g, a plant containing >150 g of dry solids in its aerial portion will develop in 4 months under favorable environmental conditions. Given enough space, the plant will grow taller than a man. Casual inspection of a row of tobacco plants growing alongside a row of corn plants gives the impression that the biomass per m^2 of space is greater for tobacco than for corn.

Current tobacco culture, e.g., for producing Burley tobacco in Kentucky, is based on growing 8000 plants/acre, which gives a harvest of 900 to 1100 dry kg/acre or, at 85% moisture, roughly 13,600 kg of fresh plants. Compared with forage crops such as alfalfa, which have been extensively studied for leaf protein production (16, 17), tobacco grown in the conventional manner does not have an impressive biomass yield. However, tobacco can be grown as a forage crop like alfalfa. In California we have been growing tobacco at a density of one plant per 0.25 ft^2 outdoors in the San Fernando Valley. During July and August 1977, plants grown for 6 weeks and cut off 4 in. above soil level had an average fresh weight of 0.125 kg each, which is equivalent to >20,000 kg fresh weight tobacco plants per acre. At 15% dry matter, a single harvest would produce about 3000 kg dry matter per acre. With harvested plants left to regenerate new shoots for a second harvest followed by replanting and two more harvests, yields approaching 60 metric tons of fresh tobacco plants per acre per year can be envisaged by conservative reckoning.

Sixty metric tons of fresh tobacco plants can be processed into 130 kg of dried crystals of F-I-p. In addition the same plants would yield two other valuable kinds of proteins and three other products, as indicated by the flow sheet in Figure 5. From the mother liquor remaining after removal of F-I-p crystals, 530 kg of dry protein can be recovered which is as nutritious as casein and nearly tasteless and odorless, and could be added to less nutritious foods to upgrade them. Extraction of the green precipitate with organic solvents yields 540 kg of a dry mixture of protein, nucleic acid, and starch having a slight but pleasant odor and a barely sweet taste. Distillation of the solvent leaves 360 kg of dry solids consisting of fats, lipids, chlorophyll, carotene, etc. Washing the Sephadex columns with water after elution of soluble proteins produces a

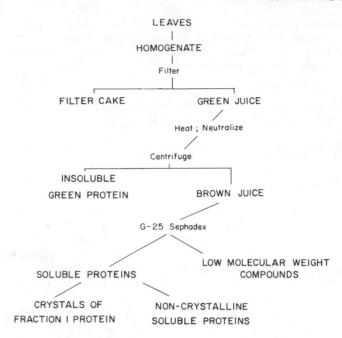

Figure 5. Flow sheet showing how few steps
are required to obtain crystalline Fraction
I protein, two other kinds of proteins, and
additional products from the aerial portion
of young, fresh tobacco plants.

solution containing 1640 kg of a dry mixture of compounds with maximum molecular weights of about 5000. This mixture contains the amino
acids, carbohydrates, vitamins, etc., that compose about 20% of the
total accumulated photosynthate (Figure 1). Finally, 2350 kg of a
dried filter cake remains, which contains the cellulose, lignin,
and protoplasmic contents of leaf and stem cells that escaped rupture during the initial blending of the plants.

From 60 metric tons of fresh tobacco plants, 1200 dry kg of
three different forms of protein can be obtained. In the U.S. mid-
west, an acre of soybean plants yields on the average 27.8 bushels
of beans containing 266 kg of protein. With the protein efficiency
ratio of casein taken as 100, soybean protein has a value of 80,
whereas the soluble proteins of tobacco leaves are nutritionally
equal to casein. Therefore, not only is the protein from tobacco
leaves obtainable in impressively greater yields than that from
soybeans, but its nutritional quality is significantly better. According to Stahmann (16), the amount of crude protein that can be
isolated per acre per year is about 1000 kg from leaves and stalks
of alfalfa, 950 kg from sorghum plants grown as forage, and 725 kg
from corn forage. Thus, tobacco plants as we have grown them can

compete in terms of protein production with the kinds of forage crops
that have been proposed as likely sources of concentrated leaf proteins for human consumption. Besides being the only plant now known
from which crystalline F-I-p can be obtained by a method as simple
as that described above, tobacco has important advantages over other
plants with regard to production and economics. Soybeans, for example, grow poorly in subtropical and tropical regions, but cultivars
of tobacco have been bred which are suited to growing from the equator to as far north as Novosibirsk, Siberia (latitude 55°).

F-I-p AS RELATED TO CROP IMPROVEMENT BY GENETIC ENGINEERING

Tobacco is a C_3 type of plant, notorious for its wastage of
photosynthate by photorespiration. Nevertheless, tobacco grown for
maximal biomass of plant product produces yields comparable to those
of other agricultural crops such as corn and sorghum. These plants
are of the C_4 type, in which photorespiration seems not to be of
significance during accumulation of photosynthate as biomass. The
estimated yield of 60 metric tons per acre per year for tobacco is
conservative. With adequate water and fertilizer, in a warm desert-
like climate with a growing season approaching 365 days of the year,
the biomass yield could be substantially improved. But what of the
possibilities of even higher yields if photorespiration could be
reduced or eliminated from the tobacco plant?

F-I-p catalyzes not only the combination of CO_2 with RuBP to
produce PGA but also the combination of O_2 with RuBP to produce PGA
plus phosphoglycolate (18), the latter being at the apex of the
photorespiration pathway leading to loss of CO_2 back into the atmosphere. The apparent indifference of F-I-p to use of CO_2 or O_2
as a substrate presents the challenge whether the techniques of
molecular biology could be used to persuade the enzyme to ignore O_2.
Perhaps mutations could be induced in the genetic information coding
for those regions in the amino acid sequences which are the binding
sites for CO_2 and O_2. In contemplating such possibilities, one should
bear in mind what is already known about some aspects of F-I-p in
higher plants in relation to coding information and previous evolution.

The amino acid sequences of the large subunit are coded by extranuclear DNA, most likely chloroplast DNA (19). Akazawa and co-workers
(20) have shown that the large subunit contains the enzymatic site.
To date, the extranuclear genes coding for the F-I-p large subunit
have been completely inaccessible to manipulation even by such sophisticated techniques as fusion of protoplasts to create new species
of plants (21).

Table 2 summarizes amino acid analyses of crystalline F-I-p
obtained from four different species of Nicotiana, with the F-I-p
from tobacco (N. tabacum) used as a basis for comparison. N. glutinosa F-I-p has three amino acids whose percentages of the total
are different from those in N. tabacum F-I-p; N. sylvestris F-I-p
has five; and N. glauca F-I-p has eleven. A subsequent study

Table 2

Comparison of Amino Acid Compositions of Cyrstalline
F-I-p in Four Species of Nicotiana

L= large subunit; S= small subunit. Cys and Try make up 5% of
the total amino acids. Data are percentages of total amino acids,
and only those significantly different from the composition of
N. tabacum F-I-p are shown. Data are from Kawashima et al. (22)

Amino Acid	N. tabacum L	N. tabacum S	N. glutinosa L	N. glutinosa S	N. sylvestris L	N. sylvestris S	N. glauca L	N. glauca S
Lys	5.04	7.90						5.92
His	2.78	0.63						
Arg	6.41	4.52				3.53		5.50
Asp	8.92	7.12						7.90
Thr	5.59	4.06			4.97		4.98	
Ser	3.17	3.58						4.40
Glu	10.00	14.63		16.70		16.10		15.80
Pro	4.32	6.52			7.53			
Gly	9.95	7.17						6.06
Ala	8.92	5.72					8.00	
Val	7.46	6.40						
Met	1.65	1.61						2.41
Ile	4.33	4.24		4.70				4.83
Leu	8.97	8.26	8.26					
Tyr	3.89	7.76						6.02
Phe	4.42	4.24				3.70		

showed no detectable differences among the specific ribulose bisphosphate carboxylase activities of these four species of F-I-p or of
two additional species for which amino acid analyses were unavailable (Table 3). In only one species (N. gossei) of the seven tested
was a difference found. That the difference in specific activity is
inherited only via the maternal line shows that it is controlled by
extranuclear DNA: the nuclear genes coding for the small subunit,
which can be controlled by plant breeders, are not the genes af-
fecting the specific activity of the enzyme--at least in this well-
defined system.
     What has happened during the evolution of this vital enzyme on
which all life depends that allows for large changes in amino acid
composition without apparent effect on the enzymatic capacity of
F-I-p? The answer is that amino acid composition can change dramatically without simultaneous mutation in the genetic code for F-I-p
as new species of plants and F-I-p evolve by amphidiploidy. Amphidiploidy is the process of interspecific hybridization followed by

Table 3

Comparison of Specific Enzymatic Activities of
Crystalline F-I-p From Different Species
and Reciprocal Interspecific Hybrids

Amino acid composition is available only for species
indicated, besides N. tabacum. Activity is in μmoles
HCO$_3$⁻ fixed per mg F-I-p per minute. Data are from
Singh and Wildman (23).

Plant	No. of differences from amino acid composition of N. tobacum F-I-p	RuBP carboxylase specific activity
N. tabacum		0.24
N. sylvestris	5	0.23
N. glutinosa	3	0.24
N. glauca	11	0.24
N. excelsior		0.23
N. suaveolens		0.24
N. gossei		0.36
N. gossei x N. excelsior		0.29
N. gossei x N. suaveolens		0.29
N. excelsior x N. gossei		0.24
N. suaveolens x N. gossei		0.24

somatic doubling of the two different sets of chromosomes contained
in the hybrid. Chromosome doubling permits the newly evolved species
of plant to complete meiosis, and alternation of generations can
ensue. Amphidiploidy is said to be responsible for creating 40% of
the half million or so different species of plants now present in
the world.

Evolution in the amino acid composition of F-I-p without simultaneous mutation in the genetic code is well illustrated by following
the origin of N. tabacum F-I-p (24). When subjected to electrofo-
cusing, the small subunit resolves into two polypeptides having different isoelectric points. The origin of these two polypeptides
has been traced. The coding information for one came from N. sylvestris (n=12) and for the other from N. tomentosiformis (n=12) during
the amphidiploid event that produced the new self-fertile N. tabacum
(n=24) species. Data from chymotryptic peptide analyses (25), amino
acid analysis, and sequence determination (26) have converged to
show four differences in amino acid composition between the two polypeptides composing the N. tabacum small subunit. It has also been

shown that at least one of these differences could not have arisen
by a single point mutation in the genetic code. This is a clear
example of how a plant protein can evolve a different amino acid
composition without simultaneous mutation of the genetic code.
N. digluta is a derivative of somatic doubling of the chromosomes in
the hybrid N. glutinosa x N. tabacum. Its F-I-p has a small subunit
composed of four polypeptides having different isoelectric points
(27). Evolution by amphidiploidy has changed the F-I-p of N. digluta
to the extent of at least eight differences in amino acid composition
from the F-I-p of N. tomentosiformis.

Several kinds of analyses of the genetic behavior of the coding
information for the small subunit of F-I-p have strongly indicated
that the coding DNA for multiple polypeptides is located on separate
chromosomes (28). It is hypothesized that polyploids may have a
selective advantage over diploids (29). Polyploids may be more ver-
satile in evolution because they contain two or more alleles at the
same gene locus. The more genes, the more opportunity for diversification by mutation. Polyploid species express multiple forms of
enzymes. Perhaps the presence of multiple polypeptides in the small
subunit of F-I-p is the molecular counterpart of polyploidy and af-
fords a mechanism for diversification whereby a crucial enzyme for
photoautotrophic life can accommodate without difficulty to what
might otherwise be a hostile cellular environment within a newly
evolved plant species. The ribulose bisphosphate carboxylase specific
activity (Table 3) of F-I-p is the same for several diploids compared
with several amphidiploids in spite of extensive differences in amino
acid composition between the different species of F-I-p.

The distribution on different chromosomes of coding information
for the small subunit appears to cause no confusion regarding biosynthesis of F-I-p. Much evidence points to the chloroplast as
being the site of synthesis of the chloroplast-DNA-coded large subunit (30). Evidence has also accumulated indicating that the small
subunit is synthesized on cytoplasmic ribosomes (31, 32). But the
genetic code for the small subunit polypeptides can be located in as
many as four different loci, all presumably transcribing and trans-
lating a somewhat different message with equal facility. How does
the enzyme distinguish between the four possibilities during its
synthesis? The answer is that it doesn't. Evolution has produced
a system for random assembly of the small subunit polypeptides with
an octamer of large subunits as the small subunit polypeptides come
off their different assembly lines (33). As shown in Figure 6, with
two kinds of small subunit polypeptides, nine types of F-I-p macromolecules constitute a population in which the most frequent macromolecules (27%) are composed of eight small subunits consisting of a
50:50 mixture of the two kinds of polypeptides. The remaining 73%
are macromolecules with every possible distribution of the polypeptides, including the two parental types but at a frequency of only
0.78%. With three sources of coding for small subunit polypeptides,
45 different kinds of macromolecules constitute the F-I-p population.

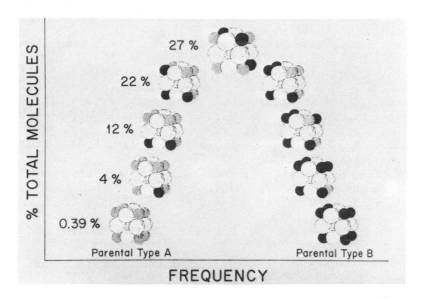

Figure 6. Histogram showing the possible kinds of
Fraction I protein molecules arising from random assembly of two kinds of small subunit polypeptides
with one kind of large subunit. Data from Hirai (33).

To paraphrase Max Delbrück's oft quoted remark: Genetic engineers
are going to have to devise some pretty crafty tricks to fool the old
bird in regard to what she has improvised after the experience of a
few billion years of F-I-p evolution!

Acknowledgements: This work was supported by a grant from the
former RANN division of the National Science Foundation. We are indebted to Drs. H. T. Huang and B. H. Ershoff for advice and encouragement. SCW extends special thanks to the Division of Biology and Medicine, now a part of the Department of Energy, for its long continued
support of the basic research that made it possible to consider tobacco as a food crop. He is also greatly indebted to a host of present and past collaborators for their efforts toward elucidating
the nature and behavior of Fraction I protein.

REFERENCES

1. Kawashima, N. and Wildman, S. G., Biochim. Biophys. Acta 229,
    240-9 (1971).

2. Chan, P. H., Sakano, K., Singh, S., and Wildman, S. G., Science
    176, 1145-6 (1972).
3. Baker, T. S., Eisenberg, D., Eiserling, F. A., and Weissman, L.,
    J. Mol. Biol. 91, 391-9 (1975).
4. Singh, S. and Wildman, S. G., Plant Cell Physiol. 15, 373-9
    (1974).
5. Sakano, K. and Wildman, S. G., Plant Sci. Lett. 2, 273-6 (1974).
6. Sakano, K., Partridge, J. E., and Shannon, L. M., Biochim.
    Biophys. Acta 329, 339-41 (1973).
7. Chollet, R., Anderson, L. L., and Hovsepian, L. C., Biochem.
    Biophys. Res. Commun. 64, 97-107 (1975).
8. Kawashima, N., Singh, S., and Wildman, S. G., Biochem. Biophys.
    Res. Commun. 42, 664-8 (1971).
9. Kwok, S. Y., Kawashima, N., and Wildman, S. G., Biochim. Biophys.
    Acta 234, 293-6 (1971).
10. Protein Requirements, FAO Nutrition Studies No. 16, UN, Rome,
    1957.
11. Block, R. J. and Bolling, D., Amino Acid Composition of Pro-
    teins and Foods, 2nd ed., Charles C. Thomas, Springfield, IL,
    1951.
12. Block, R. J. and Weiss, K. W., Amino Acid Handbook, Charles C.
    Thomas, Springfield, IL, 1956.
13. Kawashima, N., Kwok, S. Y., and Wildman, S. G., Biochim.
    Biophys. Acta 236, 578 (1971).
14. Lowe, R. H., FEBS Lett. 78, 98-100 (1977).
15. Ershoff, B. H., Wildman, S. G., and Kwanyuen, P., Proc. Soc.
    Exp. Biol. Med. 157, 626-30 (1978).
16. Oelschlegel, F. J., Schroeder, J. R., and Stahmann, M. A.,
    J. Agric. Food Chem. 17, 791-5 (1969).
17. Spencer, R. R., Mottola, A. C., Bickoff, E. M., Clark, J. P.,
    and Kohler, G. O., J. Agric. Food Chem. 19, 504-7 (1971).
18. Andrews, T. J., Lorimer, G. H., and Tolbert, N. E., Biochemistry
    12, 11-18 (1973).
19. Chan, P. H. and Wildman, S. G., Biochim. Biophys. Acta 277,
    677-80 (1972).
20. Nishimura, M., Takebe, T., Sugiyama, T., and Akazawa, T.,
    J. Biochem. Tokyo 75, 945 (1973).
21. Chen, K., Wildman, S. G., and Smith, H. H., Proc. Natl. Acad.
    Sci. USA 74, 5109-12 (1977).
22. Kawashima, N., Kwok, S. Y., and Wildman, S. G., Biochim.
    Biophys. Acta 236, 578 (1971).
23. Singh, S. and Wildman, S. G., Mol. Gen. Genet. 124, 187-96 (1973).
24. Gray, J. C., Kung, S. D., Wildman, S. G., and Sheen, S. J.,
    Nature London 252, 226 (1974).
25. Kawashima, N., Tanabe, Y., and Iwai, S., Biochim. Biophys.
    Acta 427, 70-7 (1976).
26. Strøbaek, S., Gibbons, G. C., Haslett, B., Boulter, D., and
    Wildman, S. G., Carlsberg Res. Commun. 41, 335-43 (1976).
27. Kung, S. D., Sakano, K., Gray, J. C., and Wildman, S. G.,
    J. Mol. Evol. 7, 59-64 (1975).

28. Uchimiya, H., Chen, K., and Wildman, S. G., Stadler Genet.
     Symp. 9, 83-100 (1977).
29. von Wettstein, F., Naturwissenschaften 31, 574-7 (1943).
30. Ellis, R. J., Biochim, Biophys, Acta 463, 185-215 (1977).
31. Gray, J. C. and Keckwick, R. G. O., Eur. J. Biochem. 44, 491-
     500 (1974).
32. Gooding, L. R., Roy, H., and Jagendorf, A. T., Arch. Biochem.
     Biophys. 159, 324-35 (1973).
33. Hirai, A., Proc. Natl. Acad. Sci. USA 74, 3443-5 (1977).

DISCUSSION

TOLBERT: We find the specific activity of spinach enzyme is
generally around 1.5, and we agree that the specific activity of
crystalline tobacco enzyme is much less, around 0.3 or 0.4. In our
experience, that of the tobacco enzyme before crystallization is
around 1.5. Why does the specific activity seem to decrease when
the enzyme is partially purified?

WILDMAN: Specific activity is defined as counts of CO_2 fixed
into PGA per minute per mg protein under set conditions with respect
to buffer, pH, and ionic strength of substrate, etc. I will make
one comment as to why tobacco enzyme seems to have a low specific
activity compared with spinach enzyme. Dr. Akazawa once brought us
one of his best preparations of spinach enzyme. We compared it with
the tobacco enzyme without worrying about specific activity. We put
the same amount of protein in the same buffer solution and measured
the counts, and they came out to be the same. I can't go beyond that.

McFADDEN: Have you separated the diastereomers of CRBP?
Could the nonenzymatic cleavage of ^{14}C-L-PGA preferentially be explained by predominance of one diastereomer of CRBP arising during
the cyanhydrin synthesis?

LANE: We have not separated the diastereomers. There appeared
to be some selectivity in the cyanhydrin reaction in favor of the
active inhibitory isomer. It turned out that about 92 to 93% of the
isomer obtained was in a form that bound very tightly to the enzyme
and about 7 to 8% did not bind to the enzyme at all. The ratio was
roughly 10 to 1 in actual synthesis.

PIERCE: I have a comment concerning epimeric, branched-chain
acids. We have recently used the cyanhydrin synthesis to prepare
both 2-carboxyribitol bisphosphate and 2-carboxyarabinitol bisphosphate. One of these epimers inhibits the enzyme stoichiometrically
and the other does not. The cyanhydrin synthesis, under the conditions used in Lane's original studies, gave rise to a 1 to 1 mixture
of epimeric acids. Lane's purification scheme involved DEAE chromatography, which yielded an asymmetric peak. Only fractions very
close to the peak's maximum were pooled and used for the published
inhibitor studies. Our results indicate that this narrow cut from
the DEAE column resulted in at least a 3 to 1 mixture of the two
epimers, the larger part of which corresponded to the stoichiometric

inhibitor.   That is, the 2-carboxyribitol bisphosphate reported in
the literature is not a 1 to 1 mixture of the two possible epimers.
This may explain why the hydrolysis products of oxidized 2-carboxy-
ribitol bisphosphate (i.e., 2-carboxy-3-ketoribitol bisphosphate)
were D-3-phosphoglycerate and 1-3-phosphoglycerate in approximately
equal amounts.   Had a 1 to 1 mixture of 2-^{14}C-carboxyarabinitol
bisphosphate been used, one would have expected to obtain an equal
mixture of D-3-phosphoglycerate and 1-^{14}C-D, 1-3-phosphoglycerate,
contrary to published results.

MECHANISM OF ACTION OF RIBULOSE BISPHOSPHATE CARBOXYLASE/OXYGENASE[*]

M. Daniel Lane

Department of Physiological Chemistry
The Johns Hopkins University School of Medicine
Baltimore, Maryland 21205

and

Henry M. Miziorko

Department of Biochemistry, Medical College of Wisconsin
Milwaukee, Wisconsin 53226

In 1948 Calvin and his colleagues identified D-3-phosphoglycerate as the first stable radioactive product formed during brief exposure of algae to $^{14}CO_2$ (1). This observation led to the discovery that the carboxylation of ribulose bisphosphate is the first step in the photosynthetic carbon cycle (2). An important advance that opened the way for definitive enzymatic studies was the finding both by Calvin's (3) and by Horecker's (4) groups that cell-free extracts of algae or spinach, respectively, carry out a RuBP-dependent carboxylation. It is now well established that the enzyme which catalyzes this reaction, i.e., RuBP carboxylase, is present in most if not all plants and photosynthetic microorganisms.

---

[*]The investigations reported herein were supported by Grant AM-14575 from the National Institutes of Health, United States Public Health Service. The research of HMM was supported by the Research Corporation.
    Abbreviations: RuBP, D-ribulose 1,5-bisphosphate; CRBP, 2-carboxy-D-ribitol 1,5-bisphosphate; PRR, proton relaxation rate; ϵ, enhancement of the PRR; SDS, sodium dodecylsulfate.

STRUCTURE AND MOLECULAR PROPERTIES OF RuBP CARBOXYLASE

Homogeneous preparations of RuBP carboxylase have been obtained from numerous plant sources (2); however, the enzymes from spinach leaf and tobacco leaf have been most rigorously characterized. Both carboxylases are large oligomeric structures and comprise 15 to 20% of total leaf protein. The molecular properties of the spinach enzyme are summarized in Table 1. This carboxylase has a molecular weight (5) of 557,000 and is composed of nonidentical subunits. In 1967, it was first demonstrated by Alan Rutner in my laboratory (6) that the enzyme is composed of two types of subunits which differ greatly in size and amino acid composition. Sedimentation equilibrium analysis of SDS-dissociated carboxylase indicated a high degree of weight heterogeneity. To verify the presence of more than one weight class of polypeptide, the carboxylase was subjected to gel electrophoresis in the presence of SDS; two discrete polypeptide bands were evident (Figure 1B). To isolate the subunits, native enzyme was dissociated with SDS and aminoethylated, and the heavy and light chains were resolved by gel filtration on a Sephadex G-100 column (Figure 1A). Electrophoretic analysis of the two protein fractions (I and II) revealed their correspondence to the two polypeptide species observed with SDS-dissociated native enzyme (Figure 1B). It was evident from the differences in their amino acid compositions (6) that fractions I and II were different polypeptides. The molecular weights of the heavy and light chains were found to be 55,800 and 12,000 respectively (7).

Table 1

Molecular Properties of Spinach Leaf RuBP Carboxylase
(Results from ref. 2, 5-10, 27)

Molecular weight:		
Native enzyme	557,000	
Heavy (H) chain	55,800	
Light (L) chain	12,000	
Subunit stoichiometry:	H_8 L_8	
Binding parameters:	sites/molecule	K_D
RuBP	8	10^{-6} M
2-Carboxyribitol bisphosphate	8	10^{-8} M
Mn^{2+} (in presence of CO_2)	8	10^{-5} M
Cu^{2+}	1	-
Antibody-binding sites	8	-

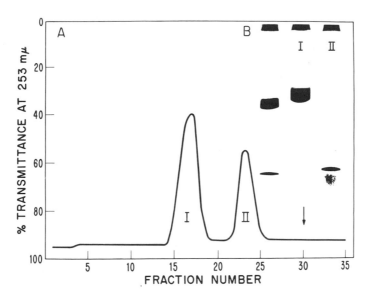

Figure 1.   (A) Elution pattern for aminoethylated RuBP
carboxylase from a Sephadex G-100 column in buffered
SDS.   (B) SDS-polyacrylamide gel electrophoresis of native carboxylase and fractions I and II from (A) above.
Results from ref. 6.

Several lines of evidence show that the native oligomeric carboxylase is composed of 8 heavy (H) and 8 light (L) chains, i.e.,
an H_8L_8 subunit stoichiometry. First, the mass ratio of the two
types of subunits in the native enzyme, taken together with their
molecular weights, indicates (7) a molar H:L ratio of 1:1. The
simplest possible structure is an oligomer composed of 8 heavy
(MW=55,800) and 8 light (MW=12,000) subunits. The finding (8) that
each carboxylase molecule has 8 binding sites for antibody RuBP,
CRBP, and Mn^{2+} (Table 1) is also compatible with an octameric structure. It appears that the catalytic center, i.e., the RuBP substrate
binding site, of the enzyme resides on the heavy subunit (11).
        Crystallization of tobacco leaf RuBP carboxylase, first accom-
plished by Wildman and associates (12), made possible direct three-
dimensional structural analysis by x-ray crystallography. X-ray
diffraction studies on several crystalline forms, combined with
electron microscopy, led to the proposal by Eisenberg's group (13-
15) that RuBP carboxylase is composed of 8 large and 8 small subunits clustered in two layers, perpendicular to a fourfold axis of
symmetry. A model compatible with available structural information
on the enzyme is shown in Figure 2.

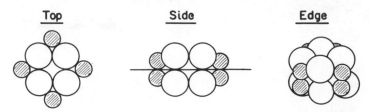

Figure 2.   Proposed model of RuBP carboxylase showing
the arrangement of the heavy and light subunits.

THE CARBOXYLATION REACTION

RuBP carboxylase catalyzes the divalent metal ion-dependent carboxylation of D-ribulose 1,5-bisphosphate to yield two molecules of 3-D-phosphoglycerate (2) as shown in Reaction 1.

$$
\begin{array}{c}
\overset{O}{\underset{\overset{\|}{C}}{\overset{\|}{C}}}{}^{\bullet} \\
\end{array}
\quad
\begin{array}{l}
CH_2{-}O{-}PO_3^{2-} \ (1) \\
\quad | \\
C{=}O \qquad\qquad (2) \\
\quad | \\
H{-}C{-}OH \qquad (3) \\
\quad | \\
H{-}C{-}OH \qquad (4) \\
\quad | \\
CH_2{-}O{-}PO_3^{2-} \ (5)
\end{array}
\;+\; HOH
\xrightarrow[\Delta G'^{\circ} = -12.4 \ kc/m]{Mg^{2+}}
\begin{array}{l}
CH_2{-}O{-}PO_3^{2-} \\
\quad | \\
{}^{-}O_2C^{\bullet}{-}CH{-}OH \\[4pt]
CO_2^{-} \\
\quad | \\
H{-}C{-}OH \\
\quad | \\
CH_2{-}O{-}PO_3^{2-}
\end{array}
\;+\; 2H^{+} \quad (1)
$$

D−Ribulose   D−(−)−3−
1,5−diphosphate   Phosphoglycerate

$H^{14}CO_3^{-}$ is incorporated exclusively into the carboxyl carbon of phosphoglycerate, whereas label from 1-^{14}C-RuBP appeared in the C3 position (2).   These results did not prove, however, into which of the two identical product molecules ^{14}C-label is incorporated; hence, the site of cleavage remained equivocal. Müllhofer and Rose (16) showed later that ^{2}H from ^{2}H$_2$O and ^{14}C from 2-^{14}C-RuBP (but not from 4-^{14}C-RuBP) are both incorporated into the 2-position of the same phosphoglycerate molecule, thus proving that cleavage takes place between C2 and C3 of RuBP.

The carboxylation reaction (Reaction 1) is essentially irreversible; all attempts to demonstrate the reverse reaction have been unsuccessful. This is consistent with the high negative free energy change (-12.4 kcal/mole) calculated for the forward reaction (17). The fact that protons are generated in the reaction at pH 8, where carboxylase activity is optimal, contributes to the high apparent equilibrium constant for the reaction.

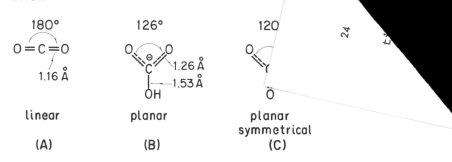

Figure 3. Structures of the major species of "CO_2."

CO_2 AS SUBSTRATE

An understanding of the enzymatic carboxylation mechanism requires knowledge of the species of "CO_2" that participates in
the reaction. At physiological pH, "CO_2" exists both as dissolved
CO_2 and as its hydrated forms, H_2CO_3, HCO_3^-, and CO_3^{2-} (see
Reaction 2),

$$CO_2 + H_2O \underset{\text{slow}}{\overset{k_1}{\rightleftharpoons}} H_2CO_3 \underset{\substack{\text{rapid} \\ pk_1 = 6.3}}{\rightleftharpoons} H^+ + HCO_3^- \underset{\substack{\text{rapid} \\ pk_2 = 10.3}}{\rightleftharpoons} H^+ + CO_3^- \qquad (2)$$

$$k_1 = 0.022 \ sec^{-1} \ \text{at } 20°C$$

the predominant species* being HCO_3^- and CO_2. Chemical considerations suggest that CO_2 is a better electrophile than HCO_3^- and therefore a superior carboxylating species. CO_2 is a linear symmetrical
molecule (Figure 3) with a C-O distance of ~1.16 Å, which is shorter than the typical carbonyl C-O distance of ~1.22 Å (18, 19). The
delocalization of electrons in CO_2 toward the two oxygen atoms renders the carbon atom particularly susceptible to nucleophilic attack
(18, 19). In contrast, bicarbonate, because of resonance stabilization, is apparently less prone to nucleophilic attack.

---

*It is not usually possible to discriminate between HCO_3^-,
H_2CO_3, and CO_3^{2-}, although at the usual assay pH (7.5 to 8.5) for
RuBP carboxylase, HCO_3^- is by far the most abundant species.

Hydration of CO_2 is accompanied by substantial angular distortion with reduction of the O-C-O bond angle (18, 19) from 180° to about 120°. The bond angles of the carboxylate anion, such as that in the product 3-phosphoglycerate, approximate those of bicarbonate. Thus, a carboxylation involving CO_2 requires a change in hybridization of the carbon orbitals from sp to sp^2. The relatively slow rate of hydration of CO_2 has been attributed to $sp \rightarrow sp^2$ rehybridization (18, 19) and suggests that a similar kinetic effect may apply in carboxylation reactions requiring CO_2 as the active species. This would be consistent with the view, discussed below, that CO_2 addition is the rate-limiting chemical step in the carboxylation mechanism.

An approach for determining the active species of "CO_2" in the RuBP carboxylase-catalyzed reaction was first employed in collaborative experiments between Cooper's laboratory and mine (20). By taking advantage of the slow rate of hydration of CO_2 (Reaction 2), this step can be made rate-limiting relative to the rate of carboxylation of RuBP (20). Thus, it was found (20) (Figure 4) that the rate of ^{14}C-phosphoglycerate formation from a $^{14}CO_2 + H^{12}CO_3^-$ mixture added initially was much faster than the rate from a mixture of $^{12}CO_2 + H^{14}CO_3^-$. ^{14}C incorporation rates were equalized by the addition of carbonic anhydrase, which causes instantaneous equilibration of ^{14}C label between all species of CO_2. Hence, molecular CO_2 appeared to be the active species in the carboxylation of RuBP.

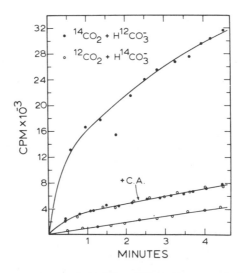

Figure 4. Kinetics of RuBP carboxylase-catalyzed labeling of 3-phosphoglycerate from initially added $^{14}CO_2 + H^{12}CO_3^-$ or $^{12}CO_2 + H^{14}CO_3^-$ in the presence or absence of carbonic anhydrase (CA). Results from ref. 20.

Recent evidence from several laboratories (21-26) suggests that CO_2 serves not only as a substrate of the carboxylase but also as an allosteric activator (in the presence of Mg^{2+}) which drastically lowers the K_m, i.e., to ~10 µM (23-25), for CO_2. This finding resolves the long-recognized discrepancy (2) between the high CO_2 fixation rates at low CO_2 concentration with intact cells or freshly prepared cell-free extracts and the high K_m for CO_2 (~500 µM) in the RuBP-catalyzed reaction using the purified carboxylase.

Prior exposure of carboxylase to CO_2 in the presence of Mg^{2+} appears to induce a conformational change which markedly increases the affinity of the enzyme for substrate CO_2. The results of Lorimer et al. (24) and Laing and Christeller (25) indicate that the enzyme reacts first with CO_2 in a rate-determining and reversible step, and then reacts rapidly with Mg^{2+} to form the active ternary complex (Reaction 3).

$$E + CO_2 \rightleftharpoons E - CO_2^- \rightleftharpoons E - CO_2^- \cdot Me^{2+} \qquad (3)$$

inactive   **active**

Furthermore, the pH dependence of the activation process suggests that CO_2 reacts with a functional group on the enzyme whose pK is alkaline, most likely an amino group. Lysyl ε-amino groups have been implicated by others (28) as essential functional groups at the active site of RuBP carboxylase. It is suggested that CO_2 reacts to form a carbamate group which is them stabilized by interaction with the divalent metal ion (Reaction 4).

$$E - \overset{..}{N}H_2 \quad \overset{O}{\underset{O}{\overset{||}{\underset{||}{C}}}} \rightleftharpoons E - NH - CO_2^- \rightleftharpoons E\text{-}NH\text{-}CO_2^- \, Me^{2+} \qquad (4)$$

                            carbamate   Mg-stabilized
                                                    carbamate

The dependence of divalent metal ion binding to RuBP carboxylase upon "CO_2" was first recognized by Miziorko and Mildvan (27), who studied the binding of Mn^{2+} to RuBP carboxylase by electron paramagnetic resonance and water proton relaxation rate (PRR) measurements. As shown in Figure 5B, tight binding of Mn^{2+} (n = 1 site per ~70,000 daltons) occurs only in the presence of CO_2; in the absence of added CO_2 (Figure 5A) only weak binding of Mn^{2+} occurs. The K_D for the dissociation of Mn^{2+} from the enzyme-CO_2-Mn^{2+} ternary complex is 640 µM in the absence of added CO_2 and 10 µM in its presence (Table 2). The number of tight Mn^{2+} binding sites, i.e., 1 site per 70,000 daltons or 8 sites per native oligomer, is consistent

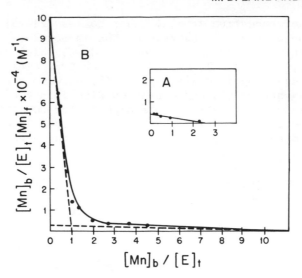

Figure 5. Scatchard plot of Mn^{2+} binding to RuBP carboxylase in the absence (A) or presence (B) of 50 mM total "CO_2". Results from EPR titration in ref. 27.

Table 2

Effect of CO_2 on Mn^{2+} Binding to RuBP Carboxylase
(Result from ref. 27)

Total CO_2 conc. (mM): $CO_2+H_2CO_3+HCO_3^-+CO_3^{2-}$	No. of tight binding sites for Mn^{2+} per 70,000 daltons	K_D (μM) of Mn^{2+} from $Enz.CO_2.Mn$ ternary complex
0.5	-	640
50	1.0±0.1	10

with the octameric paired subunit structure of the enzyme (see Table 1 and Figure 2).

The binding of Mn^{2+} to RuBP carboxylase causes a 10-fold enhancement of the effect of manganese on the longitudinal relaxation rate of water protons (27). The fact that CO_2 further enhances the PRR by 40% suggests (27) that "CO_2" is not directly coordinated to the enzyme-bound Mn^{2+}. Furthermore, the distance from bound Mn^{2+} to the carbon atom of "CO_2," 5.4 Å, determined by its ^{13}C relaxation rate suggests a second sphere complex. This would be consistent with the stabilization of CO_2, bound as a carbamate, through a salt linkage to the divalent cation (Reaction 4).

It is significant (Table 3) that an analogue (2-carboxyribitol 1,5-bisphosphate, CRBP) of the proposed 6-carbon intermediate formed as adduct between CO_2 and RuBP drastically decreases the enhancement of the water PRR. This suggests that upon binding to Mn^{2+}-enzyme, CRBP displaces water from the inner sphere of Mn^{2+} and "CO_2" from the second sphere (based on ^{13}C nuclear magnetic resonance data) of Mn^{2+}. Although direct activation of "substrate CO_2" by enzyme-bound Mn^{2+} is not strictly ruled out, Mildvan (29) points out a possible alternative. Perhaps the metal coordinates the oxygen at C2 and/or C3 of RuBP to facilitate carbanion formation at C2, an essential step in the carboxylation and oxygenation steps.

A fundamental question remains unanswered. Does CO_2 bind initially both to activate the enzyme and to serve as carboxylating substrate (as in Reaction 5a) or does a second molecule of $CO2$ function as carboxylating substrate after an initial activation (as in Reactions 5b and c)? Several lines of evidence favor the second ex-

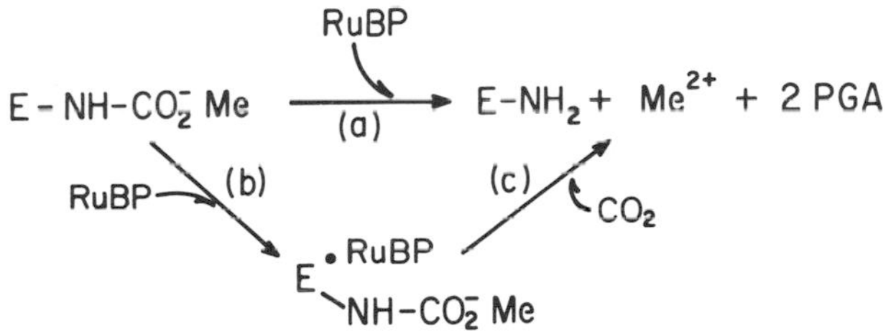

$$
\begin{array}{c}
\text{RuBP} \\
\text{E}-\text{NH}-\text{CO}_2^-\ \text{Me} \xrightarrow[\text{(a)}]{} \text{E}-\text{NH}_2 + \text{Me}^{2+} + 2\,\text{PGA} \qquad (5) \\
\text{RuBP} \searrow^{\text{(b)}} \qquad \text{(c)} \nearrow \text{CO}_2 \\
\text{E}\cdot \overset{\text{RuBP}}{\underset{\text{NH}-\text{CO}_2^-\ \text{Me}}{}}
\end{array}
$$

Table 3

Dissociation Constants and Enhancement of PRR of
Ternary and Higher Complexes of RuBP Carboxylase
(Results from ref. 27)

Ligand	K_D (mM) for ligand from complex*	ε_t
"CO_2"	7	21
CRBP	0.004	0.4
Ribitol bisphosphate	1.0	14

*At saturating Mn^{2+} (for $CO2$) or at saturating CO_2 and Mn^{2+} (for CRBP and ribitol bisphosphate).

planation:

1. With the kinetic isotope-trapping technique of Meister (30), $^{14}CO_2$ from enzyme-$^{14}CO_2$-Mg^{2+} complex was not trapped in the $^{12}CO_2$ pool during incorporation into product 3-phosphoglycerate (24). In view of the relatively long apparent half-life of the complex at 20° and pH 8, ~0.1 min (23), it should have been possible to detect "trapping" by this technique.

2. Formation of the enzyme-CO_2-Mg^{2+} complex is also required for activation of the oxygenase activity of the enzyme (26). This indicates that CO_2 has an activating rather than a substrate function, since O_2 and CO_2 compete for the same substrate site.

3. Kinetic studies by Laing and Christeller (25) indicate that catalysis, unlike activation, which requires CO_2 and Mg^{2+}, appears to be kinetically dependent only on CO_2 concentration.

Although additional studies will be required for definitive determination of whether CO_2 functions both as activator and substrate, preliminary results suggest that this is the case.

MECHANISTIC CONSIDERATIONS

On the basis of the arguments presented above, it is assumed that molecular CO_2, rather than HCO_3^- or CO_3^{2-}, is the active carboxylating species in the carboxylation reaction per se. It is evident that delocalization of electrons toward the two oxygen atoms (Reaction 6)

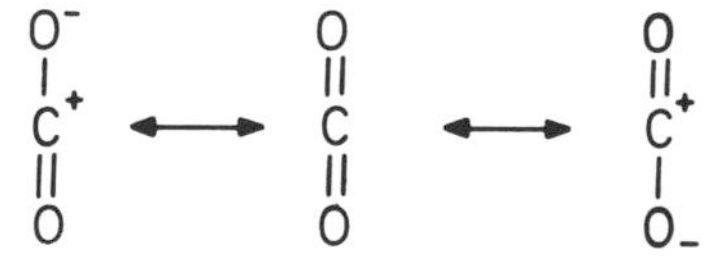

$$
\begin{array}{c}
O^- \\
| \\
\overset{+}{C} \\
\| \\
O
\end{array}
\longleftrightarrow
\begin{array}{c}
O \\
\| \\
C \\
\| \\
O
\end{array}
\longleftrightarrow
\begin{array}{c}
O \\
\| \\
\overset{+}{C} \\
| \\
O_-
\end{array}
\tag{6}
$$

renders the carbon atom particularly susceptible to nucleophilic attack by the carboxyl acceptor (Reaction 7), in this case RuBP.

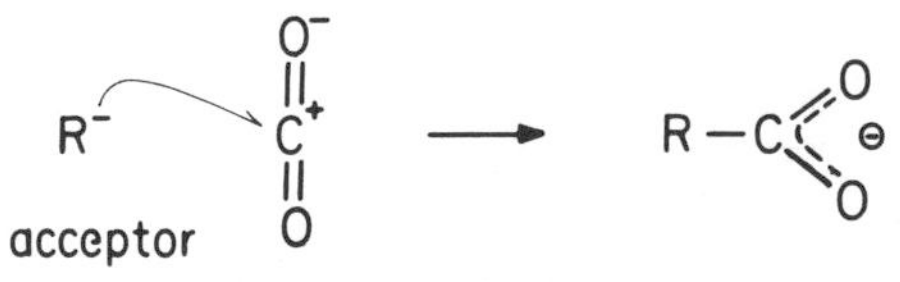

$$
R^- \quad
\begin{array}{c}
O^- \\
\| \\
\overset{+}{C} \\
\| \\
O
\end{array}
\longrightarrow
R-C\overset{O}{\underset{O_\ominus}{\diagdown}}
\tag{7}
$$

acceptor

The basic elements that are consistent with pertinent experimental findings and must be considered in formulating the carboxylation mechanism include the following:

1.   Development of a nucleophilic center at C2 of RuBP.   C2, to which CO_2 ultimately becomes bonded, is an electrophilic center (see Reaction 8)

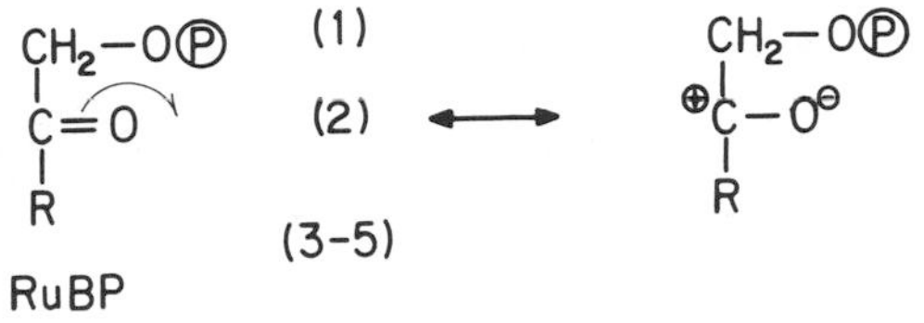

(8)

and therefore an unfavorable site for carboxylation without prior rearrangement to a center with nucleophilic character.
2.   Intramolecular dismutation in which oxidation at C3 occurs at the expense of reduction at C2.
3.   Addition of CO_2 at C2.
4.   Cleavage of the C2-C3 bond.
The first requirement would be satisfied by ene-diol formation as suggested by Calvin (31) or by isomerization to the corresponding 3-keto pentose prior to carboxylation (2) as illustrated in Reactions 9a and 9b.

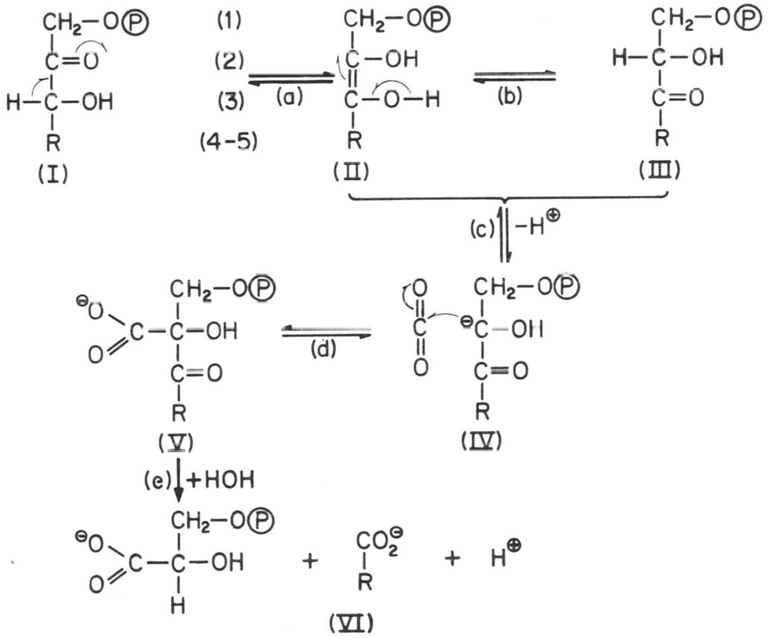

(9)

In either case the appropriate nucleophilic center at carbon-2 would be developed, and carboxylation would be expected to lead to the six-carbon intermediate (V) proposed by Calvin (31) or to a closely related transition state.   Attempts to determine the sequence of these events have been hampered by the irreversibility of the overall reac-

tion and inability to detect partial reactions by isotopic exchange
or other means. For example, the carboxylase does not catalyze isotope exchange between RuBP and ^{14}C-3-phosphoglycerate, RuBP and
$H^{14}CO_3^-$, 3-phosphoglycerate and ^3H_2O, or 3-^3H-RuBP and H_2O (2).

Although water protons do not exchange with substrate, a water
proton is incorporated at C2 of the molecule of 3-phosphoglycerate
arising from C1 and C2 of RuBP and from CO_2 (2) (Reaction 1). The
reciprocal experiment with 3-^3H-RuBP showed that ^3H was not incorporated into 3-phosphoglycerate, but rather entered solvent water; release of ^3H was dependent upon the occurrence of the overall carboxylation reaction and occurred at about the same rate. Although lack of
proton exchange between the C3 hydrogen of RuBP and water does not
rule out ene-diol or 3-keto intermediates, it does preclude their
formation in the absence of the other components (CO_2 and Mg^{2+})
needed to complete the reaction. The occurrence of a large (4- to
6-fold) kinetic isotope effect in the carboxylation reaction with
3-^3H-RuBP as substrate indicates that the enolization step (Reaction 9a) is followed by a fast and irreversible step (2). Since
there is a discrimination (2) against ^3H incorporation from ^3H_2O into
3-phosphoglycerate (Reaction 9e), this step must also be slow. The
effect with ^2H_2O on the net reaction rate is smaller than predicted
by ^3H_2O incorporation (2); thus, it appears that Reaction 9e is not
rate-determining in the overall process. The large discrimination
against 3-^3H-ribulose bisphosphate suggests (2) that enolization
(Reaction 9e) is rate-limiting in the overall reaction.

2-CARBOXYRIBITOL 1,5-BISPHOSPHATE AS AN ANALOGUE
OF THE PUTATIVE 6-CARBON CARBOXYLATED INTERMEDIATE

Support for the participation of a six-carbon carboxylated intermediate (2-carboxy-3-ketoribitol 1,5-bisphosphate, 3-keto-CRBP;
Structure V, below, and in Reaction 9) in the RuBP carboxylase-
catalyzed reaction is derived from studies with 2-carboxy-D-ribitol
1,5-bisphosphate (CRBP, Structure VII),

$H_2COPO_3^{2-}$
|
C (OH) CO_2^-
|
$C=O$
|
$H-C-OH$
|
$H_2COPO_3^{2-}$

(V)

3-keto-CRBP

$H_2COPO_3^{2-}$
|
C (OH)CO_2^-
|
$H-C-OH$
|
$H-C-OH$
|
$H_2COPO_3^{2-}$

(VII)

CRBP

an analogue of the proposed six-carbon intermediate. Labeled and
unlabeled CRBP were chemically synthesized (8) and purified chromatographically on DEAE-cellulose. Because of its structural similarity to the hypothetical intermediate, 3-keto-CRBP, it would be anticipated that CRBP would be an inhibitor and would bind more tightly
than either substrate, i.e., RuBP or CO_2, since it contains structural elements of both. As shown in Table 1, CRBP binds to RuBP
carboxylase with a $K_D \leq 10^{-8}$. This constitutes a $\geq$100-fold tighter
binding than the binding of RuBP (8). Moreover, it was found (8)
that there are 8 independent CRBP binding sites, presumably active
sites, per molecule of carboxylase.

Exposure of RuBP carboxlyase for 20 to 40 min to a 10-fold
stiochiometric excess of CRBP (a 1.25-fold excess over the 8 RuBP
substrate binding sites per molecule of enzyme), causes almost
complete loss of enzymatic activity (see Figure 6). This and the
fact that at lower CRBP/enzyme ratios the extent of inhibition at
the time end point is proportional to CRBP concentration suggest that
the inhibitor binds stoichiometrically at the active site. The presence of Mg^{2+}, an essential activator in the carboxylase-catalyzed
reaction, is also required during the preliminary incubation of enzyme with CRBP for maximal inhibition (Figure 7). On the other hand,
CO_2, the carboxylating substrate species, is not required for inhibition by CRBP. These findings agree with out earlier observation (8)
that tight binding of CRBP to the carboxylase is observed only in the
presence of divalent metal ions, e.g., Mg^{2+} or Mn^{2+}, and that enzyme
complexes formed contain equimolar amounts of inhibitor and metal ion.

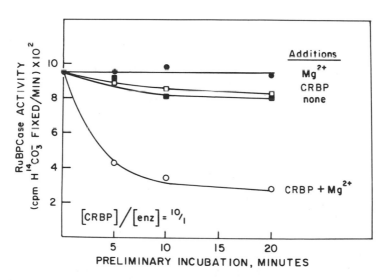

Figure 6. Dependence of the loss of carboxylase activity on CRBP
concentration during preliminary incubation. Results from ref. 10.

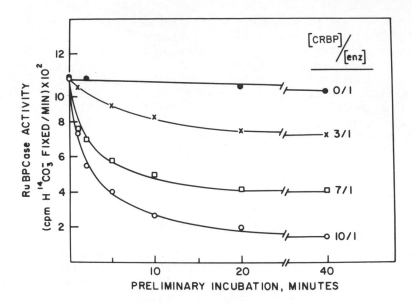

Figure 7. Dependence of inhibition by CRBP upon
the presence of Mg^{2+} during preliminary incubation.
Results from ref. 10.

It is interesting that inhibition by CRBP is a slow process requiring 20 to 40 min to reach completion (Figure 6). Thus, not only
the extent but also the rate of inhibition is dependent upon CRBP
concentration. In other experiments (10) it was shown that the rate
of loss of enzymatic activity is second order in enzyme sites and
CRBP. Direct binding experiments (8) show that CRBP binding to the
carboxylase exhibits a similar time dependence.

Once CRBP is bound to the enzyme under the conditions used for
Figure 6, its release either is extremely slow or does not occur at
all. Enzyme withdrawn for assay at any point during the preliminary incubation with inhibitor yielded reduced, but linear, carboxylation rates despite the presence of a high concentration (7 x 10^{-4} M)
of RuBP in the assay. This is particularly significant since CRBP
and RuBP appear to compete for a common binding site, apparently the
RuBP substrate site.

A common binding site is indicated by two lines of evidence (10).
First, the presence of RuBP during preliminary incubation of the
carboxylase with CRBP effectively prevents inhibition (10). Furthermore, the interaction of either RuBP or CRBP with the carboxylase
produces qualitatively identical difference spectra in the ultra-
violet region. As shown in Figure 8, both spectra exhibit sharp
spikes at 288 nm, as well as minima at 296 nm and 285 nm, and a
broad maximum at ~265 nm. The height of the 288-nm peak can be
titrated to a maximum with RuBP or CRBP; at a concentration ratio
of CRBP/enzyme of ~10, a condition which produces total inhibition

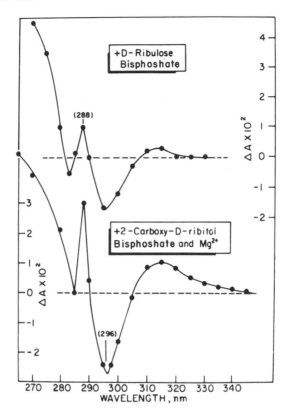

Figure 8. Difference spectra of
RuBP carboxylase promoted by RuBP
and CRBP. Results from ref. 10.

of the carboxylase (Figure 6), the 288-nm peak reaches a maximum.
The K_D ($\sim 10^{-8}$ M) for the dissociation of CRBP from the ternary complex, calculated from difference spectrum titration data, is identical to that determined by titrating catalytic activity with inhibitor (Figure 6). It is also clear that CRBP is not bound covalently
to the carboxylase, since it is released by treating the enzyme with
SDS (10).
     These results show that CRBP has the characteristics of an ana-
log of the proposed six-carbon intermediate, 3-keto-CRBP (Structure
V), in the carboxylase-catalyzed carboxylation of RuBP.

     2-CARBOXY-3-KETORIBITOL 1,5-BISPHOSPHATE (3-KETO-CRBP):
          AN INTERMEDIATE IN THE CARBOXYLATION OF RuBP

     Support for the participation of a six-carbon intermediate
(Structure V in Reaction 9) in the RuBP carboxylase-catalyzed reac-

tion was provided in the preceding discussion. However, since this
evidence is indirect, direct chemical evidence for the above proposal
was sought. Therefore, an attempt was made to synthesize 3-keto-CRBP
chemically in order to test it as a possible intermediate in the
enzymatic process (32).

CRBP differs from the proposed intermediate only in its oxidation state at C3 (compare Structures V and VII). Hence, chemical
procedures were sought which might selectively oxidize C3 of CRBP to
a carbonyl group. It was important to use mild conditions for oxidation, since the desired product, a β-keto acid, would be unstable
and decarboxylate readily. Since CRBP contains secondary hydroxyl
groups on both C3 and C4 where oxidation would occur most rapidly,
protection of C4 was attempted by formation of the γ-lactone. After
many abortive attempts (32) to oxidize specifically the lactone, it
was decided to utilize the open-chain compound, CRBP, as substrate
in the oxidation. Because of the two potential centers for oxidation
at C3 and C4, a mixture of the 3-keto and 4-keto derivatives was
expected (Figure 9).

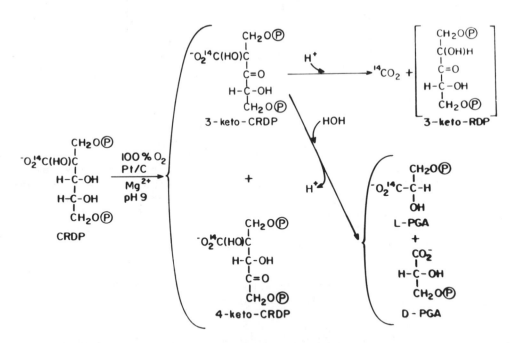

Figure 9. Catalytic oxidation of CRBP to yield 4-keto-
and 3-keto-CRBP and secondary reactions of the 3-keto-
CRBP product. PGA, 3-phosphoglycerate. From ref. 32.

A number of catalyst preparations were screened to evaluate their relative abilities to catalyze oxidation of CRBP at pH 9 and 55° in the presence of 100% oxygen (32). Extent of oxidation was assessed by reducing the filtered oxidation reaction mixture with ^{3}H-NaBH$_4$ and determining the amount of ^{3}H activity incorporated into acid-stable form. Reduction would be expected to convert the 3- or 4-keto-CRBP oxidation products to 3-^{3}H- or 4-^{3}H-CRBP, both of which are acid-stable; prior to catalytic oxidation, ^{3}H activity was not incorporated into CRBP even after extended treatment with ^{3}H-borohydride. Five percent platinum on carbon powder proved to be the most effective catalyst (32), giving rise to approximately twice as much ^{3}H incorporation into acid-stable form after a 2-hr oxidation as any of the other catalysts.

To establish conditions for optimizing the yield of keto-CRBP from CRBP by catalytic oxidation, it was necessary both to maximize oxidation rate and to minimize losses of the desired β-keto acid product, 3-keto-CRBP, via decarboxylation. Lowering the temperature to 0° during catalytic oxidation to increase the stability of the 3-keto-CRBP product decreased the rate of oxidation to an infinitesimal level (32). Further attempts to improve the yield of 3-keto-CRBP led to the discovery that magnesium ion greatly enhanced the rate of catalytic oxidation of CRBP. The presence of 10mM MgCl$_2$ led to essentially complete oxidation (91%) of CRBP in 2 hr under conditions (55°, pH 9) that supported only 17% oxidation in 4 hr in the absence of MgCl$_2$. The recovery of ~50% of the loss in acid-stable ^{14}C activity as ^{14}CO$_2$ suggested that about half of the CRBP was oxidized to the unstable 3-keto derivative. An important finding was that, in the presence of divalent magnesium ion, catalytic oxidation proceeded rapidly even at 0°; under similar conditions, but without magnesium ion, essentially no oxidation occurred. Moreover, as the temperature of oxidation in the presence of 10 mM MgCl$_2$ was lowered from 55° to 0°, the yield of 3-keto-CRBP increased dramatically (32).

Catalytic oxidation (100% oxygen, platinum on carbon catalyst, 0°) of carboxyl-^{14}C-CRBP in 10 mM MgCl$_2$ gave rise to the predicted 3-keto and 4-keto derivatives (Figure 9), which were characterized by reduction with ^{3}H-NaBH$_4$ and periodate degradation. The yield of one of these oxidation products, 3-keto-CRBP (the proposed intermediate in the enzymatic carboxylation reaction), was reduced somewhat since it is a β-keto acid and decarboxylates slowly under these conditions (Figure 9).

During the course of the synthesis of carboxy-^{14}C-3-keto-CRBP by catalytic oxidation, it was observed (32) that a ^{14}C-labeled compound which behaved chromatographically as phosphoglycerate was formed in significant quantity at 25°. The possibility that ^{14}C-phosphoglycerate arose by hydrolytic cleavage of 3-keto-CRBP between C2 and C3 yielding 2 molecules of phosphoglycerate (Figure 9) was attractive because of the analogy to the proposed enzymatic process (Reaction 9d) involving the 3-keto intermediate. We found, in fact, that chemically synthesized carboxy-^{14}C-3-keto-CRBP undergoes spontaneous hydrolytic cleavage at 25° and pH 9, giving rise to 1 molecule of D-3-phosphoglycerate and 1 molecule of L-3-phosphoglycerate, the unlabeled D-isomer arising from carbon atoms 3, 4, and 5 and the ^{14}C-labeled L-isomer from the carboxyl group and carbon atoms 1 and 2 (see Figure 9 and Table 4).

It is interesting that nonenzymatic cleavage of 3-keto-CRBP proceeds stereospecifically in that only L-phosphoglycerate is derived from the carboxyl group and carbon atoms 1 and 2 of the substrate (32). The mechanistic basis for this result is not known. However, since CRBP binds magnesium ion tightly, the divalent metal ion may participate in directing the mechanism and stereochemistry of cleavage of 3-keto-CRBP. In the enzymatic process which yields only D-3-phosphoglycerate, it is evident that the carboxylase directs the stereochemistry of the cleavage.

It was of interest to determine whether 3-keto-CRBP undergoes enzymatic cleavage. Since the 3-keto derivative is too unstable to permit resolution from CRBP, the unfractionated catalytic reaction mixture was used as the source of the intermediate. To obtain maximal yield of 3-keto-CRBP and to minimize losses via decarboxylation and C2-C3 cleavage, catalytic oxidation was conducted at 0° (32). As shown in Table 5, 12.4 nmoles of product (based on ^{14}C activity rendered acid-stable by reduction with NaBH$_4$), previously shown to be primarily 3-keto- and 4-keto-CRBP, were incubated in the presence or absence of 27 nmoles of homogeneous spinach leaf RuBP carboxylase (216 nEq of active sites). After incubation for 2 min without enzyme, all (12.4 nmoles) of the "3-keto- and 4-keto-CRBP", i.e., ^{14}C activity stabilized by borohydride reduction, remained. In contrast, when carboxylase was present, only 7.8 nmoles remained, indicating the conversion of 4.6 nmoles to product(s). Chromatography of the reaction mixture revealed that 2.9 nmoles of 3-phosphoglycerate were formed. In the reaction mixture without carboxylase, 3-phosphoglycerate was not found. The remainder (1.7 nmole) of the unstable radioactivity lost in the incubation containing enzyme appeared as

Table 4

Nonenzymatic Conversion of Carboxy-^{14}C-
3-keto-CRBP to D- and L-3-Phosphoglycerate
(Results from ref. 32)

	Phosphoglycerate formation (μmoles) based on	
	Enzymatic assay	^{14}C activity
D-3-phosphogylcerate	0.62	0.00
L-3-phosphoglycerate	0.60	0.52

Table 5

Enzymatic Conversion of Carboxy-^{14}C-3-keto-CRBP to
3-Phosphoglycerate, Ketopentose 1,5-Bisphosphate, and CO_2
(Results from ref. 32)

nmoles RuBP carboxylase	Acid-labile ^{14}C-containing compounds(s) stabilized by reduction: nmoles remaining after incubation for		nmoles 3-phosphoglycerate	nmoles CO_2
	0 min	2 min		
0.0	12.4	12.4	0.0	0.0
27.3	12.4	7.8	2.9	1.7

$^{14}CO_2$. It appears that RuBP carboxylase acted on essentially all
(4.6 nmoles) of the 3-keto-CRBP present in the mixture of oxidation
products. These results indicate that 3-keto-CRBP is a viable substrate for RuBP carboxylase, which promotes both cleavage of 3-keto-
CRBP to form 3-phosphoglycerate and decarboxylation to yield CO_2 and
ketoribitol bisphosphate (see Reaction 9).

COMPARISON OF THE CARBOXYLATION AND OXYGENATION MECHANISMS

If it is assumed that carboxylation and oxygenation take place
at the same catalytic site on RuBP carboxylase/oxygenase, the evidence presented above for a carbanion carboxylation mechanism leading to the 3-keto-CRBP intermediate (see Reaction 9) provides a
strong precedent for the mechanism of the oxygenation reaction. In
broad outline the two reactions are strikingly similar. In the oxygenation process (Reaction 10) catalyzed by purified RuBP carboxylase,
molecular oxygen, like CO_2, is added at C2 of RuBP to form one molecule of phosphoglycolate (from C1 and C2 of RuBP) and D-3-phospho-
glycerate (from C3, C4, and C5 of RuBP).

There is now compelling evidence that carboxylation and oxygenation occur at the same active site on the enzyme. Several facts
support this view: (i) Both the carboxylase and oxygenase activities
copurify during the isolation and purification of the carboxylase
(33), (ii) O_2 is a competitive inhibitor for CO_2 in the carboxylation
reaction and vice versa (34, 35), (iii) CRBP, an analogue of the
carboxylated intermediate in the carboxylation mechanism, is a potent
inhibitor of both the carboxylation and oxygenation reactions (8, 10,
35), and (iv) CO_2 (in the presence of Mg^{2+}) activates both the carboxylase and the oxygenase activities of the enzyme (24, 26).

It has been established by Tolbert and his colleagues (37) in
experiments with $^{18}O_2$ that one atom of ^{18}O is incorporated into the
carboxyl group of phosphoglycolate. ^{18}O was not incorporated, however, into the other reaction product, 3-phosphoglycerate. Evidence
was also obtained in experiments with $H_2^{18}O$ that one atom of ^{18}O was
incorporated into the carboxyl group of 3-phosphoglycerate (37); although ^{18}O was also found in the carboxyl group of phosphoglycolate,
it is presumed to have arisen via rapid $H_2^{18}O$ exchange with the keto
carbonyl oxygen of RuBP, for which there is precedent (37).

The formulation outlined in Reaction 10 accounts for all known
facts regarding the oxygenase-catalyzed reaction and the mechanistic
precedents set by our investigations of the carboxylation process.
It is visualized that the carbanion at carbon-2 of RuBP attacks O_2,
giving rise to a hydroperoxide intermediate. Presumably the enzyme
would then effect the hydrolytic cleavage of this intermediate (Reaction 10), as it does the carboxylated intermediate (Reaction 9). This
would lead to the concerted elimination of OH^- and formation of the
phosphoglycolate and phosphoglycerate products.

One aspect of this mechanism is disturbing, i.e., the apparent
lack of involvement of a transition metal. In general, the addition
of molecular oxygen to organic substrates requires the participation
of a transition metal (38). There are some exceptions, however, notably the addition of O_2 to dihydroriboflavin and its derivatives.
It should be recalled that almost 9 years ago we reported (9) the
presence of 1 gram atom of tightly-bound copper (Cu^{2+}) per mole
(560,000 g) of RuBP carboxylase. The Cu^{2+} could be removed from the
enzyme with no loss of carboxylase activity, and the copper-free enzyme re-bound Cu^{2+} readily (8). Several reports have since appeared
(37-39) which suggest that the copper-free enzyme is active in
both the carboxylation and oxygenation reactions. Unlike our study
(8), however, these studies were not done with particular care to
use conditions such that assay mixtures contained a defined amount
of metal ion less than stoichiometric with the amount of "metal-free"
enzyme added. Therefore, the question of whether a transition metal
ion is involved in the oxygenation reaction must remain open.

SUMMARY

RuBP carboxylase-oxygenase appears to catalyze carboxylation and oxygenation by homologous mechanisms. A common binding site exists on the enzyme for the acceptor substrate, RuBP. A mechanism is proposed whereby RuBP is isomerized, and a carbanion is generated at C2. Then, either CO_2 or O_2 is added as an electrophile at C2 to form the corresponding 3-keto-2-carboxy-RBP or 3-keto-2-hydroperoxy-RBP adduct. Hydrolytic cleavage at the C2-C3 bonds of these intermediates by the enzyme is envisioned to produce 2 molecules of 3-phosphoglycerate in the carboxylation sequence and 1 molecule of phosphoglycolate and 1 molecule of 3-phosphoglycerate in the oxygenation sequence. Further work will be necessary to establish the validity of the proposed mechanism.

REFERENCES

1. Calvin, M. and Benson, A. A., Science 107, 476 (1948).
2. Siegel, M. I., Wishnick, M., and Lane, M. D., in The Enzymes, 3rd ed., Vol. 6, pp. 169-92, Academic Press, New York, 1972.
3. Quayle, J. R., Fuller, R. C., Benson, A. A., and Calvin, M., J. Am. Chem. Soc. 76, 3610 (1954).
4. Weissbach, A., Smyrniotis, P. Z., and Horecker, B. L., J. Am. Chem. Soc. 76, 3611 (1954).
5. Paulsen, J. M. and Lane, M. D., Biochemistry 5, 2350-7 (1966).
6. Rutner, A. C. and Lane, M. D., Biochem. Biophys. Res. Commun. 28, 531-7 (1967).
7. Rutner, A. C., Biochem. Biophys. Res. Commun. 39, 923-9 (1970).
8. Wishnick, M., Lane, M. D., and Scrutton, M. C., J. Biol. Chem. 245, 4939-47 (1970).
9. Wishnick, M., Lane, M. D., Scrutton, M. C., and Mildvan, A. S., J. Biol. Chem. 244, 5761-3 (1969).
10. Siegel, M. I. and Lane, M. D., Biochem. Biophys. Res. Commun. 48, 508-16 (1972).
11. Nishimura, M. and Akazawa, T., Biochem. Biophys. Res. Commun. 59, 584-90 (1974).
12. Chan, P. H., Sakano, K., Singh, S., and Wildman, S. G., Science 176, 1145-6 (1972).
13. Baker, T. S., Eisenberg, D., Eiserling, F. A., and Weissman, L., J. Mol. Biol. 91, 391-9 (1975).
14. Baker, T. S., Eisenberg, D., and Eiserling, F., Science 196, 293-5 (1977).
15. Baker, T. S., Suh, S. W., and Eisenberg, D., Proc. Natl. Acad. Sci. USA 74, 1037-41 (1977).
16. Müllhofer, G. and Rose, I. A., J. Biol. Chem. 240, 1341-6 (1965).
17. Bassham, J. A., Adv. Enzymol. 25, 39-117 (1963).

18. Edsall, J. T., CO_2: Chemical, Biochemical, and Physiological
     Aspects, pp. 15-27, NASA Publ. SP-188, 1968.
19. Cooper, T. G., Filmer, D., Wishnick, M., and Lane, M. D.,
     J. Biol. Chem. 244, 1081-3 (1969).
20. Badger, M. R. and Andrews, T. J., Biochem. Biophys. Res. Commun.
     60, 204-10 (1974).
21. Andrews, T. J., Badger, M. R. and Lorimer, G. H., Arch. Bio-
     chem. Biophys. 171, 93-103 (1975).
22. Laing, W. A., Ogren, W. L., and Hageman, R. H., Biochemistry
     14, 2269-75 (1975).
23. Lorimer, G. H., Badger, M. R., and Andrews, T. J., Biochemistry
     15, 529-36 (1976).
24. Laing, W. A. and Christeller, J. T., Biochem. J. 159, 563-70
     (1976).
25. Badger, M. R. and Lorimer, G. H., Arch. Biochem. Biophys. 175,
     723-9 (1976).
26. Miziorko, H. M. and Mildvan, A. S., J. Biol. Chem. 249, 2743-
     50 (1974).
27. Norton, I. L., Welch, M. H., and Hartman, F. C., J. Biol. Chem.
     250, 8062-8 (1975).
28. Mildvan, A. S., Annu. Rev. Biochem. 43, 377-80 (1974).
29. Krishnaswamy, P. R., Pamiljans, V., and Meister, A., J. Biol.
     Chem. 237, 2932-40 (1962).
30. Calvin, M., Federation Proc. 13, 697 (1954).
31. Siegel, M. I. and Lane, M. D., J. Biol. Chem. 248, 5486-98
     (1973).
32. Andrews, T. J., Lorimer, G. H., and Tolbert, N. E., Biochemistry
     12, 11-18 (1973).
33. Bowes, G. and Ogren, W. L., J. Biol. Chem. 247, 2171-6 (1972).
34. McFadden, B. A., Biochem. Biophys. Res. Commun. 60, 312-17
     (1974).
35. Ryan, F. J. and Tolbert, N. E., J. Biol. Chem. 250, 4234-8
     (1975).
36. Lorimer, G. H., Andrews, T. J., and Tolbert, N. E., Biochemis-
     try 12, 18-23 (1973).
37. Peisach, J., Aisen, P., and Blumberg, W. E., The Biochemistry
     of Copper, pp. 339-405, Academic Press, New York, 1966.
38. Chollet, R., Anderson, L. L., and Hovsepian, L. C., Biochem.
     Biophys. Res. Commun. 64, 97-107 (1975).

MAGNETIC RESONANCE STUDIES ON RIBULOSE BISPHOSPHATE CARBOXYLASE

Henry M. Miziorko

Department of Biochemistry, Medical College of Wisconsin
Milwaukee, Wisconsin 53226

Experiments have been initiated aimed at elucidating the role of the divalent cation required in the RuBP carboxylase-catalyzed reaction. Metal ion is clearly required for activation of the enzyme. However, there is some question concerning an additional role for metal in the catalytic process. Past studies have shown that RuBP and analogous compounds decrease the enhancement of the water proton relaxation rate which arises from binding of a paramagnetic divalent cation, Mn^{2+}, to enzyme. Consider, for example, the titration of a RuBP carboxylase-Mn^{2+}-CO_2 solution with the transition state analog carboxyribitol bisphosphate. To account for the decrease in observed enhancement from an initial value of 18 to a final value of 1, displacement of water molecules bound in the inner coordination sphere of Mn^{2+} may be invoked. Such displacement could be accomplished if a ligand from CRBP or enzyme replaced a water molecule. Another explanation for the data would be that binding of CRBP results in an altered enzyme conformation which leaves Mn^{2+} with inner-sphere water ligands that can no longer rapidly exchange with the aqueous medium.

Preliminary results of recent electron paramagnetic resonance (EPR) experiments, which permit direct observation of enzyme-bound Mn^{2+}, allow the preceding list of viable explanations to be shortened. A sample of enzyme-Mn^{2+}-CRBP, separated from free Mn^{2+} and CRBP by Sephadex G-75 chromatography, yields an EPR spectrum due to bound Mn^{2+} which is quite different from the simple six-line spectrum of aqueous Mn^{2+}. Elements of multiple sextets, spread to either side of an aqueous Mn^{2+} EPR signal, are visible. This is accounted for by postulating increased zero-field splitting, which would be expected if the symmetry of enzyme-bound Mn^{2+} is distorted. In a control experiment, the EPR spectrum of an identical sample which had been denatured by 1% sodium dodecylsulfate showed none of the addi-

41

tional fine structure demonstrated above. Thus, disruption of the enzyme-Mn^{2+}-CRBP complex restores Mn^{2+} to an unstrained environment. Distortion of the symmetry of bound Mn^{2+} is most simply explained by insertion of a new ligand or of several new ligands in the inner coordination sphere of Mn^{2+}. Either CRBP or enzyme could be the ligand donor. If CRBP provides a new inner-sphere ligand, there is a strong argument for assigning the divalent cation a role in catalysis as well as in activation. If enzyme provides the ligand, the interpretation becomes less obvious, but this does not preclude a function for the cation in catalysis. Additional experimentation is in progress which will provide refined spectral data and which may allow assignment of the inner-sphere ligands of the divalent cation involved in the RuBP carboxylase-catalyzed reaction.

REGULATION OF PHOTOSYNTHETIC CARBON ASSIMILATION

D. A. Walker and S. P. Robinson

Department of Botany, University of Sheffield
Sheffield S10 2TN, England

When whole tissues or isolated chloroplasts are brightly illuminated they do not immediately commence to assimilate carbon at
maximal rates. Instead there is an initial lag or induction period
which may persist for several minutes (1). This is perhaps the best
known and most readily observed example of photosynthetic regulation.
Clearly the chloroplasts are potentially capable of rapid photosynthesis because they soon begin to evolve O_2 and fix CO_2 at high rates.
Equally clearly this potential ability is slowed or regulated during
the first few minutes of illumination. What is the nature and function of this regulation? Osterhout and Haas (2), who first observed
induction at Woods Hole in 1918, suggested two possible causes.
Either the lag represented a period during which substrates were
built up to the level required for full activity or else the catalysts concerned might be activated in the light. Sixty years later
there is little that can be added to these statements in principle,
except of course the inevitable notion that it might be both. This
article deals with these possibilities in regard to ribulose bisphosphate (RuBP) carboxylase and in particular to the role of orthophosphate in metabolic regulation.

LIGHT ACTIVATION

The concept of light activation or dark deactivation of RuBP
carboxylase springs from experiments such as those by Bassham and
Jensen (3) which showed that RuBP decreased in the dark in Chlorella
and in isolated spinach chloroplasts but not at a rate commensurate
with the quantity of RuBP present or to the extent predicted by the
$\Delta F'$ value (about -8 kcal) of the carboxylase reaction. These findings would be readily explained if the enzyme changed its character-

istics in the dark towards those often reported in the past, i.e.,
if it assumed a more or less inactive form with such a low affinity
for CO_2 that it was even difficult to see how it could fill its
postulated role in vivo (4). Today there seems little doubt that
the affinity of the fully activated carboxylase for CO_2 and its
maximum velocity are adequate to maintain the rates of photosynthesis
displayed by the parent tissue (5). What is less certain is the extent to which the carboxylase is fully activated in vivo (6) and the
precise manner in which the activity of the enzyme is governed. Full
activation in vitro is achieved by preincubation with Mg^{2+} and CO_2
(7, 8). In C_3 plants, the $[CO_2]$ within the stroma is not likely to
be less in the dark than in the light, and the major changes in the
stroma upon illumination appear to be increased pH, increased $[Mg^{2+}]$,
and a more reducing environment (8). Here again there are many uncertainties including the concentration and degree of binding of
residual RuBP and transported Mg^{2+}. Nevertheless in experiments
with the reconstituted chloroplast system it is possible to achieve
70 to 90% of full activity by increasing exogenous $[Mg^{2+}]$ from 1 mM
to 2 to 5 mM (8). Similarly CO_2 fixation by intact chloroplasts is
inhibited by an ionophore that facilitates Mg^{2+} exchange across the
envelope and is restored by the addition of exogenous Mg^{2+} (9).

Lorimer and his colleagues have shown (7) that Mg^{2+} activation
of carboxylase preincubated with bicarbonate is a very rapid process
and so, of course, is the influx of protons into the thylakoid compartment and the associated efflux of Mg^{2+} (10). Therefore, if Mg^{2+}
and increased pH are the major factors in light activation, it seems
improbable that the lags in CO_2 fixation normally observed (1) and
the very long lags that may be observed in the presence of high
$[P_i]$ (1) can be associated with light activation of the carboxylase.
Note that high $[P_i]$ inhibits the action of the carboxylase (11, 12)
but not its activation (12) (i.e., high $[P_i]$ would be more likely
to depress the final rate of photosynthesis than to extend the lag).
On the other hand, P_i does not enter the chloroplast freely, except
by exchanging with compounds such as triose phosphate via the P_i
translocator (13), and even in the presence of high external $[P_i]$
the increase in stromal $[P_i]$ is perhaps unlikely, in itself, to
cause prolonged lag extension by inhibition of the carboxylase.
Thus McNeil (12) has shown that a $[P_i]$ of 20 mM will still permit a
rate >100 μmoles CO_2 per mg chlorophyll (Chl) per hr in the standard assay (14), whereas it would normally bring about complete inhibition of photosynthesis by isolated chloroplasts (1).

If dark deactivation of carboxylase activity is real and complete (3, 6), then its function could be conservation of RuBP.
This compound does not cross the chloroplast envelope. Its persist-
ence, even at a relatively low concentration in the dark, would ensure prompt resumption of photosynthesis in the light. In the absence of separate control mechanisms, synthesis of RuBP by reactions
other than the reductive pentose phosphate pathway (e.g., from
starch) followed by carboxylation in the dark could also prove to
be a nonproductive drain on ATP. [Involvement of P_i in light ac-

tivation of the carboxylase has recently been investigated by Heldt
and Lorimer and is discussed elsewhere in this Symposium (15).]

ACCUMULATION OF INTERMEDIATES

     In our view neither induction itself, nor extension of the lag
by P_i, nor reversal of P_i inhibition by compounds such as triose
phosphates (1), can be explained by light activation of the carboxylase as such, even though this process might be a contributory factor.   Conversely, RuBP apparently falls to a relatively low level
in the dark (3) and increases in a significant fashion (Figure 1)
during the lag period.   It is therefore reasonable to ask whether
or not photosynthetic carbon assimilation and its associated O_2
evolution will start very rapidly if RuBP is supplied at an adequate
concentration.   If we accept that the reconstituted chloroplast system is, at least in many regards, similar to an immense chloroplast
with an immediately accessible stromal compartment, then the answer
is "yes," provided there is also a favorable ATP/ADP ratio (Figure 2).   This qualification, we believe, may go to the heart of induction and the crucial role of P_i in photosynthetic regulation (8).

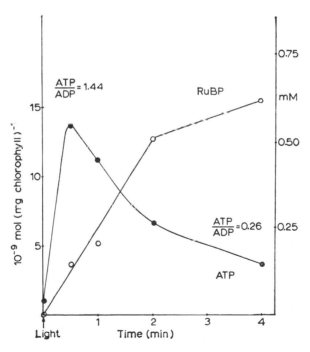

Figure 1.   Change in [ATP] and increase in [RuBP]
during induction.   From Lilley et al. (16).

THE ROLE OF ORTHOPHOSPHATE

It has been proposed (17) that the chloroplast is not a self-sufficient photosynthesizing organelle capable of catalyzing the process summarized by equation 1,

$$CO_2 + H_2O \xrightarrow{h\nu} CH_2O + O_2 ,$$
(1)

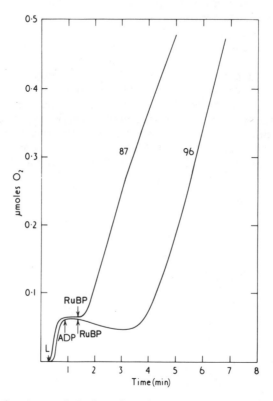

Figure 2. Simulation of induction in reconstituted chloroplast system, showing very short lag with RuBP as a substrate (despite need to build up PGA) and pronounced lag extension by additional ADP. Each reaction medium contained 330 mM sorbitol, 1 mM EDTA, 10 mM KCl, 50 mM Hepes (pH 8.0), 1 mM P_i, 10 mM $MgCl_2$, 10 mM $NaHCO_3$, 4 mM isoascorbate, 1 mM dithiothreitol, 0.1 μmole NADP, 0.2 μmole ADP, 220 unit catalase, 200 μg spinach ferredoxin, and broken chloroplasts plus stromal extract equivalent to 100 μg chlorophyll in a total volume of 1 ml. After the added NADP was reduced, 1 μmole of RuBP was added. In the lower trace, 1 μmole of ADP was added prior to RuBP, resulting in a lag of 3 min before the maximum rate of oxygen evolution was achieved. Rates of oxygen evolution (μmoles per mg Chl per hr) are given alongside the traces.

but rather a P_i-consuming system which produces triose phosphate according to equation 2:

$$3CO_2 + 3H_2O + P_i \rightarrow \text{triose phosphate} + 3O_2 . \tag{2}$$

Conversely, starch synthesis within the stroma is adequately summarized by equation 1, but this process (in which P_i is recycled) accounts for only a fraction (often <25%) of maximal CO_2 fixation. This concept is based on the fact that the major products of photosynthetic carbon assimilation in isolated C3 chloroplasts are sugar phosphates (16, 18) (plus some PGA) and that photosynthesis soon falls to the low level commensurate with starch synthesis if no P_i is added (17, 19). Photosynthesis is then restored by the addition of P_i (or some compound such as PP_i that yields P_i on external hydrolysis), and in the short term small quantities of exogenous P_i bring about the evolution of oxygen in accord with the stoichiometry of equation 2 (17, 19). The isolated chloroplast has a pool of phosphate amounting to 0.4 to 1.5 μmole/mg Chl (16, 20). The low direct permeability of the chloroplast envelope to most phosphorylated compounds and the strict counterexchange nature of the phosphate transporter effectively prevent major changes in the chlo-

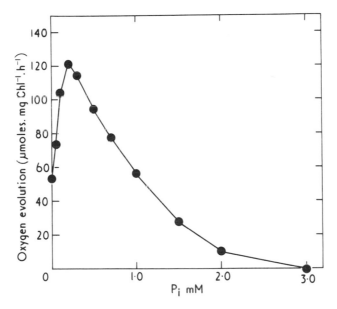

Figure 3. Photosynthesis by isolated pea chloroplasts as a function of $[P_i]$. Similar curves can be obtained with chloroplasts from spinach, wheat, and sunflower. Oxygen evolution was measured under standard conditions (21).

roplast phosphate pool (Figure 4). Only counterexchange of polyphos-
phates--e.g., PP_i/P_i or ATP/ADP (21, 22)--or unidirectional transport
of phosphates could alter the size of the chloroplast phosphate pool,
and both these processes appear to be relatively slow. Similarly,
it is now accepted that triose phosphate is normally the major transport metabolite exported from the chloroplast via the P_i translocator
(13) and that equation 2 also represents the principal exchanges a-
cross the envelope (i.e., one P_i enters for every triose phosphate
exported). The extension of the lag by high $[P_i]$ was interpreted
in this way even before the P_i translocator was formulated. Thus
it was proposed (23) that "A direct obligatory exchange between
orthophosphate (outside) and sugar phosphates (inside) could account
for the inhibition of photosynthesis by orthophosphate and its reversal by sugar phosphates."

     If photosynthesis is slowed by too little P_i or by too much
(Figure 3), it is also reasonable to ask how these changes in $[P_i]$
are "perceived" within the stroma and what relevance they have, if
any, to in vivo regulation. Our present view (which relates to that
expressed in the preceding section) is that the controlling sequence
is the reduction of PGA to DHAP. The first stage of this process is
the freely reversible reaction (catalyzed by PGA kinase) which, by
mass action, is readily inhibited by ADP (8, 24, 25). The relevance
of this inhibition to induction and regulation of levels of cycle
intermediates is discussed below.

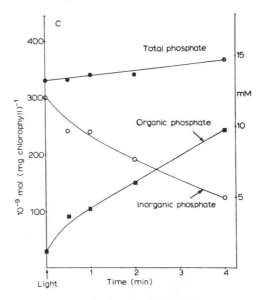

Figure 4.   Inverse relationship between $[P_i]$ and [organic phosphate]
     in spinach chloroplast stroma.   From Lilley et al. (16).

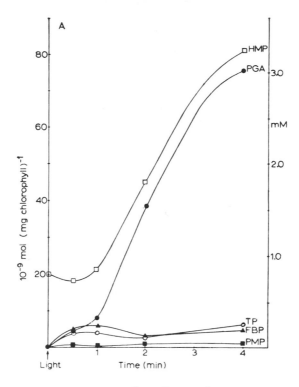

Figure 5. Increase in stromal [PGA] and [hexose monophosphates]
during induction. From Lilley et al. (16). (In this experiment
and in those of Figures 1 and 4, the induction period lasted about
3 min.) TP, triose phosphate; FBP, fructose 1,6-bisphosphate; PMP,
pentose monophosphates.

INDUCTION AND REGULATION IN RELATION TO ATP/ADP RATIOS

Early measurements of the distribution of $^{14}CO_2$ between cycle
intermediates (in reaction mixtures containing chloroplasts in either
the inductive or steady-state phase of photosynthesis) suggested that
pentose monophosphates might be the "pace-setter" in induction (only
pentose phosphates displayed kinetics similar to those for CO_2 fixation). Since 1967, however, it has been possible to prepare more
active chloroplasts and to investigate the distribution of metabolites between chloroplast and medium by using ^{32}P. The results (16)
indicate that there is a sharp decline in P_i during induction and a
corresponding rise in organic phosphates, whereas the total phosphate
within the chloroplast remains more or less constant (Figure 4). The
largest rises occurred in hexose monophosphates, 3-phosphoglycerate
(Figure 5), and ribulose 1,5-bisphosphate (Figure 1), although the
absolute concentration of RuBP was much smaller than that of the

other two.   The ratio [ATP]/[ADP] first rose steeply to 1.44 and then
declined to 0.26 in the steady state (16).   It seems clear, therefore, that during the lag the CO_2 acceptor is built up to its full
steady-state concentration.   During this period ATP is formed faster
than it is consumed so that the [ATP]/[ADP] ratio rises abruptly, a
change that would favor the conversion of PGA to DPGA, triose phosphate, and pentose monophosphate.   As more ribose 5-phosphate (R5P)
becomes available for conversion to RuBP, however, the rate of ATP

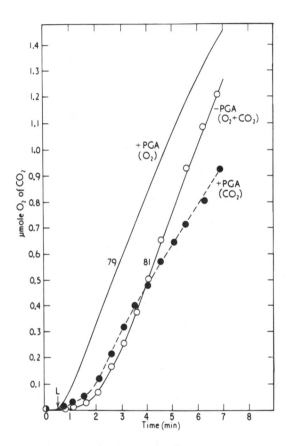

Figure 6.   Simultaneous measurement of oxygen evolution and CO_2
fixation by intact spinach chloroplasts in the presence and absence
of PGA.   The reaction medium contained 330 mM sorbitol, 2 mM EDTA,
1 mM $MgCl_2$, 1 mM $MnCl_2$, 50 mM Hepes (pH 7.6), 0.5 mM P_i, 220 unit/
ml catalase, 10 mM $NaH^{14}CO_3$ (15 Ci/mole), and chloroplasts equivalent to 200 μg chlorophyll in a total volume of 2 ml.   One reaction
mixture also contained 2 mM PGA.   Oxygen evolution is shown by continuous lines and incorporation of $^{14}CO_2$ into acid-stable products
by circles.   Although the lag in oxygen evolution is virtually eliminated by PGA, the lag in CO_2 fixation is only slightly decreased.

consumption rises and the [ATP]/[ADP] ratio falls, and this tends
to slow the conversion of PGA to pentose monophosphate. As these
trends continue, more and more PGA is needed to overcome the unfavorable equilibrium position of the PGA → DPGA reaction.

This interpretation is favored by several lines of evidence.

(a) Although PGA virtually eliminates the lag in O_2 evolution,
a lag in CO_2 fixation can still be demonstrated, which indicates
that a finite time must elapse before PGA reduction can fill the
RuBP pool (Figure 6).

(b) If the lag is artificially extended by high P_i, subsequent
addition of PGA causes an immediate onset of CO_2 fixation (26). This
would be expected if there had been some buildup of RuBP during this
time and any light activation of enzymes had gone to completion.
(Excessive prolongation of induction may involve some inhibition of
RuBP carboxylase by high P_i which would be reversed on addition of
PGA as a result of P_i efflux via the translocator).

(c) Feeding R5P to whole chloroplasts in the absence of exogenous P_i inhibits CO_2-dependent and PGA-dependent O_2 evolution, and

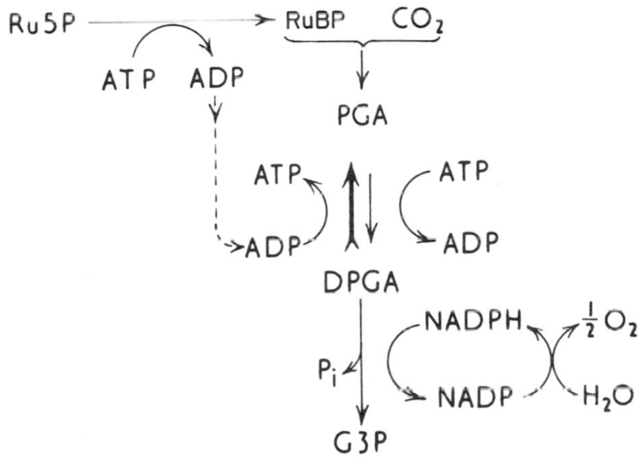

Figure 7. Diagramatic representation of inhibition of O_2 evolution
by an ATP sink. In this instance ribulose 5-phosphate → RuBP
consumes ATP and the resulting ADP inhibits PGA → 2,3-diphosphoglyc-
erate (DPGA).

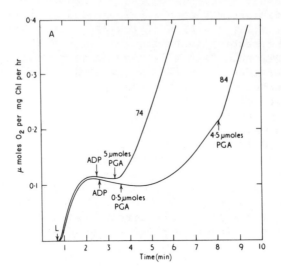

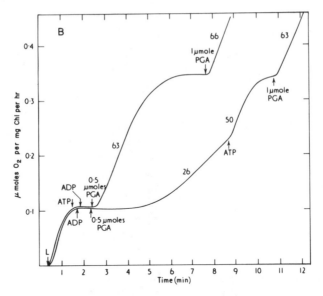

Figure 8.   Oxygen evolution in reconstituted chloroplasts by PGA.
Reaction mixture as for Figure 2, with NaHCO$_3$ omitted. P$_i$ was
2 mM and NADP, 0.2 mM.   (A) Reversal of ADP inhibition by increased
concentrations of PGA.   ADP (1 mM) added as indicated.   (B) Reversal
of ADP inhibition by ATP.   ATP (5 mM) was added for upper trace and
ADP (1 mM) for both traces as indicated.

this inhibition is reversed by addition of P_i (27). Such an inhibition would result if conversion of R5P to RuBP produced a sufficiently unfavorable [ATP]/[ADP] ratio to block conversion of PGA
to DPGA (Figure 7).

(d) In the reconstituted chloroplast system, addition of ADP,
or a sink for ATP such as R5P (Figure 7) or glucose plus hexokinase,
inhibits PGA-dependent O_2 evolution (8, 25). Such inhibition ter-
minates spontaneously as a consequence of photophosphorylation or
upon the addition of an ATP-generating system such as creatine phosphate plus its kinase (8, 25).

(e) Phosphoglycerate is probably exported as PGA^{2-}, and the
alkaline pH of the illuminated stroma therefore allows the accumulation of high concentrations of PGA despite the ease with which
this compound will traverse the envelope at more acid pH values (16).

(f) The rate of PGA reduction by chloroplast extracts is dependent on the ratio [PGA] [ATP]/[ADP] within the physiological
range of concentrations of these metabolites. During PGA-dependent
oxygen evolution by reconstituted chloroplasts, the lag produced
by addition of ADP can be largely eliminated if this ratio is sufficiently increased by the addition of PGA (Figure 8A) or ATP (Figure 8B). The activity of chloroplast PGA kinase in the direction
of DPGA formation seems to be controlled by mass action (cf. refs.
8, 24, 25), i.e., by the concentrations of its substrates and products. In the presence of NADPH, DPGA is reduced to triose phosphate and the steady-state concentration of DPGA is likely to be
low. Thus it is the concentrations of PGA, ATP, and ADP which determine the rate of PGA reduction. Figures 9 and 10 show the results
of an experiment with the reconstituted chloroplast system designed
to mimic the conditions of the intact chloroplast during induction.
After the added NADP had all been reduced, ATP, ADP, and PGA were
added simultaneously to give varying concentrations of PGA and varying ATP/ADP ratios with a fixed total concentration of adenine nucleotides (1.5 mM). As shown in Figure 9, the maximum rate of oxygen evolution achieved was dependent on the concentration of PGA and
on the ATP/ADP ratio. As this ratio was decreased, higher concentrations of PGA were required to reach a given rate of oxygen evolution.
The time taken to reach the maximum rate is plotted as a function of
PGA in Figure 10. When all of the adenine nucleotides were added as
ATP (ATP/ADP > 100), the lag was <0.2 min. The lag increased as the
ATP/ADP ratio was decreased, but for a given ratio it was dependent
on the PGA concentration. Thus both the rate of oxygen evolution
and the time taken to reach that rate were dependent on the ratio
[PGA] [ATP]/[ADP].

It is therefore suggested that at the onset of illumination
there is a rapid activation of the carboxylase and PGA starts to
accumulate. This leads to increased formation of triose phosphate
and, as a result of the regenerative reactions of the cycle, RuBP
climbs to its steady-state level and induction ends. During steady-
state photosynthesis, triose phosphate is rapidly exported to the
cytoplasm in exchange for P_i, which is normally made available by

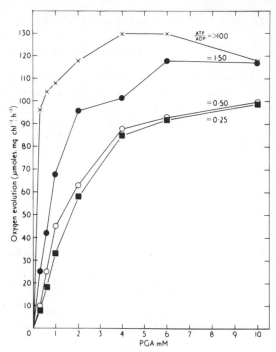

Figure 9. Effect of ATP/ADP ratio on the rate of PGA-dependent oxygen evolution by reconstituted chloroplasts at different PGA concentrations. Reaction mixtures as for Figure 8 except that ADP was initially moitted. ATP, ADP, and PGA were added together, once the NADP had been reduced. The rate of oxygen evolution was dependent on ATP/ADP ratio and on PGA concentrations.

processes such as sucrose synthesis. Should sink activity be diminished and the supply of P_i curtailed, however, less triose phosphate would be exported (8). The combination of high [triose phosphate], high [PGA], and low [P_i] is precisely what is required for allosteric activation of ADP glucose pyrophosphorylase (28) so that peak activity of this key enzyme in starch synthesis coincides with maximal availability of starch precursors (29). [There is now ample evidence that low [P_i] favors starch synthesis by isolated chloroplasts and that sequestration of cytoplasmic P_i by mannose favors starch synthesis in vivo (8)]. Curtailment of P_i, however, also tends to slow assimilation because, if ribulose 5-phosphate accumulates, it will inhibit its own formation (as noted above) by acting as an ATP sink and thereby slowing PGA reduction (Figure 7). Our view of these facts is that photosynthesis in low [P_i] is not curtailed by diminution of photophosphorylation (which would largely be independent of [P_i] except at very low concentrations) but by the ADP inhibition of triose phosphate formation. We do not regard [P_i] as a physiological control mechanism, but its inhibitory action

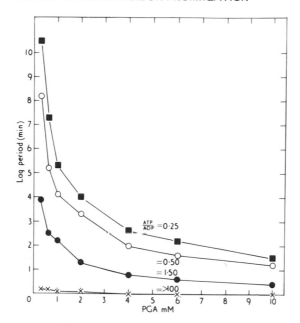

Figure 10. Time taken to reach maximum rate as a function of PGA concentration and ATP/ADP ratio. Data from
the experiment of Figure 9. The lag is increased by low
ATP/ADP and lower PGA concentration.

in vitro illustrates what happens when the delicate balance between
export and internal consumption (and therefore the equally delicate
balance between organic phosphate and P_i within the stroma) is upset.
Enforced export of intermediates ensues, and photosynthesis is inhibited. Restoration of activity can then be achieved only by the
addition of intermediates which replace those lost to the exterior
or prevent further loss by competing with P_i for uptake via the P_i
translocator (8). In vivo the control by $[P_i]$ may be further modulated by cytoplasmic metabolites (PGA, DHAP, etc.) just as PP_i ame-
liorates P_i inhibition in vitro (8).

SUMMARY

     It may be concluded that the conversion of PGA to DPGA plays
a key role in induction and in the regulation of cycle activity. The
high concentrations of PGA in actively photosynthesizing chloroplasts
reflect this role and the control exerted by adenylate ratios. Thus
the cycle can operate at its maximum rate only in the presence of
high PGA and low ribulose 5-phosphate concentrations.

Once induction is complete, the reductive pentose phosphate pathway will continue to function at its maximum rate if sink activity within the cytoplasm makes available sufficient P_i to support rapid export of triose phosphate. If triose phosphate tends to build up in the stroma, it will favor pentose monophosphate accumulation. A relative excess of ribulose 5-phosphate would, in turn, inhibit PGA reduction (and hence its own formation) by drawing too heavily on the available ATP.

REFERENCES

1. Walker, D. A., in The Intact Chloroplast, pp. 235-78, J. Barber, Editor, Elsevier, Amsterdam, 1976.
2. Osterhout, W. J. V. and Haas, A. R. C., J. Gen. Physiol. 1, 1-16 (1918).
3. Bassham, J. A. and Jensen, R. G., in Harvesting the Sun, pp. 79-110, A. San Pietro et al., Editors, Academic Press, New York, 1967.
4. Walker, D. A. and Lilley, R. McC., in Proc. 50th Annu. Meet. Soc. Exp. Biol., Cambridge, pp. 189-98, 1974.
5. Lilley, R. McC. and Walker, D. A., Plant Physiol. 55, 1087-92 (1975).
6. Bahr, J. T. and Jensen, R. G., Arch. Biochem. Biophys. 185, 39-48 (1978).
7. Lorimer, G. H., Badger, M. S., and Andrews, T. J., Biochemistry 15, 529-36 (1976).
8. Walker, D. A., Curr. Top. Cell. Regul. 11, 203-41 (1976).
9. Portis, A. R. Jr. and Heldt, H. W., Biochim. Biophys. Acta 449, 434-46 (1976).
10. Hind, G., Nakatani, H. Y., and Izawa, S., Proc. Natl. Acad. Sci. USA 71, 1484-8 (1974).
11. Paulson, J. M. and Lane, M. D., Biochemistry 5, 2350-7 (1966).
12. McNeil, P. H., Unpublished.
13. Fliege, R., Flügge, U-I., Werdan, K., and Heldt, H. W., Biochim. Biophys. Acta, in press (1978).
14. Lilley, R. McC. and Walker, D. A., Biochim, Biophys. Acta 386, 226-9 (1974).
15. Lorimer, G. H. et al., See paper in this Symposium.
16. Lilley, R. McC., Chon, C. J., Mosbach, A., and Heldt, H. W., Biochim. Biophys. Acta 460, 259-72 (1977).
17. Walker, D. A. and Herold, A., Plant Cell Physiol. Special Issue, 295-310 (1977).
18. Walker, D. A., in Proc. NATO Adv. Study Inst. Aberystwyth, 1965, Vol. 2, pp. 53-69, T. W. Goodwin, Editor, Academic Press, New York, 1966.
19. Cockburn, W., Baldry, C. W., and Walker, D. A., Biochim. Biophys. Acta 131, 594-6 (1967).
20. Hall, D. O., pp. 135-70 in ref. 1.
21. Robinson, S. P. and Wiskich, J. T., Arch. Biochem. Biophys. 184, 546-54 (1977).

22. Robinson, S. P. and Wiskich, J. T., Plant Physiol. $\underline{59}$, 422-7
     (1977).
23. Walker, D. A. and Crofts, A. R., Annu. Rev. Biochem. $\underline{39}$, 386-
     428 (1970).
24. Slabas, A. R. and Walker, D. A., Biochem. J. $\underline{154}$, 185-92 (1976).
25. Slabas, A. R. and Walker, D. A., Biochim. Biophys. Acta $\underline{430}$,
     154-64 (1976).
26. Cockburn, W., Walker, D. A., and Baldry, C. W., Biochem. J.
     $\underline{107}$, 89-95 (1968).
27. Cockburn, W., Baldry, C. W., and Walker, D. A., Biochim.
     Biophys. Acta $\underline{143}$, 614-24 (1967).
28. Priess, J. and Kosuge, T., Annu. Rev. Plant Physiol. $\underline{21}$, 433-66
     (1970).
29. Heldt, H. W., Chon, C. J., Maronde, D., Herold, A., Stankovic,
     Z. S., Walker, D. A., Kraminer, A., Kirk, M. A., and Heber, U.,
     Plant Physiol. $\underline{59}$, 1146-55 (1977).

DISCUSSION

JENSEN: The values for the amounts of intermediates, determined by Dr. Heldt, were obtained by measuring the radiolabel in
the intermediates, were they not?

WALKER: Yes. The point you are about to make, which I concede
immediately, is that Figure 1, which shows the increase in RuBP,
because of the nature of the measurements suggests that the RuBP
starts out at a low level, close to zero. I am sure that this is
not the case, and that the chloroplast contains residual RuBP.
There is a lot of uncertainty about this in relation to the amount
of binding, etc. Lorimer will address this subject later. I don't
want to say that light activation of the carboxylase isn't involved
in the normal induction period--it may contribute to it--but certainly
many observations associated with induction can't be explained by
light activation of catalysts but can be explained by accumulation
of substrate. If one extends the lag period artificially by high
orthophosphate and then does precisely the sort of experiments that
you have done on light activation of the carboxylase, one finds
that activation is complete in the same time as for controls. We
find that the presence of high orthophosphate extends the lag but
does not affect the activation or slow the enzyme. Lorimer has seen
the same sort of thing in chloroplasts that showed a really extended
induction period (due to building up of intermediates) because they
were prepared from leaves that had not been preilluminated, but
showed fairly rapid carboxylase activation.

L. ANDERSON: The K_i(ADP) of chloroplast PGA kinase is tenfold
lower than the K_m(ADP). Don't you think then that mass action is
helped along by the peculiar catalytic properties of the enzyme?

WALKER: Yes I do. Everything that we have done leads us to
the conclusion that mass action is the principal factor in intermediate buildup. Certainly the rates can be changed considerably,

either raised by increasing the PGA or ATP concentration, or diminished by adding ADP, and so on. I see your point, but it certainly
is not resolved. Our view is that there is certainly a mass action
effect rather than an effect on catalysis as such.

L. ANDERSON: In our experiments we had saturating magnesium,
so that in terms of enzymology they were clearly sound. Did you
include dithiothreitol in the assay media in the reconstituted
chloroplast system? I think the effect of light activation is to
eliminate the lag period, perhaps in tandem with the system you
invoke here.

WALKER: We normally have dithiothrietol present, but in these
particular experiments we did not. Dithiothrietol certainly makes
a big difference in many of the reconstituted chloroplast experiments.

L. ANDERSON: Would you shorten the lag more if you included
dithiothrietol? Is the lag due both to an ADP-ATP effect and to a
light-activation effect? I ask because, in the old experiments of
Kalo and Gibbs, in which they inhibited with arsenite, the lag
seemed extremely long, and they never continued for hours to see
whether the rates became reasonable because by that time the chloroplasts were dead anyway. I wonder whether the role of light activation is to reduce the lag period, but perhaps other factors also
enter in.

WALKER: I would like to reiterate our main conclusion. If
you want to study induction, you have to ask what actual effect you
are observing, and you find immediately that the effect is not constant because the length of the induction period depends on a great
many factors such as composition of the reaction mixture and the
prehistory of the leaf. Material from leaves that have been actively
photosynthesizing until the point at which their chloroplasts are
isolated, incubated in a low orthophosphate medium, may show virtually
no induction, but material from leaves kept in the dark, so that sugar
phosphate pools are depleted, show long induction periods (up to 35
min). Similarly, one can artificially extend the induction period
with high phosphate. I believe that under most circumstances,
although not enough data are available for a definite conclusion,
the activation of the carboxylase as such will be shown to be a
more or less constant factor. Certainly when we extend the induction period with high phosphate, we find no corresponding slowing
of the activation. I am not disputing that light activation may
contribute to the first small induction period, but I still think
the major controlling device is the orthophosphate level in relation
to the sugar phosphate level: the way these interact through transport and the way they affect the PGA to triose phosphate step.

JENSEN: The measurement of photosynthesis in intact chloroplasts by O_2 evolution is a measure of PGA available for reduction.
Addition of inorganic phosphate would cause loss of PGA from the
chloroplast with an increase in the dark-to-light initial lag of
O_2 evolution. However, if sufficient RuBP is available for carboxyl-

ation, CO_2 fixation would be expected to operate as soon as the enzyme became active in the light. Are these expectations borne out in your experiment when you compare $^{14}CO_2$ fixation with O_2 evolution during the initial lag phase?

WALKER: The addition of high $[P_i]$ enforces export of intermediate (including PGA) from the chloroplast and, as a consequence, induction (i.e., both CO_2 fixation and O_2 evolution) can be extended more or less indefinitely as a function of $[P_i]$. If PGA is added at the outset of the illumination period, the P_i inhibition is reversed but O_2 evolution starts in advance of CO_2 fixation, which reaches its maximum rate only after 1 or 2 min. If, however, PGA is added after several minutes of illumination, there is an immediate and simultaneous onset of CO_2 fixation and O_2 evolution. McNeil at Sheffield has measured activation of RuBP carboxylase in intact illuminated chloroplasts by techniques similar to those used by Jensen and Bahr. He finds that high $[P_i]$ does not affect the activation of the carboxylase. Although carboxylase activation might conceivably contribute to induction, it seems clear that induction cannot be explained on this basis alone, and lag extension by high $[P_i]$ appears to be more or less independent of carboxylase activation.

PHOTORESPIRATION AND THE EFFECT OF OXYGEN ON PHOTOSYNTHESIS[*]

David T. Canvin

Department of Biology, Queen's University
Kingston, Ontario K7L 3N6, Canada

Photorespiration is the light-dependent, oxygen-sensitive CO_2 evolution from green leaves that originates from the metabolism of compounds in the glycolate pathway. It has been termed "an inevitable consequence of the existence of atmospheric oxygen" (1) and has been attributed to the ability of ribulose bisphosphate carboxylase to act also as an oxygenase catalyzing the reaction of oxygen with ribulose bisphosphate (RuBP) to yield phosphoglycolate (2), the precursor of the substrate for the glycolate pathway. In this reaction, oxygen not only produces the substrate for photorespiration but also competitively prevents the fixation of CO_2 (3, 4). It has been known for some time that the inhibition of photosynthesis by oxygen is comprised of two components, inhibition of true photosynthesis and stimulation of photorespiration (5, 6). With the discovery of the oxygenase activity of RuBP carboxylase, both these components were attributed to the effect of oxygen on the enzyme (4, 7), and the joint action of oxygen and CO_2 on the enzyme was proposed to be responsible for the regulation of soybean net photosynthesis (8). The oxygen concentration around a leaf--and in it, as the oxygen concentration is similar, (9)--can be quickly changed; and, if oxygen acts only on the oxygenase, one might expect a change in oxygen concentration to be accompanied by a corresponding rapid change in CO_2 fixation to a new steady rate. Few published results show the pattern of change of photosynthesis when the oxygen concentration is changed, but it seems to be widely believed that the change in CO_2 fixation after an oxygen concentration change is rapid and the new rate is steady. With a change from 21% O_2 to 2% O_2, the

_____

[*]Supported in part by the National Research Council of Canada.

expectation, depending on CO_2 concentration, would be either no change in the rate of CO_2 fixation or a rapid change to a stable higher rate. However, several reports (10-14) showed that, at high CO_2 concentrations, CO_2 fixation was inhibited when the oxygen concentration was lowered from 21% down to 1 to 3%. Some of the characteristics of inhibition of photosynthesis by low oxygen concentration have been identified and are reported here.

MATERIALS AND METHODS

Sunflowers (<u>Helianthus</u> <u>annuus</u> L. var. CM90RR) were grown as previously described (15) at light intensities of either 400 or 800 μeinstein·m^{-2}·sec^{-1}. Soybean [<u>Glycine</u> <u>max</u> (L.) Merr. var. Wayne] and cowpea [<u>Vigna</u> <u>unguiculata</u> (L.) Walp.] were grown as nodulated plants in sand culture with N-free mineral nutrient solution (16). Conditions in the growth chamber were temperatures of 25° (day) and 20° (night), light intensity of 400 μeinstein·m^{-2}·sec^{-1}, and day length of 16 hr. Castor oil plants (<u>Ricinus</u> <u>communis</u> L. var. Baker 296 Dwarf) were grown in soil in the glasshouse with temperatures ranging from 20° to 25° and natural light supplemented for 16 hr,

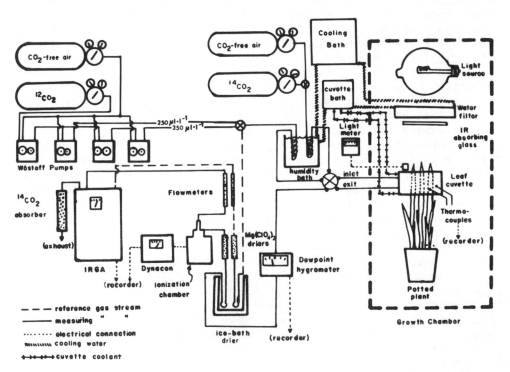

Figure 1.   A schematic of the gas circuit.   From McCashin (18).

including the day period, with 300 μeinstein·m^{-2}·sec^{-1} light from fluorescent lamps. Spinach (Spinacia oleracea L. var. Longstanding Bloomsdale Dark Green) was grown on vermiculite with nutrient solution (16). Conditions in the growth chamber were temperatures of 20° (day) and 15° (night), light intensity of 400 μeinstein·m^{-2}·sec^{-1}, and day length of 10 hr.

Measurements of photosynthesis were made in an open gas-exchange system (Figure 1) slightly modified from that described previously (15, 17). The plant chamber was fitted with water jackets, and circulating water from a constant-temperature bath was used to help control leaf temperature. Humidity was measured by a dew-point hygrometer (Model 880, EG&G, Environmental Equipment Division, Waltham, MA). Light was supplied to the leaf from a Sylvania 1000 W/BU-HOR metal arc lamp through a 4-cm water filter and an infrared filter (Corning glass 4602, Corning Glass Works, Corning, NY). Quantum flux densities for all experiments except the light intensity study was 650 μeinstein·m^{-2}·sec^{-1}. Other conditions are specified below with the results.

Gas mixtures of known composition were supplied to the leaf by precision gas-mixing pumps (19) from cylinders of CO_2 and of various oxygen concentrations in nitrogen. The latter cylinders were interconnected with Tygon tubing so that the oxygen concentration to the pumps could be instantaneously changed by closing the valve on one tank and opening that on another. With flow rates of 0.96 l·min^{-1} through the measuring circuit and 0.5 l·min^{-1} through the reference circuit, 30 to 40 sec were required for complete flushing of the system and detection of CO_2 fixation at the new oxygen concentration.

When studying the effect on CO_2 uptake immediately after a change in oxygen concentration, it is essential to avoid generation in the CO_2 measurements of perturbations or artifacts due to the changing procedure. Artifacts due to variations in flow rate to the leaf or in CO_2 concentration did not occur because both these factors were held constant by the precision gas-mixing pumps. Artifacts due to different oxygen concentrations in the measuring cell and the reference cell of the infrared gas analyzer (20) were avoided by adjusting the flow rate in the reference circuit so that both the measuring and the reference cells of the analyzer were flushed simultaneously with the new oxygen concentration. In the three upper traces of Figure 2, low flow rates were used to exacerbate effects that could arise from changes in oxygen concentration. When the new oxygen concentration flushes the reference cell first (trace 1) an apparent evolution of CO_2 occurs upon switching to 2% O_2 and an apparent uptake occurs upon switching to 50% O_2. The change due to 50% O_2 can be reversed by flushing the measuring cell first (trace 2). When the cells are flushed simultaneously (trace 3), and especially at the flow rates used in this study (trace 4), no change occurs in the trace when the oxygen concentration is changed.

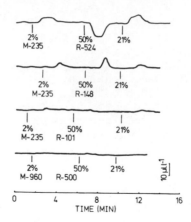

Figure 2. Recorder tracing of the effect of flow rate to the measuring and reference cells of the infrared gas analyzer on the CO_2
measurements when the oxygen concentration was changed. Traces referred to in descending order with trace 1 at top. All traces
start with 21% O_2. Inlet CO_2 was constant at 750 $\mu l \cdot l^{-1}$. Oxygen
concentrations were changed to the new values at times indicated.
M and R refer to the measuring and reference gas circuits, and the
numbers after them are the flow rates of gas in $ml \cdot min^{-1}$.

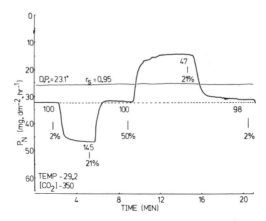

Figure 3. Recorder tracing of the effect of oxygen on photosynthesis of a sunflower leaf. Temperature was 29.2°, inlet $[CO_2]$ 350 $\mu l \cdot$
l^{-1}, and stomatal resistance, r_s, 0.95 $sec \cdot cm^{-1}$. The rate of photosynthesis, p_N, was initially 31.6 mg CO_2 $dm^{-2} \cdot hr^{-1}$ in 21% O_2, and
the oxygen concentration was changed to the new values at the times
shown by a vertical line with a percent figure. Other numbers next
to the trace given the rate of photosynthesis in that oxygen concentration as a percentage of the initial rate. The trace labeled D.P.=
23.1° is the dew point of the exit gas from the leaf chamber. Inlet
dew point was 8.6°C.

RESULTS

In sunflowers, rapid changes to a new steady rate of photosynthesis are observed after changes in the oxygen concentration at
29.2^o and 350 $\mu l \cdot l^{-1}$ CO_2 (Figure 3). When 2% O_2 replaced 21% O_2,
the change to a new steady rate, which was 145% of that in 21% O_2,
was almost a square-wave and was complete within 90 sec. Although
the change to a new steady rate was somewhat slower (requiring about
3.5 min) when the oxygen concentration was changed to 50%, the pattern was similar. The dew-point trace shows that transpiration or
stomatal aperture was not affected by change in oxygen concentration, which confirms earlier results (21, 22).

In the castor oil plant, however, at 20.5^o and 750 $\mu l \cdot l^{-1}$ CO_2,
a change in the oxygen concentration from 21% to 2% resulted in 30%
inhibition of photosynthesis (upper trace, Figure 4). This inhibition was rapidly reversed by 21% O_2 and could be repeated. The inhibition by 2% O_2 was greater than by 50% O_2, and a stimulation of
photosynthesis was seen upon switching from 2% to 50% O_2 (middle
trace, Figure 4). Although the inhibition was rapid, it was not
permanent, and the rate of photosynthesis slowly recovered. At
20° in 2% O_2, the photosynthesis rate was equal to that in 21% O_2
after 15 min, the final rate achieved after 27 min was 107% of the
rate in 21% O_2. At 25° (lower trace, Figure 4), switching to 2% O_2

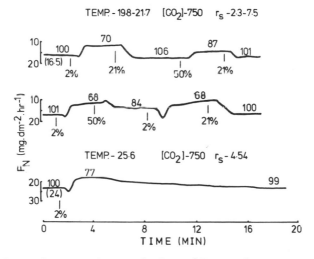

Figure 4. Recorder tracings of the effect of oxygen on photosynthesis of a leaf of a castor oil plant. $[CO_2]$ was 750 $\mu l \cdot l^{-1}$. The two
upper traces were done on one leaf at 20.5^o, and the initial rate of
photosynthesis was 16.5 mg CO_2 $dm^{-2} \cdot hr^{-1}$. The lower trace was done
on another plant at 25.6^o, and the initial rate of photosynthesis
was 24 mg CO_2 $dm^{-2} \cdot hr^{-1}$. Times of changing the oxygen concentrations are shown. Other numbers by the traces show the rate of photosynthesis as a percentage of the initial rate in 21% O_2.

caused a 23% inhibition, but after 16 min the rate was almost equal
to the initial rate. The variation in temperature and stomatal resistance appears to be due to rhythmic movements of the stomata with,
at 20° to 22°, a peak-to-peak period of about 2 hr.

Changes were made, as shown in the lower trace of Figure 4, from
21% to various other oxygen concentrations, and the rates 2 to 4 min
after the change and the final steady rates were calculated (Figure
5). A change to an oxygen concentration <21% always resulted shortly afterward in an inhibition of photosynthesis to a degree dependent
on the oxygen concentration. In all cases the inhibition was reversed within 90 sec after the oxygen concentration was returned to
21%. The final steady rates of photosynthesis in 2% and 8% O_2 were
about 107% of the initial rate in 21% O_2. Concentrations >21% always caused a rapid inhibition of photosynthesis, and the rate did
not change with time.

At 20° and 650 μeinstein$\cdot$m$^{-2}\cdot$sec^{-1} no inhibition was observed
in the photosynthesis rate of a castor oil leaf upon changing the
oxygen concentration from 21% to 2% if the CO_2 concentration was
$\leq$ 100 μl$\cdot$l^{-1} (Figure 6). At a CO_2 concentration $\geq$200 μl$\cdot$l^{-1} the
initial rate was always considerably less than the final rate, and
at $\geq$400 μl$\cdot$l^{-1} the initial rate was less than the rate in 21% CO_2.
At an inlet CO_2 concentration of 50 μl$\cdot$l^{-1} a change in O_2 concentration from 21% to 2% resulted in a jump in the photosynthesis rate
to 438% the rate in 21% O_2 (0.97 mg CO_2 dm$^{-2}\cdot$hr^{-1}). The compensation point of this plant in 21% O_2 was 37 μl$\cdot$l^{-1}.

In the castor oil leaf, the final percentage stimulation of
photosynthesis by 2% O_2 decreased as the light intensity increased.

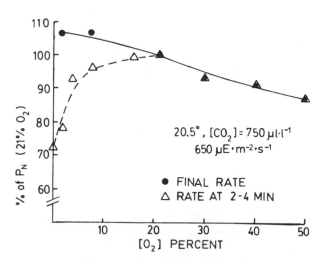

Figure 5. The effect of changing the oxygen concentration from 21%
to a different value on the photosynthesis rate of the leaf of a
castor oil plant. Initial rate of P_N (21% O_2) corresponding to
100 was 16.5 mg CO_2 dm$^{-2}\cdot$hr^{-1}.

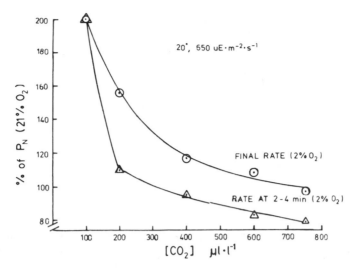

Figure 6. The effect of CO_2 concentration (inlet) on the photosynthesis rate of the leaf of a castor oil plant when the oxygen concentration was changed from 21% to 2%. The rate in 21% O_2 for each
CO_2 concentration was as follows: 3.9 mg CO_2 $dm^{-2} \cdot hr^{-1}$ for 100 $\mu l \cdot$
l^{-1} CO_2; 8.95 for 200; 14.8 for 400; 15.6 for 600; 14.8 for 750.

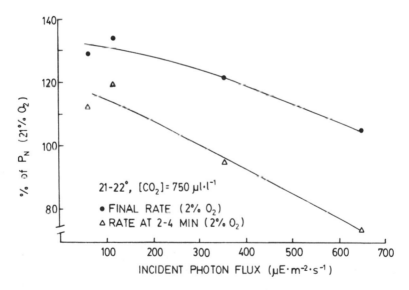

Figure 7. The effect of light intensity on the photosynthesis rate
of a castor oil plant leaf when the oxygen concentration was changed
from 21% to 2%. Rates in 21% O_2 were 4 mg CO_2 $dm^{-2} \cdot hr^{-1}$ for
63 $\mu einstein \cdot m^{-2} \cdot sec^{-1}$; 6.5 for 115; 13.5 for 360; 14.8 for 650.

Inhibition of photosynthesis upon changing the oxygen concentration from 21% to 2% was observed also in sunflowers, the rate of photosynthesis showing a number of cyclical changes of decreasing amplitude (upper trace, Figure 8). A rate of photosynthesis in 2% O_2 equal to that in 21% O_2 was not reached until about 14 min after the change, and the final rate was 108% of that in 21% O_2. The inhibition that occurred for 13 min after the change from 21% O_2 to 2% O_2 was rapidly reversed by returning the oxygen concentration to 21% (lower trace, Figure 8).

At 30° or above and 750 $\mu l \cdot l^{-1}$ CO_2 no transients or inhibition of photosynthesis occurred in sunflower when the oxygen concentration was changed from 21% to 2% (Figure 9), but at 15° the cyclical variations in photosynthesis and the inhibition were more marked (Figure 10) than at 20°.

The effect of changing the oxygen concentration from 21% to 2% on the rate of photosynthesis was measured on leaves of various ages on a sunflower plant (Figure 11). The fluctuations and degree of inhibition were greater in younger leaves. In the oldest leaf measured (leaf 2), except for a very short period, no inhibition of photosynthesis was observed. Leaves were counted from the bottom, excluding cotyledons. Leaf pairs one through five were fully expanded; leaf pair six, about one-half; and leaf pair seven, one-fifth.

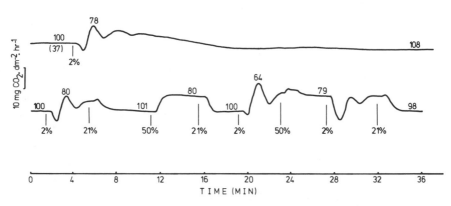

Figure 8. Recorder tracings of the rate of photosynthesis of a sunflower leaf at 20° during changes in oxygen concentration. Both tracings begin in 21% O_2 with an initial photosynthesis rate of 37 mg CO_2 $dm^{-2} \cdot hr^{-1}$. Times of oxygen changes are shown. Numbers above the trace give the rate of photosynthesis as a percentage of the initial rate in 21% O_2. The inlet CO_2 concentration was 750 $\mu l \cdot l^{-1}$, and the quantum flux density incident on the leaf was 650 $\mu einstein \cdot m^{-2} \cdot sec^{-1}$.

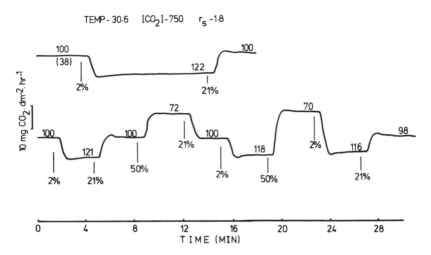

Figure 9.   Recorder tracings of the photosynthesis rate of a sunflower leaf at 30° during changes in oxygen concentration.   Initial
rate in 21% O_2 was 38 mg CO_2 $dm^{-2} \cdot hr^{-1}$.   Other explanation as for
Figure 8.

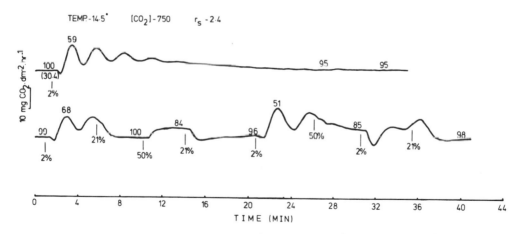

Figure 10.   Recorder tracing of the photosynthesis rate of a sunflower leaf at 15° during changes in oxygen concentration.   Initial
rate in 21% O_2 was 30.4 mg CO_2 $dm^{-2} \cdot hr^{-1}$.

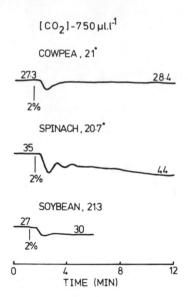

Figure 11. Recorder tracing of the photosynthesis rate of sunflower leaves of various ages when the oxygen concentration was changed from 21% to 2%. Leaves were counted in pairs from the bottom of the plant, and one leaf of each pair was measured. Rates (in mg CO_2 $dm^{-2} \cdot hr^{-1}$) are shown on the trace. Final degree of stimulation can be seen from final rate.

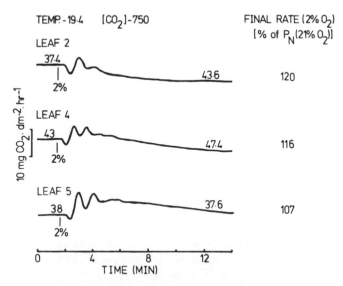

Figure 12. Recorder tracings of the rate of photosynthesis in cowpea, spinach, and soybean leaves when the oxygen concentration was changed from 21% to 2%. Tracings all start in 21% O_2. Rates (in mg CO_2 $dm^{-2} \cdot hr^{-1}$) are given above the traces.

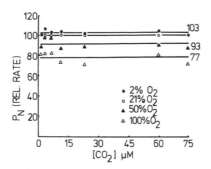

Figure 13. The effect of oxygen on photosynthesis by <u>Chlorella</u> <u>py-</u>
<u>renoidosa</u> Chick. (strain 252) at various CO_2 concentrations. Cells
were suspended in MES buffer at pH 5 on filter paper. Quantum flux
density was 360 μeinstein·m^{-2}·sec^{-1} and temperature was 25°.
(Unpublished data of B. Shelp).

The effect of changing the oxygen concentration from 21% to 2%
on photosynthesis of fully expanded leaves of cowpea, spinach, and
soybean was investigated at 21° and 750 μl·l^{-1} CO_2 (Figure 12).
Inhibition was not observed in any of the leaves, although cyclical
variation and a low initial rate were seen in spinach. Except for
and initial larger increase in photosynthesis, no effect was observed in cowpea. In soybean, the rate of photosynthesis changed
smoothly to an increased steady rate.

The effect of changing the oxygen concentration on photosynthesis by <u>Chlorella</u> <u>pyrenoidosa</u> Chick. (strain 252) was investigated
with the artificial leaf system described earlier (23), except that
cells were suspended in MES* buffer, pH 5.0. At 20°, 750 μl·l^{-1}
CO_2, and 650 μeinstein·m^{-2}·sec^{-1} the rate of photosynthesis changed
smoothly to a new steady rate when the oxygen concentration was
changed. The rate of photosynthesis in 21% O_2 was 92 μmoles·hr^{-1}
per mg chlorophyll. Rate of photosynthesis expressed as a percentage of the rate in 21% O_2 was 107 for 2% O_2, 82 for 50% O_2, and 54
for 100% O_2.

In a more extensive study done at 25°, the effect of oxygen on
photosynthesis by <u>Chlorella</u> (Figure 13) was found to be independent
of the CO_2 concentration. A 3% stimulation of photosynthesis was
observed in 2% O_2, a 7% inhibition in 50% O_2, and a 23% inhibition
in 100% O_2.

* 2-(N-morpholino)ethanesulfonic acid.

DISCUSSION

The inhibition of photosynthesis due to lowering the O_2 concentration from 21% to 2% (Figures 4, 8, 10) and the reversal of this inhibition by restoring it to 21% (Figures 4, 8, 10) appear to occur only at saturating CO_2 concentrations. The leaves of the castor oil plant, under our experimental conditions, were saturated with CO_2 at about 400 $\mu l \cdot l^{-1}$ CO_2 (Figure 6), and also the leaves of the sunflower at 15° (Figure 10). At low CO_2 concentrations no transients occurred when the oxygen concentration was lowered from 21% to 2%, but at CO_2 concentrations approaching saturation, transients occurred, and the initial rate in 2% O_2, though not lower than the previous rate in 21% O_2, was considerably lower than the final rate achieved in 2% O_2 (Figure 6). Relatively high light intensities were required for inhibition of photosynthesis upon changing from 21% O_2 to 2% O_2 (Figure 7). The light intensity used in these studies was close to that required for light saturation with the leaf of the castor oil plant (Figure 7). Viil et al. (10) reached similar conclusions regarding the effects of CO_2 concentration and light on inhibition of photosynthesis by 0.8% O_2, although they used very high concentrations of CO_2. Temperature affects the appearance of the inhibition of photosynthesis by 2% O_2 (Figures 8, 9, 10): as the temperature is increased, higher CO_2 concentrations are required to achieve CO_2 saturation of photosynthesis. At very high CO_2 concentration and low temperature, also, the inhibition by 2% O_2 may not be observed (13). Although data on this point are lacking, it is also possible that leaves of different ages respond differently to a change from 21% to 2% O_2 (Figure 11) because of differences in the CO_2 concentration required to saturate photosynthesis.

The inhibition of photosynthesis observed upon changing the gas composition around the leaf from 21% O_2 to 2% O_2 cannot be attributed to stomatal changes because stomatal aperture is not affected by O_2 concentration (Figure 3) (13, 21, 22) and the changes in photosynthesis are much more rapid than any change in stomatal aperture (24). The inhibition of photosynthesis also cannot be attributed to a direct effect of oxygen on RuBP carboxylase because lowering the oxygen concentration would be expected to stimulate CO_2 fixation (3, 4, 8). Thus, the effect of low oxygen on photosynthesis must be due to an effect of oxygen on the generation of RuBP. These exact conclusions have been previously presented by Viil et al. (13).

Viil et al. (13) argued that photorespiration could not be involved in the inhibition of photosynthesis at low oxygen because photorespiration should not be operating at the 0.3% CO_2 concentrations used in their experiments. In our experiments, the possible involvement of photorespiration cannot be so easily dismissed, since photorespiration can be observed at CO_2 concentrations of 200

to 400 $\mu l \cdot l^{-1}$ (25), and the inhibition of photosynthesis upon changing from 21% to 2% O_2 can also be observed at these CO_2 concentrations (Figure 6). It does not seem likely, however, that the sup-
pression of photorespiration on changing to 2% O_2 (25) would limit
the generation of RuBP because in 2% O_2 more 3-phosphoglyceric acid
would be available than was previously available in 21% O_2 (8).

It should be quite clear that the inhibition of photosynthesis
seen upon changing from 21% to 2% O_2 was not permanent. Rates of
photosynthesis slightly less than, equal to, or greater than that in
21% O_2 were eventually reached in 2% O_2 (Figures 4 to 8, 10). This
must mean that the oxygen-dependent system used by the leaf to maintain RuBP levels can be replaced by an oxygen-independent system.
It is not known whether the capacity of the oxygen-independent system would allow a rate of photosynthesis which would accurately reflect the capacity of the system that might be expected if the removal of any oxygenase activity occurred concurrently with the lowering of oxygen; in other words, the capacity of the oxygen-independent system to generate RuBP may be less than that of the oxygen-
dependent system.

The oxygen-independent system seems to be always present at a
low level because no restriction of CO_2 fixation occurred when the
O_2 was changed from 21% to 2% if the CO_2 concentration was low
(Figure 6). How this oxygen-independent system for RuBP generation
is maintained at an adequate level in low light intensities (Figure 7) is not known. At high light intensities and high CO_2 concentrations, the capacity of the oxygen-independent system was not
adequate, but it could and did increase with time. The system developed rather slowly, however, requiring in the case of the castor
oil leaf, 15 min to reach the same apparent capacity as the oxygen-
dependent system at 20°, and 16 min at 25° (Figure 4) and reaching
maximum capacity at 20° after 27 min.

In sunflower, the capacity of the oxygen-independent system became equal to that of the oxygen-dependent system after 12 min at 20°
(Figure 8), and it reached its maximum after about 26 min at 20° (Figure 8) and after 10 min at 25°. In sunflower at 15° maximum capacity
of the oxygen-independent system was reached after about 22 min but it
was only 95% the capacity of the oxygen-dependent system (Figure 10).
In sunflower at 30° and above the capacity of the oxygen-independent
system was always sufficient to handle the rate of CO_2 fixation (Figure 9), or else its development was so rapid that it could not be detected. The cyclical variations in the photosynthesis of a sunflower
leaf on changing the oxygen concentration from 21% to 2% had a peak-
to-peak time of 135 to 157 sec at 15° and 127 to 135 sec at 20°. This
limited amount of data seems to indicate that the development of the
oxygen-independent system for RuBP generation is not very sensitive to
temperatures between 15° and 20° but that above 20° it is more rapid.

Under the conditions used in this study (20°, 750 $\mu l \cdot l^{-1}$ CO_2,
650 $\mu einstein \cdot m^{-2} \cdot sec^{-1}$), a stimulatory effect of oxygen could be

demonstrated in spinach but not in cowpea, soybean, or <u>Chlorella</u>
(Figure 12). Probably the conditions needed to show this effect
vary for each species of plant and possibly for different varieties
or strains. In a study of photosynthesis by 73 sunflower varieties
at 1050 $\mu l \cdot l^{-1}$ CO_2, 39 showed lower rates of photosynthesis in 2% O_2
than in 21% O_2 (14). It is not known whether the growth conditions
of the plants have any influence on whether or not the effect can
be seen.

In the above discussions one of the possible effects of oxygen
was mentioned that does not account for oxygen inhibition in 2% O_2.
It is not so easy to mention an effect or mechanism that does account for it; however, the fact that photosynthesis rate is perturbed
and inhibited when the oxygen concentration is lowered from 21%
(Figure 5) suggests that oxygen normally acts on some site that regulates RuBP generation. The oxygen concentration that seems to
give the best rates of RuBP generation seems, surprisingly, to be
around 21% (Figure 5). Above 21% O_2 it appears that inhibition of
CO_2 fixation at the oxygenase site of RuBP carboxylase always results
in lower photosynthesis rates.

Viil et al. (13) discussed the possibility that the apparent
requirement for oxygen is due to the necessity of pseudocyclic phosphorylation to maintain ATP levels. Heber and French (26) found,
with spinach chloroplasts, that oxygen uptake did have "some positive feedback role in oxygen evolution." They further established
that, under conditions precluding photosynthesis, oxygen uptake by
chloroplasts was saturated at about 8% O_2. They pointed out, however, that if photosynthesis were allowed to occur, the point of
saturation of oxygen uptake would be shifted to a higher oxygen
concentration, and Huber and Edwards (27) observed increases in
pseudocyclic phosphorylation in C_4 mesophyll preparations up to
100% O_2. Thus the operation of pseudocyclic phosphorylation remains as one possible explanation of the stimulatory effects of
oxygen observed in this work. Certainly, even if this explanation
is not correct, there is a site of action of oxygen on photosynthesis, other than the oxygenase site, that may play a role in regulating the rate of photosynthesis. When the oxygen concentration
around a leaf is varied, the final rate of photosynthesis achieved
may depend on the interaction of this effect with the effect of
oxygen on the carboxylase.

REFERENCES

1. Lorimer, G. H. and Andrews, T. J., Nature London <u>243</u>, 359-60
   (1973).
2. Bowes, G., Ogren, W. L., and Hageman, R. H., Biochem. Biophys.
   Res. Commun. <u>45</u>, 716-22 (1971).
3. Ogren, W. L. and Bowes, G., Nature New Biol. <u>230</u>, 159-60 (1971).
4. Bowes, G. and Ogren, W. L., J. Biol. Chem. <u>247</u>, 2171-6 (1972).

5. Forrester, M. L., Krotkov, G., and Nelson, C. D., Plant Physiol. 41, 422-7 (1966).
6. D'Aoust, A. L. and Canvin, D. T. Can. J. Bot. 51, 457-64 (1973).
7. Chollet, R. and Ogren, W. L., Bot. Rev. 41, 137-79 (1975).
8. Laing, W. A., Ogren, W. L., and Hageman, R. H., Plant Physiol. 54, 678-86 (1974).
9. Samish, Y. B., Photosynthetica 9, 372-5 (1975).
10. Viil, J., Laisk, A., Oja, V., and Pyarnik, T., Dokl, Akad. Nauk SSSR (Engl. transl. of bot. sci. sect.) 204, 86-8 (1972).
11. Jolliffe, P. A. and Tregunna, E. B., Can. J. Bot. 51, 841-53 (1973).
12. Chmora, S. N., Slobodskaya, G. A., and Nichiporovish, A. A., Soviet Plant Physiol. Engl. Transl. 23, 745-50 (1976).
13. Viil, J., Laisk, A., Oja, V., and Pyarnik, T., Photosynthetica 11, 251-0 (1977).
14. Lloyd, N. D. H. and Canvin, D. T., Can. J. Bot. 55, 3006-12 (1977).
15. D'Aoust, A. L. and Canvin, D. T., Photosynthetica 6, 150-7 (1972).
16. Hewitt, E. J., Sand and Water Culture Methods Used in the Study of Plant Nutrition, Commonwealth Agriculture Bureau, Farnham Royal, Bucks. England, 1966.
17. Ludwig, L. J. and Canvin, D. T., Can. J. Bot. 49, 1299-319 (1971).
18. McCashin, B. G., M.Sc. Thesis, Queen's University, Kingston, Ontario, 1978.
19. Bate, G. C., D'Aoust, A. L., and Canvin, D. T., Plant. Physiol. 44, 1122-6 (1969).
20. D'Aoust, A. L., Bate, G. C., and Canvin, D. T., Can. J. Bot. 49, 317-19 (1971).
21. Gauhl, E. and Bjorkman, O., Planta 88, 187-91 (1969).
22. Bull, T. A., Crop Sci. 9, 726-9 (1969).
23. Lloyd, N. D. H., Canvin, D. T., and Culver, D. A., Plant Physiol. 59, 936-40 (1977).
24. Meidner, H. and Mansfield, T. A., Physiology of Stomata, McGraw-Hill, New York, 1968.
25. Ludwig, L. J. and Canvin, D. T., Plant Physiol. 48, 712-19 (1971).
26. Heber, U. and French, C. S., Planta 79, 99-112 (1968).
27. Huber, S. and Edwards, G., Biochem. Biophys. Res. Commun, 67, 28-35 (1975).

DISCUSSION

SLOVACEK: Have you tried perfusing inhibitors into leaves, such as antimycin or known uncouplers, to test their effect on the inhibited rate?

CANVIN: I have discussed such experiments with my associate, Dr. Woo, but we have not done them. The difficulty with getting agents into whole leaves is that often the results are ambiguous and cannot be interpreted. With antimycin A, for example, I assume but cannot prove (and many people may disagree with me) that the dark respiration electron flow is shut down; therefore, with antimycin A in the light, the only thing affected would be the photosynthetic electron flow. If my assumption is not correct, and respiratory electron flow is still performing some useful function, then it also will be affected, and I will see a perturbation whose cause I don't know.

SLOVACEK: I brought this up because in the chloroplast one also observes the inhibition of photosynthesis when going to low oxygen, and this appears to be due to an excessive level of ATP.

CANVIN: The state of the leaves is important. Different leaves have different oxygen-independent abilities. The stage they are in affects the system one is testing. For example, a leaf undergoing photosynthesis in 21% oxygen has a pseudocycling system, and its oxygen-independent system is absent or minimal. If it is placed in the light in very low oxygen, its chloroplasts have to develop this oxygen-independent system to build up their ATP. I find that this takes 27 min, which is close to Walker's 30 min.

IN VIVO CONTROL MECHANISM OF THE CARBOXYLATION REACTION[*]

James A. Bassham, Sheryl Krohne, and Klaus Lendzian[**]

Lawrence Berkeley Laboratory, University of California
Berkeley, California 94720

INTRODUCTION

It is hardly surprising that both the synthesis and the activity of the enzyme ribulose 1,5-bisphosphate (RuBP) carboxylase are
highly regulated. This most abundant enzyme on earth catalyzes the
entry of CO_2 into the reductive pentose phosphate pathway (Calvin
cycle) (1), the pathway leading to the reduction of CO_2 to sugar
phosphates in all green plants (2), including those with a preliminary C_4 cycle (3) for CO_2 accumulation. Such first reactions are
often the sites of important metabolic regulation. The carboxylation reaction is one of four steps in the Calvin cycle unique to
that cycle and not found in the oxidative pentose phosphate cycle
(the other such reactions are the ones converting fructose and sedoheptulose bisphosphates to their respective monophosphates and the
reaction converting ribulose 5-phosphate to RuBP) (Figure 1). All
four of these reactions are inactivated or are less active in the
dark, when the oxidative pentose phosphate cycle and the glycolytic
pathway operate. The inactivation in the dark of these four reactions unique to the reductive cycle is required to prevent the operation of futile cycles.

In the light, the rates of the reactions catalyzed by RuBP
carboxylase and the bisphosphatases are balanced in order to keep
the concentrations of Calvin cycle intermediates within acceptable
ranges as carbon is withdrawn for biosynthesis from pools of both

_____

[*]This research was supported by the Division of Biomedical and
Environmental Research of the U. S. Department of Energy.
[**]Present address: Institut für Botanik und Mikrobiologie der
Technischen Universität München, D-8 München 2, Germany.

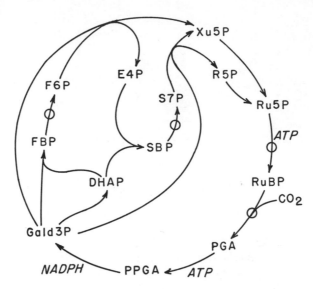

Figure 1.   Rate-limiting steps in the Calvin cycle.   RuBP, ribulose
1,5-bisphosphate; Ru5P, ribulose 5-phosphate; Xu5P, xylulose 5-
phosphate; S7P, sedoheptulose 7-phosphate; SBP, sedoheptulose 1,7-
bisphosphate; E4P, erythrose 4-phosphate; F6P, fructose 6-phos-
phate; FBP, fructose 1,6-bisphosphate; DHAP, dihydroxyacetone
phosphate; Gald3P, glyceraldehyde 3-phosphate; PPGA, phosphoryl 3-
phosphoglycerate; PGA, 3-phosphoglycerate.

triose phosphates and hexose phosphates.   The pools of these compounds are very small compared with the flux of carbon through
the cycle, so very precise regulation is required.
        A further regulatory requirement is placed on the RuBP carbox-
ylase by its oxygenase activity.   Given air levels of CO_2 and O_2,
with the level of CO_2 further decreased inside the chloroplasts
(at least in C_3 plants) by the high rate of photosynthesis in
bright light, O_2 binds competitively with CO_2 at the active site
of the enzyme, after which O_2 reacts with RuBP, giving as one
product, phosphoglycolate, a substrate for photorespiration (4-8).
This means that although a low K_m for CO_2 is desirable for the
efficient operation of the enzyme with air level CO_2, it would also
be desirable (in order to minimize photorespiration) if the K_m for
CO_2, and hence for O_2, would rise in the absence of CO_2.
        It appears that the activity of the enzyme RuBP carboxylase
responds to all of these requirements in one way or another.
Besides having a fast metabolic response, RuBP carboxylase
increases and decreases in amount in response to genetic and

hormonal control, physiological adaptation, etc. The control of
synthesis and degradation of the protein will doubtless be covered
elsewhere in this Symposium and will not be further discussed in
this paper. Principally, our discussion will be of results related
to metabolic regulation obtained over the years in our laboratory.

FREE ENERGY CHANGE AND LACK OF REVERSIBILITY;
STROMA CONCENTRATIONS

The carboxylation of RuBP and hydrolytic redox cleavage of the
six-carbon intermediate to give two molecules of 3-phosphoglycerate
(PGA) is one of the least reversible steps to occur along any important biochemical pathway. The Gibbs free energy change under physiological standard conditions

$$RuBP^{4-} + CO_2 + H_2O \rightarrow 2PGA^{3-} + 2H^+ \qquad \Delta G' = -8.4 \text{ kcal}$$

(all reactants with unit activity except H^+ with activity 10^{-7}) is
calculated to be $\Delta G' = -8.4$ kcal (9). With Chlorella pyrenoidosa
photosynthesizing under air and saturating light, chloroplast concentrations of metabolites were estimated to be 1.4 mM for PGA and
2.04 mM for RuBP.

The method of estimating concentrations was to allow the algae
to photosynthesize with $^{14}CO_2$ under conditions of steady-state
photosynthesis with constant levels of CO_2 and specific radioactivity until the intermediate compounds of the Calvin cycle were fully
labeled with ^{14}C. Samples of the algae were then killed, and the
metabolites were separated by two-dimensional paper chromatography.
The ^{14}C content of each compound could then be determined and used
to calculate its concentration in the cells. This requires some
assumption about the effective soluble volume of the chloroplasts
or space in which the metabolite is dissolved. In the original
calculations it was assumed that the stroma region of the chloroplast occupied about 1/4 the total volume of the algae, and that
Calvin cycle metabolites were located only in that space. Also,
a pH of 7.5 was assumed. With the calculated concentrations used
as activities, steady-state ΔG of -9.8 kcal was calculated. At
pH 8, the value would be $\Delta G^s = -11.2$ kcal. It should be noted, for
later discussion, that the algae were grown for some days in air,
not in CO_2-enriched air.

More recently (J. S. Paul et al., private communication),
isolated mesophyll cells from Papaver somniferum were used in similar experiments to determine steady-state concentrations of metabolites. The amounts of metabolites were measured with respect to
chlorophyll concentration, which averaged 10 mg per ml packed cell
volume in two experiments. If a stroma volume of 20 µl per mg
chlorophyll was assumed, the concentration of RuBP was calculated
to be only 0.07 mM, or 70 µM! This was with cells that gave photo-

synthesis rates comparable with the rates of the leaves from which
the cells were isolated (10). Note that in this calculation,
stroma space, considered to be the exclusive location of RuBP in
green cells, is taken to be 0.2 ml, or 1/5 the total cell volume.
The concentration of PGA, much of which might be outside the stroma,
was very high, 0.307 μmoles per mg chlorophyll. Possibly the inability of isolated mesophyll cells to export photosynthate, which
results in an accumulation of sucrose in the cells (J.S. Paul, private communication), also causes a buildup of PGA in the cytoplasm.
Calculated stroma concentration of other sugar phosphate intermediates of the Calvin cycle and of glucose-6-phosphate were generally
two to ten times as high in the mesophyll cells as in Chlorella.
If we arbitrarily assume that only 1/3 of this PGA is in the stroma,
a concentration in the stroma of 5 mM is calculated. At these concentrations, a $\Delta G^S = -8.38$ kcal is obtained for the carboxylation
reaction at pH 8. The more important point is the great difference
in RuBP concentration between Chlorella and the mesophyll cells
from poppy.

Of the chemical free energy expended by conversion to heat in
the Calvin cycle, about 40% is used in the carboxylation reaction
and another 40% in the other three regulated steps, and the remaining 20% is distributed among the reversible reactions (9).

LIGHT-DARK REGULATION IN VIVO AND IN CHLOROPLASTS

Although in Chlorella photosynthesizing in air the level of
RuBP is high when the light is turned off, the concentration declines
rapidly for the first two minutes and then reaches a concentration
about 5% of that in light, from which it declines very slowly
(Figure 2) (11). Since the K_m for RuBP for the fully activated
enzyme is about 0.035 mM (12), and the $\Delta G'$ for the carboxylation
reaction is -8.4 kcal, this failure of the reaction to continue
after two minutes of darkness means that the enzyme activity has
greatly declined.

The light-dark inactivation of RuBP carboxylase is also evident with isolated spinach chloroplasts (13), in which, following
a period of photosynthesis with $^{14}CO_2$, the level of RuBP in the
dark declined to about one-half the value in light and then remained constant (Figure 3). When the light was again turned on,
the level of RuBP rose very rapidly for 30 sec and then declined
to a steady-state level. We attribute such transients to dark
inactivation of the carboxylase, followed by light reactivation
requiring 30 sec.

When the drop in RuBP level in the isolated spinach chloroplasts was prevented by addition of ATP to the suspending medium
just after the light was turned off, very little uptake of $^{14}CO_2$
occurred as long as the light was off (even though the chloroplasts
contained as much RuBP as in the light). When the light was turned
on again, high rates of $^{14}CO_2$ uptake resumed (14). Although the

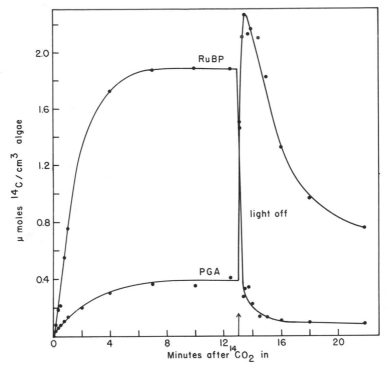

Figure 2.   Changes in RuBP concentration in <u>Chlorella</u> <u>pyrenoidosa</u> in light and dark periods (from ref. 11).

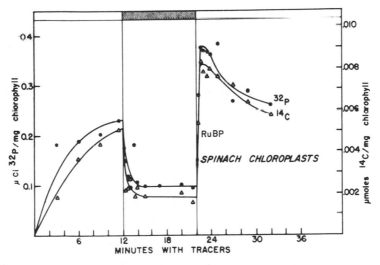

Figure 3.   Changes in RuBP concentration in isolated spinach chloroplasts in light and dark periods (from ref. 13).

rate of entry of ATP into whole chloroplasts may be low compared
with the requirements of photosynthesis (15-17), this low rate is
apparently sufficient to maintain the level of RuBP when it is not
being consumed, once the RuBP carboxylase is inactivated.

LIGHT-DARK REGULATION OF RuBP CARBOXYLASE BY Mg^{2+} AND OTHER FACTORS

The primary mechanism for light-dark regulation of the activity of RuBP carboxylase appears to be via changes in Mg^{2+} ion concentrations and pH. Other metabolites, particularly NADPH and
6-phosphogluconate, may also contribute both to light-dark regulation and perhaps to regulation in the light. Isolated RuBP carboxylase is activated by preincubation with CO_2 or bicarbonate plus
high levels of Mg^{2+} (e.g., 10 mM), before the enzyme is exposed to
RuBP (18-21). Preincubation with physiological levels of RuBP in
the absence of either bicarbonate or Mg^{2+} results in conversion of
the enzyme to an inactive form with high K_m values for CO_2, and the
enzyme does not recover its activity for many minutes upon subsequent exposure to physiological levels of bicarbonate and Mg^{2+}
(19,20). Full activation of the isolated purified enzyme requires
that the preincubation with CO_2 and Mg^{2+} be carried out also in
the presence of either NADPH or 0.05 mM 6-phosphogluconate, each
at physiological levels (19-21).

With respect to light-dark regulation, it seems clear that the
changes in Mg^{2+} levels and pH in the chloroplasts result in changes
in RuBP carboxylase activity, with the light-induced increases in
pH and Mg^{2+} resulting in increased enzyme activity. The pH optimum
of the isolated enzyme shifts towards the pH actually found in
chloroplasts in the light (about 8) with increased Mg^{2+}, and the
value of K_m for CO_2 is lower at pH 8 than at pH 7.2 (22-24).

The activation of the isolated enzyme by NADPH seems to be
another part of the light-dark regulation, but the activation by
6-phosphogluconate is at first surprising, since this compound
appears in the dark. Kinetic studies show that the 6-phosphoglu-
conate is still present during the first two minutes of light
after a dark period (13), and it may be that a useful activation
occurs then, while the level of NADPH is still being built up.
In the dark, 6-phosphogluconate would not activate the carboxylase
since the optimal conditions of pH and Mg^{2+} levels would not be met.

Presumably, any light activation of the enzyme via NADPH
would involve the transfer of electrons to $NADP^+$ via ferredoxin.
This does not explain how oxidized ferredoxin could further increase the activity of the isolated RuBP carboxylase, as reported
by Vaklinova and Popova (25) and confirmed by Popova in our laboratory (unpublished work).

CAN THE LOW $K_m(CO_2)$ FORM BE MAINTAINED OUTSIDE CHLOROPLASTS?

Although it appeared for many years that $K_m(CO_2)$ for RuBP carboxylase is too high to support the reductive pentose phosphate cycle, a number of laboratories have found evidence in recent years that the $K_m(CO_2)$ is sufficiently low. In particular, Bahr and Jensen (26) found that a low $K_m(CO_2)$ form of the enzyme obtained from freshly lysed spinach chloroplasts could be stabilized with dithioerythritol, ATP, MgCl$_2$, and R5P. Lilley and Walker (27) have shown that the activity and $K_m(CO_2)$ for the enzyme from spinach chloroplasts are more than adequate to support photosynthesis.

A common problem with some of the reported studies on the biochemical constants of RuBP carboxylase outside the intact chloroplast, whether in the form of crude protein extract or of purified enzyme, is the changing value of $K_m(CO_2)$ and fixation rate during the time of the enzyme assays. Even with the purified enzyme preincubated with Mg^{2+} and bicarbonate in the presence of NADPH or 6-phosphogluconate as effectors, the most activated rate of reaction was always during the first 3 to 5 min, followed by a decline to a slower rate. In some of the studies reported from other laboratories, linear fixation was obtained for <2 min, and K_m determinations were reported using only the rate during the first minute after the addition of RuBP to preincubated enzyme. Although it may be common practice among biochemists to look only at initial rates of a reaction when calculating kinetic constants, it would seem that subsequent behavior may be telling us something in the case of a large protein molecule with complex regulation including evident slow changes in state. We have therefore endeavored, with limited success, to discover conditions under which the enzyme, outside the intact chloroplast, might be able to exhibit prolonged activity and K_m values required for in vivo photosynthesis.

Lysed and subsequently reconstituted spinach chloroplasts can be made, under carefully chosen conditions, to carry on photosynthetic CO_2 fixation at substantial rates linear up to 30 min (28). It should be mentioned that even after several years' experience with this system, prolonged high rates cannot be guaranteed on any given day, presumably because of variability in biological material, minor impurities in reagents, or other uncontrolled variables. Nevertheless, we have succeeded often enough to be able on good days to carry out investigations on the kinetics of CO_2 fixation.

An early result was that, even though Mg^{2+} is maintained at 20 mM and pH at 8.0 (the chloroplast values in light), CO_2 fixation in the originally reported reconstituted system is strictly light dependent. Since dithiothreitol (1 mM) and glutathione (50 mM) are used, the enzymes of the rate-limiting reactions are fully active, but of course in the dark there is no reduction of PGA and hence no cyclic regeneration of RuBP. Two kinds of experiments have been done in the past year: determination of the $K_m(CO_2)$ in light and

determination of $K_m(CO_2)$ in the dark with soluble enzymes only, in
the presence and absence of various additional effectors. In both
types of experiments we make use of the gas-handling steady-state
apparatus (29) which has been adapted to allow connection to small
round-bottom flasks used previously in our studies with isolated
chloroplasts (30) and reconstituted chloroplasts (28). Each 15-ml
flask has been fitted with 2-mm-i.d. inlet and outlet tubes connected by small flexible tubing to gas manifold tubes attached to
the steady-state apparatus. We can thus control and monitor CO_2
pressure, O_2 pressure, and ^{14}C content during the course of the
experiment. The $^{12}CO_2$ and $^{14}CO_2$ are supplied to the closed system
by the pressurized cylinder and regulating system previously described (31).

        For experiments with reconstituted chloroplasts, isolation of
chloroplasts (30) and reconstitution were as described earlier (28)
except that, instead of the 14:1 ratio of soluble components to
lamellae, a ratio of 7:4 was employed in order to boost the concentration of RuBP to levels such that the determined $K_m(CO_2)$ would be
meaningful. At this ratio there is 14.5 mg soluble protein per mg
chlorophyll, and, for reproducibility, we adjust the reconstituted
system to this ratio rather than relying on volumetric measurement
of thylakoid fractions. The rest of the assay mixture consists of
4 mM NADP, 2 mM ADP, 1 mM PGA, 0.05 mg ferredoxin, 4 mM Na isoascorbate, and solution Z (28), all in a total volume of 0.5 ml in 15-ml
round-bottom flasks.

        Each pair of flasks was run at a separate gas concentration
starting at the highest concentration. Five different concentrations were used to get the maximum number of data points possible
by using all the manifold inlets and outlets. Flasks not being
used were clamped off until needed. They were unclamped and opened
to the gas at time zero shown in the results. All flasks were in a
nitrogen atmosphere before the assay, and the assay mixture was put
into the flask just before each separate experiment. Samples (50 μl)
were taken during the assay and killed in 450 μl methanol. From
the resulting 500 μl of mixture, aliquot samples were taken and
counted in scintillation vials after acidification and drying.
Carbon dioxide fixation in the 7:4 reconstituted system was linear
from 5 to 20 min after the introduction of $^{14}CO_2$ at levels ranging
from 0.013 to 0.128% (Figure 4). Some time is required during the
first 5 min after $^{14}CO_2$ introduction to replace the gas initially
in the flasks and to equilibrate gas and liquid phases.

        When the reciprocals of the rate vs. CO_2 concentrations are
plotted (Figure 5) we obtain a value for $K_m(CO_2)$ of 0.023% of 230
ppm. This is well below the air level of 0.032% CO_2. Moreover,
this K_m plot, unlike many that have been published, shows a truly
linear character over a meaningful range of CO_2 pressures. The $1/v$
intercept at $1/s = 0$ is 0.0175 and therefore V_{max} = 60 μmoles CO_2
per mg Chl per hr, or 4.2 μmoles CO_2 per mg soluble protein per hr
($\sim$8.5 μmoles CO_2 per mg RuBP carboxylase per hr). The rate at

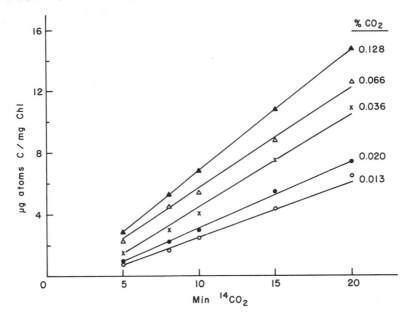

Figure 4. Steady-state CO_2 fixation by reconstituted spinach chloroplasts at five levels of CO_2. Conditions described in text. Plot shown only for 5 to 20 min, since an initial equilibration time of 2 min is required.

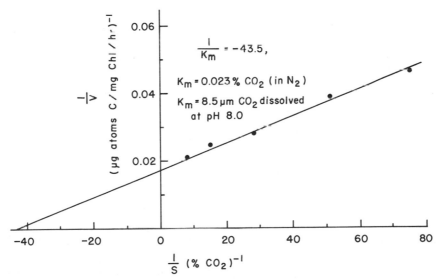

Figure 5. Lineweaver-Burk plot: $K_m(CO_2)$ for RuBP. Plot based on data in Figure 4.

0.036% CO_2 in air (slightly more CO_2 than in ambient air) was 60% of V_{max}. A Hill plot of the data (Figure 6) gives a slope of n = 1 which shows noncooperativity of the CO_2 binding sites, as previously reported by Bahr and Jensen (26).

The in vivo rate at air level should be about 75 μmoles CO_2 fixed per mg RuBP carboxylase per hr. Thus, although we have obtained physiological $K_m(CO_2)$ values that remain unchanged over 10 min, the reaction rates are only about 7% of the in vivo rates. One reason for the lower in vitro rates may be the fact that all the soluble components are considerably diluted in the reconstituted system as compared with in vivo concentrations. With the system used (14.5 mg protein per mg Chl) this dilution is about 14:1. Although this should not affect the usual assumptions made in enzyme kinetics as long as CO_2 and RuBP concentrations are maintained, it might affect the state of this enzyme, which we know is subject to complex regulation.

Experiments similar to those just described were performed, but with ^{14}C-labeled PGA in which the specific radioactivity of each carbon position in the molecule was the same as that of the $^{14}CO_2$ employed (44.7 μCi per μgram-atom). Small samples were withdrawn periodically during the experiment, killed in 80% alcohol, and subsequently analyzed by two-dimensional paper chromatography and radioautography (11). Since the Calvin cycle intermediates were fully labeled with ^{14}C of known specific radioactivity (except for a little dilution by unlabeled intermediates initially present), it was possible to calculate the concentration of RuBP in the flasks at each CO_2 concentration (Table 1). Ideally for a $K_m(CO_2)$ determination one would like saturating but not inhibitory concentrations of RuBP. In these experiments the concentrations of RuBP at the four lower CO_2 pressures (0.0116 to 0.0613%) are all well above the 35 μM value for the $K_m(RuBP)$ of the enzyme. Once again, a low $K_m(CO_2)$ value was obtained (0.027% CO_2).

One may at least conclude from these results that RuBP carboxylase can be maintained outside the intact chloroplast in the low $K_m(CO_2)$ form, given cyclic regeneration of RuBP driven by cofactors from the illuminated thylakoids. The next question was whether the low $K_m(CO_2)$ form could be maintained without thylakoids. To date we have not succeeded in this, but we have learned some interesting properties. We can obtain considerable stimulation of the fixation rate by the addition of several metabolites, including NADPH and ferredoxin (Figure 7).

The experiments are performed by preincubating the soluble enzymes with Mg^{2+} and $^{14}CO_2$ for 10 min before addition of RuBP. Without effectors, we can obtain a "$K_m(CO_2)$" of 0.031% by using the 1-min points as rates, but an examination of the kinetics (Figure 8) shows how misleading this is. Always, with the stroma enzymes and enough RuBP present initially to avoid using it up (10 mM in the case shown in Figure 8) there is a very rapid slowing of the rate. We attribute this to allosteric inactivation of the enzyme (18-20).

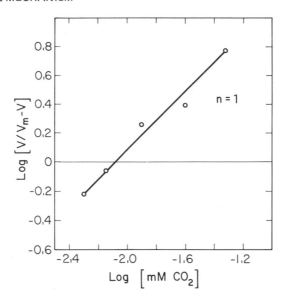

Figure 6. Hill plot of RuBP carboxylase activity with varying CO_2 pressure. Plot based on data in Figure 4.

Table 1

Steady-State Levels of RuBP in Reconstituted Spinach Chloroplasts

Concentrations of RuBP were determined by ^{14}C content at steady state with $^{14}CO_2$ and $^{14}C-U-PGA$ (see text).

% CO_2	mM RuBP	Rate: μmoles CO_2 fixed per mg Chl per hr
0.1428	0.0224	90.0
0.0613	0.0644	68.4
0.0266	0.1262	50.4
0.0202	0.2061	39.6
0.0116	0.2998	25.2

Just as with the isolated enzyme, this inactivation can be partially overcome with added effectors such as 6-phosphogluconate (Figure 9) or NADPH, but not to a sufficient extent at these high concentrations of RuBP to give sustained linear rates characteristic of the low $K_m(CO_2)$ form.

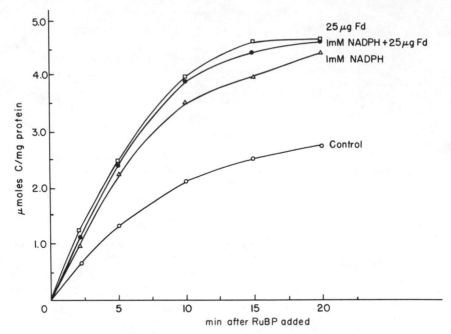

Figure 7. Stimulation of CO_2 fixation in diluted stroma enzymes by NADPH and ferredoxin. Whole isolated spinach chloroplasts were lysed in Hepes buffer (25 mM, pH 8) and thylakoids were removed by centrifugation at 20,000 x g for 10 min. The supernatant solution was diluted in Hepes buffer to give 50 μg soluble protein in each 0.5 ml of solution in a reaction flask. Reaction conditions: $NaHCO_3$, 2 mM; RuBP, 4 mM; and as described in text.

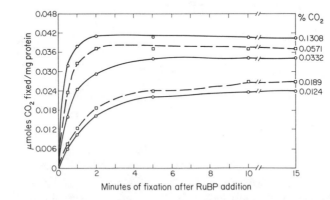

Figure 8. CO_2 fixation by stroma enzymes at five levels of CO_2. Whole isolated spinach chloroplasts were lysed in solution Z (28). Each 0.5 ml of reaction mixture in flask contained 940 μg soluble protein. Reaction conditions as in text.

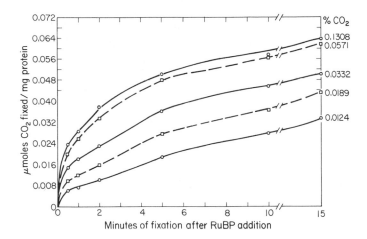

Figure 9. CO_2 fixation by stroma enzymes at five levels of CO_2 with added 6-phosphogluconate.   Conditions same as for Figure 8.

     This inactivation by RuBP is probably the reason for the finding
often reported by various workers that higher rates can be obtained
when the RuBP is supplied by a RuBP-generating system such as ribose
5-phosphate and ATP plus enzymes rather than by a high initial concentration of RuBP.   As reported earlier, levels of RuBP in whole
photosynthesizing poppy leaf cells are quite low: around 70 μm, or
about double the K_m(RuBP).   Judging by the data in Table 1, however,
a concentration as high as 300 μM RuBP is not seriously inhibitory
with stroma enzymes at concentrations only severalfold less than in
intact chloroplasts.
     If inhibition by RuBP concentrations of 0.5 mM or more is the
cause of our failure to obtain linear kinetics with stroma enzymes,
it might be possible to dilute the enzyme concentration and work
with much lower RuBP concentrations, say 0.1 mM, without seriously
depleting the RuBP as substrate during the course of the experiment.
We have done such experiments.   Although a closer approach to linearity was obtained, some falling off in rate with time was still
encountered (Figure 10).   Perhaps all that can be concluded at
present is that even relatively low concentrations of RuBP decrease
the activity of the RuBP carboxylase in diluted stroma enzymes over
a period of time.
     If this premise is accepted, we are left with the question of
how <u>Chlorella</u> tolerates the high RuBP concentrations (2.0 mM) found
in the measurements cited earlier.   The reason for the inhibitory

regulation of RuBP carboxylase by RuBP in leaves is presumably to minimize photorespiration resulting from the competitive binding of O_2 at the CO_2 binding site. RuBP binding allosterically is thought to raise the K_m for both CO_2 and O_2. Perhaps this protection is simply not so important for Chlorella. In shallow ponds and lakes there is less light, more dissolved CO_2, no water stress, lower O_2, and consequently little chance for photorespiration. When Chlorella are exposed to high O_2, low CO_2, and high light in the laboratory, huge amounts of glycolate are produced. Under such drastic conditions, the enzyme in Chlorella may be inactivated by a decrease in K_m for CO_2 and O_2, since the rate of phosphoglycolate formation appeared to decrease after about 30 sec even though RuBP concentration was still substantial (32). The regulatory behavior of RuBP carboxylase from Chlorella might be worth further study.

CONCLUSION

The activity of RuBP carboxylase is strongly regulated in light and dark, becoming minimal in vivo in the dark as part of the mechanisms of avoiding wasteful reactions. Changes in pH, Mg^{2+}, NADPH and other metabolites account for this change in activity. Probably these effectors also account for some fine control of

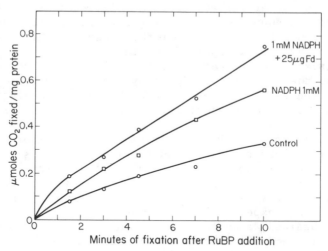

Figure 10. CO_2 fixation with diluted stroma enzymes and 0.1 mM RuBP. Whole isolated spinach chloroplasts were lysed in Hepes buffer (25 mM, pH 8) and thykaloids were removed by centrifugation at 20,000 x g for 10 min. The supernatant solution was diluted in Hepes buffer to give 20 μg soluble protein in each 0.5 ml of solution in a reaction flask. Reaction conditions: RuBP, 0.1 mM; $^{14}CO_2$, 0.033%; and as described in text.

RuBP carboxylase activity in the light, permitting the rate of carboxylation to be balanced against rate of utilization of photosynthate. A major form of regulation of RuBP carboxylase in leaves is conversion of the enzyme to a high $K_m(CO_2)$ form in the presence of high ratios of total RuBP to enzyme binding sites and in the presence of very low levels of CO_2. This regulation provides a measure of protection against photorespiration resulting from the attack of O_2 on RuBP leading to phosphoglycolate formation. Steady-state concentrations of RuBP in the unicellular alga Chlorella are much higher than would appear to be tolerable in leaf cells, which suggests the possibility of different degrees of inhibition by RuBP in algae and in higher plants.

REFERENCES

1. Bassham, J. A., Benson, A. A., Day, L. D., Harris, A. Z., Wilson, A. T., and Calvin, M., J. Am. Chem. Soc. 76, 1760-70 (1953).
2. Norris, L., Norris, R. E., and Calvin, M., J. Exp. Bot. 6, 64-74 (1955).
3. Hatch, M. D. and Slack, C. R., Biochem. J. 101, 103-11 (1966).
4. Bassham, J. A. and Kirk, M. R., Biochem. Biophys. Res. Commun. 9, 376-80 (1962).
5. Bowes, G. Ogren, W. L., and Hageman, R. H., Biochem. Biophys. Res. Commun. 45, 716-22 (1971).
6. Bowes, G., Ogren, W. L., and Hageman, R. H., Plant Physiol. 56, 630-3 (1975).
7. Lorimer, G. H., Andrews, T. J., and Tolbert, N. E., Biochemistry 12, 18-23 (1973).
8. Richardson, K. E. and Tolbert, N. E., J. Biol. Chem. 236, 1285-90 (1961).
9. Bassham, J. A. and Krause, G. H., Biochim. Biophys. Acta 189, 207-21 (1969).
10. Paul J. S. and Bassham, J. A., Plant Physiol. 60, 775-8 (1977).
11. Pedersen, T. A., Kirk, M., and Bassham, J. A., Physiol. Plantarum 19, 219-31 (1966).
12. Chu, D. K., Biochemical and Physical Studies of Ribulose 1,5-Disphosphate Carboxylase From Spinach Leaves, Ph.D. Thesis, University of California, Berkeley, 1974.
13. Bassham, J. A. and Kirk, M., in Comparative Biochemistry and Biophysics of Photosynthesis, pp. 365-78, K. Shibata et al., Editors, University of Tokyo Press, 1968.
14. Jensen, R. G. and Bassham, J. A., Biochim. Biophys. Acta 153, 227-34 (1968).
15. Heber, V. W. and Santarius, K. A., Z. Naturforsch. 25b, 718-28 (1970).
16. Heldt, H. W., Sauer, F., and Rapley, F., in Proc. 2nd. Int. Congr. Photosynth. Res., Stresa, 1971, pp. 1345-55, G. Forti et al., Editors, Junk, The Hague, 1972.

17.   Stokes, D. M. and Walker, D. A., in <u>Photosynthesis and Respir-</u>
      <u>ation</u>, pp. 226-31, M. D. Hatch et al., Editors, Wiley, New
      York, 1971.
18.   Pon, N. G., Rabin, B. R., and Calvin, M., Biochem. Z. <u>338</u>,
      7-9 (1963).
19.   Chu, D. K. and Bassham, J. A., Plant Physiol. <u>52</u>, 373-9 (1973).
20.   Chu, D. K. and Bassham, J. A., Plant Physiol. <u>54</u>, 556-9 (1974).
21.   Chu, D. K. and Bassham, J. A., Plant Physiol. <u>55</u>, 720-6 (1975).
22.   Bassham, J. A., Sharp, P., and Morris, I., Biochim. Biophys.
      Acta <u>153</u>, 901-2 (1968).
23.   Sugiyama, T., Nakayama, N., and Akazawa, T., Biochem. Biophys.
      Res. Commun. <u>30</u>, 118-23 (1968).
24.   Lorimer, G. H., Badger, M. R., and Andrews, T. J., Biochemistry
      <u>15</u>, 529-36 (1976).
25.   Vaklinova, S. and Popova, L., pp. 1869-74 in ref. 16.
26.   Bahr, J. T. and Jensen, R. G., Plant Physiol. <u>53</u>, 39-44
      (1974).
27.   Lilley, R. McC. and Walker, D., Plant Physiol. <u>55</u>, 1087-92
      (1975).
28.   Bassham, J. A., Levine, G., and Forger, J. III, Plant Sci.
      Lett. <u>2</u>, 15-21 (1974).
29   Bassham, J. A. and Kirk, M. R., Biochim. Biophys. Acta <u>43</u>,
      447-64 (1960).
30   Jensen, R. G. and Bassham, J. A., Proc. Natl. Acad. Sci. USA
      <u>56</u>, 1095-101 (1966).
31.   Platt, S. G., Plaut, Z., and Bassham, J. A., Plant Physiol.
      <u>57</u>, 69-73 (1976).
32.   Bassham, J. A. and Kirk, M., Plant Physiol. <u>52</u>, 407-41 (1973).

DISCUSSION

BLACK: How do you calculate concentrations of intermediates
in <u>Papaver</u> vs. <u>Chlorella</u>: do you take into consideration factors
such as the volume of the vacuole, since these two types of cells
are obviously quite different?

BASSHAM: When we did the <u>Chlorella</u> work several years ago,
we didn't have the figures, later published by Heldt and others,
about the volume of stroma space per milligram of chlorophyll.
In <u>Chlorella</u>, the chloroplast comprises about half the volume of
the cell, and about half of it is stroma; thus, the stroma is a
fourth of the packed volume. We can alternatively assume a certain
concentration of chlorophyll, and we come out with about the same
answer. In <u>Papaver</u> we measured 10 mg chlorophyll per ml packed
cells, multiplied this by 20 to get 200 μl, which is a fifth of the
packed volume; that is the basis for our calculation. It may be in
error because the stroma volume reported for other plants may not
apply to <u>Papaver</u>, but that's the approximation we use. We also
made an estimate by looking at the volume under the microscope;
visually the chloroplast volume looks smaller--more like a tenth.

PAECH: Regarding the inhibitors you mentioned, not only commercial RuBP but also homemade RuBP always contains inhibitors; they are inherent in its preparation, unless it is made by the coupled enzyme system with the kinase and isomerase directly from R5P.

BASSHAM: I think it is very important to clear up whether RuBP itself has any part in regulating the enzyme or whether the action is due only to inhibitors, and I trust this will be done.

REGULATION OF RIBULOSE 1,5-BISPHOSPHATE

CARBOXYLASE IN THE CHLOROPLAST[*]

Richard G. Jensen, Richard C. Sicher Jr.

Departments of Biochemistry and Plant Sciences
University of Arizona, Tucson, Arizona 85721

and

James T. Bahr

Mobil Chemical Company, Edison, New Jersey 08817

INTRODUCTION

Carbon dioxide is incorporated by the action of ribulose 1,5-bisphosphate carboxylase/oxygenase (RuBP carboxylase) during photosynthesis. Curves relating photosynthesis rates to illuminance for many species, especially C_3 plants, show that the photosynthesis rate approaches a limiting value asymptotically at high radiation (1, 2). CO_2 availability becomes a primary limiting factor to photosynthetic CO_2 assimilation under these conditions (2). However, more capacity for CO_2 assimilation per leaf area can be induced. When the photosynthetic product demand was increased by partial defoliation or shading of plant leaves, increased photosynthesis rates were observed within days (3, 4). Increased levels of RuBP carboxylase were noted (4), suggesting that under saturated light conditions in the field the photosynthesis rates were limited by the activity of the carboxylase.

---

[*]This research was supported in part by National Science Foundation Grant PCM 75-23240. University of Arizona Agricultural Experiment Station Paper 265.
    Abbreviations: CCCP, carbonyl cyanide m-chlorophenylhydrazone; Ches, cyclohexylaminoethanesulfonic acid; Chl, chlorophyll; DCMU, 3-(3,4-dichlorophenyl)-1,1-dimethylurea; DTE, dithioerythritol; FBP, fructose 1,6-biphosphate; PGA, glycerate 3-phosphate; R5P, ribose 5-phosphate; RuBP, ribulose 1,5-bisphosphate; Ru5P, ribulose 5-phosphate.

Carboxylation by RuBP carboxylase requires the presence of both CO_2 and RuBP in the chloroplast. Too low levels of RuBP will not sustain carboxylation, and too high levels might inhibit it. The enzyme itself becomes more active in response to light-induced changes in the chloroplast stroma (5). Recent information on the kinetic properties of the carboxylase suggests that the amount of enzyme as well as its degree of activation in the chloroplast should be considered as regulating CO_2 assimilation during photosynthesis in isolated chloroplasts (6).

Since RuBP carboxylase is found in all photosynthetic organisms and often comprises more than 50% of the soluble leaf protein, it is the most abundant protein in nature (7). It is a protein of high molecular weight (550,000) existing as a spherical aggregate of two types of subunits, large and small (L_8S_8) (6, 8). The molecular weight of the larger subunit is 51,000 to 58,000 and that of the smaller one is 12,000 to 18,000. The large subunit is catalytically active even in the absence of the small one (9). The function of the small subunit might well be regulatory (10), but how this is accomplished in the molecule remains to be determined. Baker et al. (11) have proposed that the structure of crystalline tobacco RuBP carboxylase consists of a two-layered structure, each layer having four large and four small spherical masses. Physical and chemical studies also have provided good evidence for eight copies of the large subunit (9, 12, 13), and there are a maximum of eight binding sites for RuBP in the molecule (14, 15).

CATALYTIC PROPERTIES

Because of the central role of RuBP carboxylase in photosynthetic CO_2 fixation, considerable effort has been aimed at clarifying its regulatory properties, with special emphasis on the physiological conditions necessary for activity. Earlier kinetic studies showed that the activity of the isolated enzyme, as exhibited by its apparent affinity for CO_2, was too low to account for the observed rates of photosynthetic CO_2 fixation (16-19). As first reported, the $K_m(CO_2)$ for purified RuBP carboxylase was quite high, between 70 and 600 μM (6, 20). In intact isolated spinach chloroplasts, separated leaf cells, and intact leaves, the apparent $K_m(CO_2)$ for CO_2 fixation is in the range of 10 to 20 μM (2, 6, 17, 19, 20). Water, in equilibrium with 1 atm air with 0.03% CO_2, contains 10 μM CO_2 at 25°C. Precise agreement between the $K_m(CO_2)$ value for steady-state CO_2 fixation by plants and the value for purified carboxylase would not be expected because during steady-state CO_2 fixation in the chloroplast conditions are not necessarily optimal for maximal CO_2 fixation. Nevertheless, the activities of purified RuBP carboxylase reported previously (20, 21) were quite inadequate to explain the observed rates of photosynthetic CO_2 fixation in plants.

The question of how RuBP carboxylase operates at air levels of CO_2 in vivo is now of considerable research interest. Several years ago we reported on the kinetic and regulatory properties of RuBP carboxylase released from intact spinach chloroplasts (17, 22) by dilution of intact chloroplasts into a hypotonic medium containing all the components for assay of carboxylase or oxygenase activity. The carboxylase activity of the enzyme assayed in this manner was equal to or slightly greater than the rate of photosynthetic CO_2 fixation by intact chloroplasts at the same CO_2 concentration. The apparent $K_m(CO_2)$ of 11 to 18 μM at pH 7.8 for the released RuBP carboxylase was comparable to that for light-dependent CO_2 fixation by intact chloroplasts (23). At low CO_2 concentrations, the kinetics of the carboxylase were not stable after release from the chloroplast. During assays observed to 10 min, a lower steady-state rate was obtained after 3 min which displayed a $K_m(CO_2)$ value of about 500 μM, comparable with that seen in most previous work with the purified carboxylase. Incubation of the enzyme in buffer alone or in buffer plus RuBP before addition of CO_2 and Mg^{2+} hastened the decline in activity.

Incubation of the enzyme in Mg^{2+} and CO_2 increased the apparent activity (V_{max}) and gave an apparent $K_m(CO_2)$ of about 60 μM (pH 7.8) (17), but increased activity was unstable during long-term assays, like that of the enzyme immediately after release from the chloroplast. It now appears that the enzyme released from intact chloroplasts was partially activated and could be further activated by incubation with Mg^{2+} and CO_2. The RuBP carboxylase, while still in the intact chloroplast, is not fully activated, and its degree of activation can be altered by incubating chloroplasts with various CO_2 concentrations (5).

As the rate of carboxylation of RuBP in the plant defines the rate of gross photosynthesis, the rate of RuBP oxygenation is the major determinant of glycolate production for photorespiration (24). O_2 has been shown to be a competitive inhibitor of carboxylation as CO_2 is of oxygenation (16, 22, 25, 26). Within experimental error, the apparent K_m values for CO_2 or O_2 are nearly equal to the apparent K_i values for O_2 or CO_2 respectively (16, 26). The relative rates of the two reactions are regulated by the concentration of O_2 and CO_2. When both reactions have been measured under the same conditions, these metabolites have been found to activate or inhibit both reactions to the same extent (27-29).

Two conditions, temperature and pH, might provide differential regulation of two reactions. The apparent $K_m(CO_2)$ and $K_m(O_2)$ values vary differently with temperature (26). The pH optima of the two reactions at constant O_2 and CO_2 are broad and similar, but may differ by as much as 0.2 to 0.3 pH units (22, 30). The apparent pH optimum of the carboxylase is dependent on CO_2 concentration even when the enzyme is activated at constant conditions because the $K_m(CO_2)$ for the catalytic reaction is pH sensitive (31).

ENZYME ACTIVATION AND CATALYSIS

The time-dependent and order of addition-dependent kinetics of RuBP carboxylase result from the activating effects of Mg^{2+} and CO_2 and the inactivating effects of RuBP in addition to their roles as substrates (17, 28, 30, 32-34). The response of the initial activity of the enzyme to the concentration of CO_2 and Mg^{2+} during preincubation indicates the reversible formation of an active enzyme-CO_2-Mg^{2+} complex. Kinetic analyses have indicated that the enzyme is activated by a slow but reversible initial binding between enzyme and CO_2 followed by a rapid reaction with Mg^{2+} (27, 35, 36). The amount of activation is a function of both CO_2 and Mg^{2+} concentration, and the final degree of activation at fixed CO_2 and Mg^{2+} is sharply pH dependent (27, 35) with a distinctly alkaline pK_a. The relationship between CO_2 and Mg^{2+} concentrations and pH during activation suggests that the protonated enzyme does not react with CO_2. When the enzyme is first incubated with RuBP and the reaction is initiated with Mg^{2+} and CO_2, a marked lag is observed in the course of product formation (17, 36, 37). The process of activation and the reactions competing with it may be written as follows:

$$\text{Activating process:} \quad E_{(inactive)} + CO_2 \xrightleftharpoons{\text{slow}} E\text{-}CO_{2(inactive)}$$

$$E\text{-}CO_{2(inactive)} + Mg^{2+} \xrightleftharpoons{\text{fast}} E\text{-}CO_2\text{-}Mg^{2+} \text{(active)}$$

$$\text{Competing processes:} \quad E + H^+ \rightleftharpoons EH^+$$

$$E + RuBP \rightleftharpoons E\text{-}RuBP$$

Catalysis leading to CO_2 fixation occurs by binding of the active $E\text{-}CO_2\text{-}Mg^{2+}$ complex with RuBP and a second CO_2 to form two molecules of 3-phosphoglycerate (PGA). Catalysis leading to oxygenation occurs when the $E\text{-}CO_2\text{-}Mg^{2+}$ complex binds RuBP and O_2 to form one molecule each of PGA and phosphoglycolate. Mg^{2+} is required for the activating process but apparently is not required for catalysis (36).

It has often been observed that RuBP carboxylase generally displays substrate inhibition at high concentrations of RuBP. A regulatory role for RuBP has been suggested on the basis of the inability of metabolite effectors to stimulate, once the reaction has begun (33, 38), and on the basis of a much lower steady-state rate obtained at longer times of assay after Mg^{2+} and CO_2 activation (17, 33, 34) or when the assay is initiated with Mg^{2+} and CO_2 after RuBP preincubation (17, 33). RuBP appears to reduce the maximal velocity of the enzyme or to increase the concentration of CO_2 required for maximal activity, or both.

ACTIVATION OF RuBP CARBOXYLASE IN THE CHLOROPLAST

The RuBP carboxylase activity of intact spinach chloroplasts prepared and stored at pH 6.7 and 0° to 4°C remains reasonably con- stant for at least 3 to 4 hr. Rates of about 200 µmole CO_2 fixed per mg Chl per hr are typically observed upon lysis of the plastids to release the carboxylase (5). The rate of fixation is linear for at least 60 to 120 sec, which indicates a lack of significant changes in enzyme activation during this time. Assays based on the initial 30 to 60 sec of reaction after lysis of the chloroplast have been used to indicate the degree of activation of RuBP carboxylase in the chloroplast (5). The degree of activation as measured by this technique corresponds to the fraction of the enzyme existing as the $E-CO_2$ and $E-CO_2-Mg^{2+}$ complexes because the formation of $E-CO_2$ by CO_2 binding to the enzyme is a slow process whereas the binding of Mg^{2+} to $E-CO_2$ to form $E-CO_2-Mg^{2+}$ is rapid.

When intact chloroplasts were incubated at 25°C, changes in the degree of activation of the carboxylase were observed. These changes were not due to chloroplast breakage or to the release of carboxylase into the medium. Incubation in the absence of CO_2 resulted in a

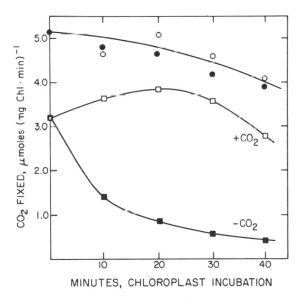

Figure 1. Reactivation of RuBP carboxylase after release from chloro- plasts. Chloroplasts were incubated in the dark at pH 7.8 and 25°C for the indicated times in the presence (o, □) or absence (•, ■) of 280 µM CO_2 (11 mM $NaHCO_3$). The activity was determined im- mediately upon release from the chloroplasts (□, ■) or after full activation (o, •) with 25 mM $MgCl_2$ and 20 mM $NaH^{14}CO_3$ (pH 7.8). Assay times were 60 sec. (See ref. 5.)

gradual decline in the initial activity of the released chloroplast
RuBP carboxylase (Figure 1) at a rate that varied somewhat between
preparations and had a half-time of 15 to 40 min. For a given preparation, the rate of decline was similar regardless of the pH of
the suspension media between 6.6 and 8.4. Rechilling the chloroplast suspension halted further loss in carboxylase activity but
did not restore it to the original value. Addition of CO_2, as
$NaHCO_3$, to intact chloroplasts incubated in the dark, stabilized or
increased the activity of the chloroplast RuBP carboxylase (Figure 1). With 11 mM $NaHCO_3$, an activation 50 to 60% higher than
the initial carboxylase activity was observed after 10 min of chloroplast incubation.

The carboxylase could also be reactivated by CO_2 after its release from the chloroplast. The released RuBP carboxylase was incubated for 4 min with $NaHCO_3$ and Mg^{2+} before initiation of fixation
by addition of RuBP (Figure 1, top curve). These conditions are
similar to those used for maximal activation of purified RuBP carboxylase (4-7, 17, 27, 35, 36). The rate determined in this manner for
fully Mg^{2+}-CO_2-stimulated RuBP carboxylase was independent of the
presence of CO_2 during incubation of the intact chloroplasts and of
the level of activation prior to lysis of the chloroplasts. This
again indicated that the decline in activity of chloroplast carboxylase in the absence of CO_2 can be reversed.

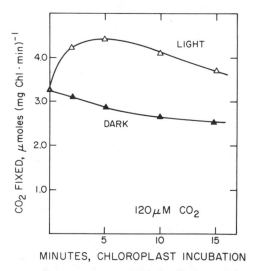

Figure 2. Stabilization of chloroplast RuBP carboxylase activity
during incubation with CO_2. Chloroplasts (33 μg Chl in 575 μl)
were incubated at pH 7.8, 25°C, in the dark (▲) or light (Δ). Assay
time was 60 sec. $NaHCO_3$ at 4.5 mM was added to the chloroplast incubation medium (calculated CO_2 concentration was 120 μM). (See ref. 5.)

In the presence of CO_2 the activity is always higher for RuBP
carboxylase from illuminated chloroplasts than for that from dark
chloroplasts (Figure 2). The activation in the light in the presence of CO_2 was studied with several photosynthetic inhibitors.
The electron transport inhibitor DCMU and the uncoupler CCCP both
abolished the light-dependent activation in the presence of CO_2.
Partial inhibition of the light-dependent activation was seen with
35 mM acetate, which has been shown to equilibrate the pH of the
medium and the chloroplast stroma (39). Partial inhibition was
observed also with D,L-glyceraldehyde.

These responses of the degree of activation of RuBP carboxylase
located within intact chloroplasts fit the kinetics of the purified
enzyme. The chloroplast carboxylase is inactivated in the dark under conditions such that the $E\text{-}CO_2$ complex can dissociate in a CO_2-
deficient medium. Addition of CO_2 results in activation of the carboxylase. The light activation and dark inactivation can be explained by light-dependent changes of Mg^{2+} concentration in the
chloroplast, which have been measured at 1 to 3 mM (40).

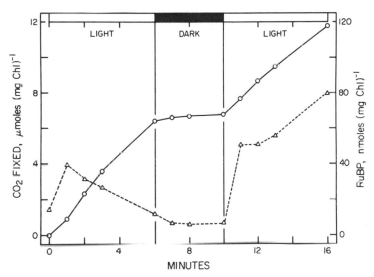

Figure 3.   Comparison of CO_2 fixation with levels of RuBP in isolated
intact spinach chloroplasts. CO_2 fixation (o—o) was determined in
solution C plus PP_i (4.0 mM), $NaH^{14}CO_3$ (10 mM, 0.5 µCi µmole^{-1}), and
chloroplasts (145 µg Chl ml^{-1}). Samples for RuBP (Δ---Δ) were taken
from a similar reaction mixture containing $NaH^{12}CO_3$ and diluted tenfold into 25 mM Hepes-NaOH (pH 8.0), 20 mM $MgCl_2$, 0.4 mM 2,6-dichlo-
rophenolindophenol, 4 µM carbonylcyanide m-chlorophenylhydrazone, and
10 mM $NaH^{14}CO_3$ (5.0 µCi µmole^{-1}). This technique allows for a rapid
measurement (<10 min) of chloroplast RuBP without interference from
other metabolites. (Sicher, Bahr, and Jensen, in preparation.)

At air levels of CO_2 and a stroma pH of 8.0 there is a significant amount of active carboxylase in the chloroplast. With lower stroma pH values, as expected in the dark, the binding of CO_2 to form the E-CO_2 complex is reduced and the carboxylase is less active (5). The inactivation by lower pH values and the drop in Mg^{2+} concentration in the stroma appear to be too slow to fully explain the more rapid light-to-dark cessation of CO_2 fixation seen with intact chloroplasts (41). Most probably, the lower pH reached in the dark affects catalysis before it affects activation of the enzyme. At longer times, however, the activation of the enzyme does decrease in response to the conditions in the chloroplast in the dark. This degree of activation (measured in the E-CO_2 and E-CO_2-Mg^{2+} forms) is usually about 20 to 50% that of the fully activated carboxylase.

LEVELS OF RuBP AND PHOTOSYNTHETIC CO_2 FIXATION

CO_2 fixation depends on the operation of the Calvin cycle to produce RuBP, which is required as a substrate for RuBP carboxylase but can also have a negative effect on the enzyme's activity. We have studied the RuBP level in isolated chloroplasts and its effect on the rate of photosynthesis and activation of RuBP carboxylase, and we have determined the level of RuBP necessary to saturate CO_2 fixation.

Metabolic studies with both algae (42) and intact isolated chloroplasts (41) indicate that RuBP carboxylase is active in the light and inactive in the dark. Recent measurements of RuBP in intact chloroplasts confirm these observations (Figure 3). Upon turning on the light there was a brief initial lag in CO_2 fixation, during which the chloroplast RuBP pool increased or "burst," followed by a steady decrease in RuBP accompanying linear fixation of CO_2. When the lights were turned off, photosynthetic CO_2 fixation ceased, and the amount of RuBP decreased but remained significant. CO_2 fixation resumed in the light and was accompanied by a dramatic and continued increase in the chloroplast RuBP level. The higher RuBP levels during the second period of illumination indicate that RuBP availability is not the rate-determining factor for CO_2 fixation.

Previous investigations have established that pyrophosphate (PP_i) added to isolated chloroplasts stimulates CO_2 fixation (23, 43-45). During CO_2 fixation, PP_i sustains a greater chloroplast level of RuBP and a greater rate of CO_2 fixation (Figure 4). This is consistent with the hypothesis that PP_i increases the level of Calvin cycle intermediates. During the first minute of illumination there was a burst in chloroplast RuBP, which in the absence of PP_i subsequently dropped to a steady-state level of 8 to 10 nmoles RuBP per mg Chl. Chloroplasts treated with PP_i accumulated RuBP, especially as the rate of CO_2 fixation declined. In light, isolated and intact chloroplasts can accumulate large amounts of RuBP, especially in the absence of CO_2 and O_2. Under N_2 and with 2 mM PP_i, >320 nmole RuBP per mg Chl has been measured after 40 min light.

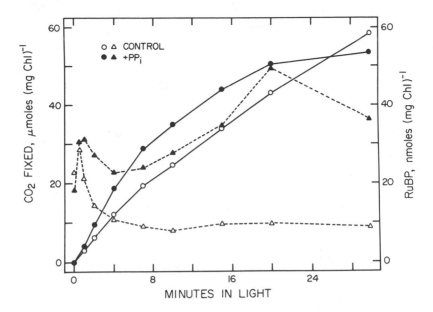

Figure 4.   Effect of pyrophosphate on CO_2 fixation and chloroplast
levels of RuBP.   In the presence (●, ▲) and absence (o, Δ) of PP_i
(1.25 mM), CO_2 fixation (——) was determined with solution C plus
10.4 mM $NaH^{14}CO_3$ (0.5 μCi μmole[-1]) and chloroplasts (79 μg Chl
ml[-1]).   Chloroplast RuBP (---) was determined from similar reaction
mixtures (see Figure 3).

LIMITING LEVELS OF RuBP

        CO_2 fixation by isolated chloroplasts was completely inhibited
by 15 mM D,L-glyceraldehyde, with a corresponding drop in RuBP to
almost zero during the first 2 min of illumination (Figure 5).
Chloroplasts treated with D,L-glyceraldehyde lose the ability to
generate RuBP, consistent with the hypothesis that D,L-glyceraldehyde
inhibits the transketolase reaction (46).   When treated with D,L-
glyceraldehyde at levels that only partially inhibit (1.0 mM),
chloroplasts produced only 2 to 3 nmoles less RuBP per mg Chl than
did the untreated control; however, CO_2 fixation was inhibited by
>40%.   The RuBP level of the untreated chloroplasts (15 to 17 nmoles
per mg Chl) appears to be near or at the point where the rate of CO2
fixation becomes limiting.   In our experience with many preparations,
RuBP levels become limiting for photosynthetic CO_2 fixation in
chloroplasts at 15 to 25 nmoles RuBP per mg Chl.

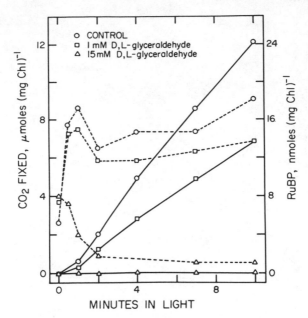

Figure 5. The effect of D,L-glyceraldehyde on CO_2 fixation and levels of chloroplast RuBP. CO_2 fixation (——) determined with solution C plus PP_i (1.0 mM), $NaH^{14}CO_3$ (8.3 mM, 0.5 µCi µmole^{-1}), chloroplasts (105 µg Chl ml^{-1}), and either no (o), 1.0 mM (□), or 15.0 mM (Δ) D,L-glyceraldehyde. RuBP(---) was determined from similar media (see Figure 3).

LIGHT INTENSITY AND RuBP LEVELS

At a limiting light intensity (25 µeinstein·m^{-2}·sec^{-1}) the initial rate of CO_2 fixation was 18.9 µmoles per mg Chl per hr (Figure 6), compared with an initial rate of 98.3 for the same chloroplasts at saturating light intensity (800). Although the kinetics of the light-on RuBP burst were modified at the lower light intensity, the chloroplast RuBP pool size was similar during the 4 to 10-min period in both experiments. This strongly suggests that the reduced fixation of CO_2 at the limiting light intensity was not caused by the lack of availability of RuBP. However, if iso-lated spinach chloroplasts are initially deficient in Calvin cycle intermediates they may not generate sufficient RuBP at low light in-tensities, causing decreased rates of CO_2 fixation (Figure 7). When the pool of Calvin cycle intermediates is maintained by sup-plying the chloroplasts with triose phosphates from fructose bis-phosphate (FBP), the RuBP level measured after 10 min of CO_2 fixa-tion (a time sufficient for subsidence of the light-on RuBP burst) actually increased at the lower light intensities (Figure 8). Even

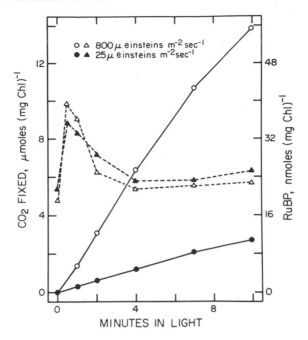

Figure 6. The effect of light intensity on CO_2 fixation and levels
of RuBP. The light intensity was either 800 (o, Δ) or 25 (●, ▲) μ-
einstein·m^{-2}·sec^{-1}. CO_2 fixation (——) reaction mixtures contained
solution C plus NaH$^{14}CO_3$ (9.2 mM, 0.5 μCi μmole^{-1}), PPi (0.9 mM),
and chloroplasts (117 μg ml^{-1}). RuBP (---) was determined from similar solutions containing NaH$^{12}CO_3$ (see Figure 3).

though CO_2 fixation responds to light and is less at lower intensities, this may be due not necessarily to a limitation on RuBP
synthesis but possibly to limitations on activity of the
RuBP carboxylase
        Because he found that "catalytic amounts" of R5P shortened the
initial lag in CO_2 fixation, Walker (45) attributed the lag effect
to a lack of intermediates in the Calvin cycle. We have found that
RuBP levels rapidly "peak" during the lag period, which indicates
that the synthesis of RuBP exceeds its consumption. This implies
that the lag can be caused by light-dependent activation of the
carboxylase and of CO_2 fixation, especially when the chloroplast
has enough RuBP. At longer times, beyond 10 to 15 min, when CO_2
fixation in isolated chloroplasts begins to drop, the level of
RuBP again rises. This indicates that photosynthesis is not limited by lack of RuBP, but that carboxylation by RuBP carboxylase is
limiting production of PGA.
        Although not measured in these experiments, the drop in RuBP
carboxylase activity is most likely due to a decrease in the chloroplast Mg^{2+} or pH gradient between the thylakoid membranes and the

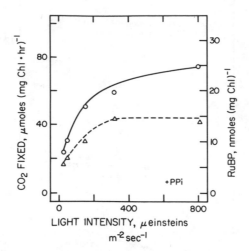

Figure 7. CO_2 fixation and spinach chloroplast RuBP levels as a
function of light intensity. CO_2 fixation (o—o) reaction mixtures
contained solution C plus $NaH^{14}CO_3$ (10.4 mM, 0.5 μCi μmole^{-1}), PP_i
(1.25 mM), and chloroplasts (89 μg Chl ml^{-1}). Rates of $^{14}CO_2$ incorporation were determined between 0 and 10 min. Samples for RuBP
(Δ---Δ) were taken from similar reaction mixtures (see Figure 3).
The chloroplast RuBP level was measured at 10 min in the light.

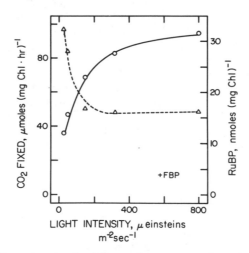

Figure 8. CO_2 fixation and chloroplast RuBP levels as a function of
light intensity in the presence of fructose bisphosphate. Conditions
were the same as for Figure 7 except that FBP (1.0 mM) and 1.25 units
ml^{-1} of aldolase were added to provide an excess of carbon to support the Calvin cycle.

stroma. Obviously electron transport and photophosphorylation producing NADPH and ATP are not limiting, as the RuBP levels are sufficient. However, the role of electron transport in maintaining
the proper pH and Mg^{2+} gradient across the thylakoid membrane must
be considered a major factor in the regulation of RuBP carboxylase
activity in the chloroplast (47).

The lower limit of RuBP capable of saturating the carboxylation
reaction (15 to 25 nmoles per mg Chl) corresponds to a concentration
in the chloroplast stroma (volume, 25 µl per mg Chl) of 0.6 to 1.0
mM RuBP. With a maximum of eight binding sites per RuBP carboxylase
molecule, there would be 3 to 4 mM binding sites for RuBP present in
the chloroplast (6). As the K_m(RuBP) is 20 to 30 µM, most of the
chloroplast RuBP would be bound to the RuBP carboxylase. It might
be expected that catalysis by RuBP carboxylase would be limited by
RuBP if only part of the binding sites were occupied. No evidence
has been found for significant interaction between RuBP binding sites
(15). Perhaps not all the binding sites are available to RuBP on
the RuBP carboxylase in the chloroplast. Wishnick et al. (15) noted
that, at high ionic strength, RuBP carboxylase has a decreased number of binding sites. In 0.25 M Tris-Cl, RuBP carboxylase was estimated to have 4 $\pm$ 1 RuBP binding sites per molecule.

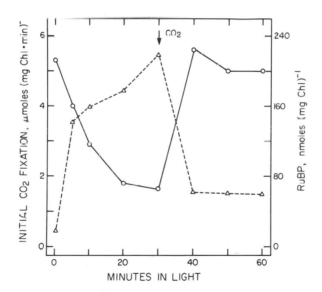

Figure 9. Activity of RuBP carboxylase and levels of RuBP in chloroplasts in the light. Chloroplasts (36 µg Chl ml^{-1}) were illuminated in solution C plus PP$_i$ (4.0 mM) which had been flushed with
N_2 to remove CO_2 and O_2. At 30 min, 3.9 mM NaH$^{12}CO_3$ was added.
Initial CO_2 fixation (o—o) of the chloroplast RuBP carboxylase was
measured immediately upon lysis, as for Figure 1 (5). RuBP (Δ---Δ)
was determined as for Figure 3.

High levels of RuBP in the chloroplast might affect RuBP carboxylase activity. Besides serving as a substrate, RuBP inactivates
the purified enzyme (6). If chloroplasts are illuminated in N_2 in
the absence of CO_2 and with PP_i, then RuBP levels increase. Figure
9 shows that RuBP production rose to 220 nmoles per mg Chl in 30
min while RuBP carboxylase activation, measured as initial activity
upon lysis, declined 70%. Subsequent addition of CO_2 dropped RuBP
to 60 nmoles per mg Chl and restored the RuBP carboxylase activity
in the intact chloroplasts. The drop in RuBP carboxylase activity
in the chloroplast was probably attributable more to the lack of
CO_2 (5). The degree of inactivation by high levels of RuBP is difficult to determine in this experiment.

A CO_2 fixation experiment with intact chloroplasts in a special
medium in the dark (48) gave results suggesting that RuBP levels in
the chloroplast could be involved in the inactivation of RuBP carboxylase. For maximal CO_2 fixation (still at rates only 10 to 15%
those observed in the light) a reducing agent such as dithioerythritol (DTE) is necessary. FBP and aldolase are also included, to produce triose phosphates that enter the chloroplast to provide carbon

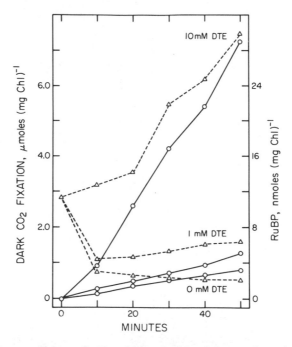

Figure 10. Comparison of CO_2 fixation and RuBP levels as influenced
by dithioerythritol (DTE) in intact chloroplasts in the dark. CO_2
fixation (o—o) was measured with intact chloroplasts (73 µg Chl
ml^{-1}) in 0.33 M sorbitol, 0.1 M Ches-KOH, pH 8.9, 5 mM FBP, 2 mM
oxalacetate, 10 mM $NaH^{14}CO_3$ (0.5 µCi µmole^{-1}), 10 mM $KHPO_4$, pH 8.9,
in dark (25oC) with DTE as given. RuBP (Δ---Δ) was determined from
similar solutions containing $NaH^{12}CO_3$ (see Figure 3).

for RuBP and ATP synthesis.   In Figure 10, $^{14}CO_2$ fixation is compared
with the levels of RuBP produced in the chloroplast in the dark.
With little (1.0 mM) or no DTE, RuBP levels dropped to <7 nmoles
per mg Chl and CO_2 fixation was low.   With 10 mM DTE, maximal CO_2
fixation was noted at 9.2 μmoles CO_2 per mg Chl per hr with a steady
rise in RuBP.   The only difference between the three runs was the
amount of DTE, which activated Ru5P kinase to produce RuBP (48).
Each run had the same levels of CO_2 and other ingredients, yet the
higher the DTE level the greater the rate of RuBP carboxylase inactivation in the chloroplast (Figure 11).   DTE itself has no effect
on activation of RuBP carboxylase, which suggests that the inactivation observed could have been related to the buildup of RuBP.
     The evidence that saturating levels of RuBP can exist in iso-
lated chloroplasts when CO_2 fixation is limited by carboxylase
activity suggests a change in emphasis on what controls the rate of
photosynthesis.   It has been observed that the activities of several
Calvin cycle enzymes are enhanced in the light: Ru5P kinase (49),
FBP and sedoheptulose 1,7-bisphosphate (SBP) phosphatases (50-52),
and glyceraldehyde 3-phosphate (G3P) dehydrogenase (53).   Electron
transport and photophosphorylation in the thylakoids utilize light
energy to form NADPH and ATP, which are used to reduce PGA and
phosphorylate Ru5P.   However, when RuBP levels are saturating the

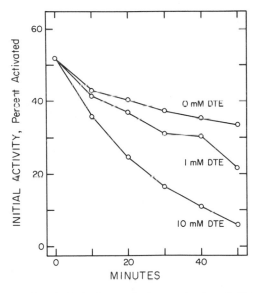

Figure 11.   Change in initial activity of the RuBP carboxylase as
influenced by DTE during chloroplast CO_2 fixation in the dark.   In
the experiment of Figure 10, initial activity was measured upon
chloroplast lysis (see Figure 1) but with samples from the dark
fixation media having $NaH^{12}CO_3$.

chloroplast, then these factors, essential for RuBP synthesis, would not be involved in limiting photosynthetic CO_2 fixation. The activity of the chloroplast RuBP carboxylase as it responds to "activation" and "catalysis" by CO_2 and other conditions in the stroma environment (e.g., pH, Mg^{2+}) must be considered as the factors controlling the rate of photosynthesis.

REFERENCES

1. Black, C. C. Jr., Annu. Rev. Plant Physiol. 24, 253-86 (1973).
2. Gaastra, P., Meded. Landbouwhogesch. Wageningen 59, 1-68 (1959).
3. Thorne, J. H. and Koller, H. R., Plant Physiol. 54, 201-7 (1974).
4. Wareing, P. F., Khalifa, M. M., and Treharne, K. J., Nature London 220, 453-7 (1968).
5. Bahr, J. T. and Jensen, R. G., Arch. Biochem. Biophys. 185, 39-48 (1978).
6. Jensen, R. G. and Bahr, J. T., Annu. Rev. Plant Physiol. 28, 379-400 (1977).
7. Kung, S. D., Science 191, 429-34 (1976).
8. McFadden, B. A. and Tabita, F. R., Biosystems 6, 93-112 (1974).
9. Nishimura, M. and Akazawa, T., Biochem. Biophys. Res. Commun. 54, 842-8 (1973).
10. Nishimura, M., Takabe, T., Sugiyama, T., and Akazawa, T., J. Biochem. 74, 945-54 (1973).
11. Baker, T. S., Eisenberg, D., and Eiserling, F. A., Science 196, 293-5 (1977).
12. Kawashima, N. and Wildman, S. G., Biochem. Biophys. Res. Commun. 41, 1463-8 (1970).
13. Rutner, A. C., Biochem. Biophys. Res. Commun. 39, 923-9 (1970).
14. Siegel, M. I. and Lane, M. D., J. Biol. Chem. 248, 5486-98 (1973).
15. Wishnick, M., Lane, M. D., and Scrutton, M. C., J. Biol. Chem. 245, 4939-47 (1970).
16. Badger, M. R. and Andrews, T. J., Biochem. Biophys. Res. Commun. 60, 204-10 (1974).
17. Bahr, J. T. and Jensen, R. G., Plant Physiol. 53, 39-44 (1974).
18. Lilley, R. M. and Walker, D. A., Plant Physiol. 54, 1087-92 (1975).
19. Walker, D. A., New Phytol. 72, 209-35 (1973).
20. Siegel, M. I., Wishnick, M., and Lane, M. D., in The Enzymes, pp. 169-92, P. D. Boyer, Editor, Academic Press, New York, 1972.
21. Kawashima, N. and Wildman, S. G., Annu. Rev. Plant Physiol. 21, 325-58 (1970).
22. Bahr, J. T. and Jensen, R. G., Arch. Biochem. Biophys. 164, 408-13 (1974).
23. Jensen, R. G. and Bassham, J. A., Proc. Natl. Acad. Sci. USA 56, 1095-101 (1966).

24. Krause, G. H., Thorne, S. W., and Lorimer, G. H., Arch. Bio-
    chem. Biophys. 183, 471-9 (1977).
25. Bowes, G. and Ogren, W. L., J. Biol. Chem. 247, 2171-6 (1972).
26. Laing, W. A., Ogren, W. L., and Hageman, R. H., Plant Physiol.
    54, 678-85 (1974).
27. Badger, M. R. and Lorimer, G. H., Arch. Biochem. Biophys. 175,
    723-9 (1976).
28. Chollet, R. and Anderson, L. L., Arch. Biochem. Biophys. 176,
    344-51 (1976).
29. Ryan, F. J., Barker, R., and Tolbert, N. E., Biochem. Biophys.
    Res. Commun. 65, 39-46 (1975).
30. Andrews, T. J., Badger, M. R., and Lorimer, G. H., Arch. Bio-
    chem. Biophys. 171, 93-103 (1975).
31. Bowes, G., Ogren, W. L., and Hageman, R. H., Plant Physiol.
    56, 630-3 (1975).
32. Chu, D. K. and Bassham, J. A., Plant Physiol. 54, 556-9 (1974).
33. Chu, D. K. and Bassham, J. A., Plant Physiol. 55, 720-6 (1975).
34. Laing, W. A., Ogren, W. L., and Hageman, R. H., Biochemistry
    14, 2269-75 (1975).
35. Lorimer, G. H., Badger, M. R., and Andrews, T. J., Biochemistry
    15, 529-36 (1976).
36. Laing, W. A. and Christeller, J. T., Biochem. J. 159, 563-70
    (1976).
37. Pon, N. G., Rabin, B. R., and Calvin, M., Biochem. Z. 338,
    7-19 (1963).
38. Chu, D. K. and Bassham, J. A., Plant Physiol. 52, 373-9 (1973).
39. Werden, K., Heldt, H. W., and Milovancev, M., Biochim. Bio-
    phys. Acta 396, 276-92 (1975).
40. Portis, A. R. Jr. and Heldt, H. W., Biochim. Biophys. Acta 449,
    434-46 (1976)
41. Jensen, R. G. and Bassham, J. A., Biochim. Biophys. Acta 153,
    227-34 (1968).
42. Pedersen, T. A., Kirk, M., and Bassham, J. A., Physiol. Plant.
    19, 219-31 (1966).
43. Jensen, R. G. and Bassham, J. A., Biochim. Biophys
    219-26 (1968).
44. Stankovic, Z. S. and Walker, D. A., Plant Physiol. 59, 428-
    32 (1977).
45. Walker, D. A., New Phytol. 72, 209-35 (1973).
46. Slabas, A. R. and Walker, D. A., Biochem. J. 153, 613-19 (1976).
47. Werdan, K., Heldt, H. W., and Milovancev, M., Biochim. Biophys.
    Acta 396, 276-92 (1975).
48. Latzko, E., Garnier, R. V., Gibbs, M., Biochem. Biophys. Res.
    Commun. 39, 1140-4 (1970).
49. Avron, M. and Gibbs, M., Plant Physiol. 53, 136-9 (1974).
50. Anderson, L. E., in Proc. 3rd Int. Congr. Photosynth., pp. 1393-
    405, M. Avron, Editor, Elsevier, Amsterdam, 1974.
51. Buchanan, B. B., Schurmann, P., and Kalberer, P. P., J. Biol.
    Chem. 246, 5952-9 (1971).

52.   Schurmann, P. and Buchanan, B. B., Biochim. Biophys. Acta 376,
      189-92 (1975).
53.   Ziegler, H. and Ziegler, I., Planta 65, 369-80 (1965).

DISCUSSION

BASSHAM: How did you assay for RuBP?

JENSEN: We use an enzymatic assay in which we take the chloroplast itself right out of the reaction mixture and break it open in
the presence of inhibitors that stop the carbon from going from R5P
to RuBP; that is the control point in the whole assay. The chloroplast uses its own carboxylase to consume the RuBP, with saturating
CO_2 and magnesium. We spent a number of months working on that
control. It allows the assay to be done rapidly, without going
through the perchloric acid treatment, but one must know the pit-
falls. For example, if there is too much RuBP present, the enzyme
becomes inactivated and one must add more carboxylase. We compared
this assay with the acid treatment and found it similar but a lot
faster and easier for us.

L. ANDERSON: The RuBP levels you see are remarkably constant.
You attribute this only to modulation of carboxylase activity. But
in a cyclic steady-state system such as the Calvin cycle, don't you
also have to postulate control of the two drains on the cycle,
namely, starch formation and triose phosphate export?

JENSEN: You are right, but you have to understand the context
in which I am making these suggestions. I am simply saying that
when RuBP is saturating, there are other processes that are not
saturated. What I wanted to show is that the chloroplast has the
capability of running at low light intensity; fixation does drop
but this is due not to lack of substrate but clearly to inactivation
of carboxylase.

WALKER: A point I was trying to make this morning is that one
must move away from the initial induction period into what is clearly
the steady-state situation. For example, your slides (Figures 7 and
8) showing the effects of RuBP and fructose 1,6-bisphosphate on CO_2
fixation showed no stimulation, but you had barely any induction
in Figures 3 to 6.

JENSEN: That's right.

WALKER: If you changed your conditions slightly so that you
had a prolonged induction, you could certainly see effects of intermediates. If you put in high phosphate, you can get no photosynthesis,
and then you can reverse the situation with R5P. In weighing the
relative contributions of enzyme activity and other control mechanisms
(and I think Dr. Anderson's point about the sinks is very important)
we must separate the initial induction period.

JENSEN: Would you suggest that it is good to follow an experiment for 8 to 10 min, as we were doing when studying fixation vs.
RuBP levels?

WALKER: Yes.

BIOCHEMICAL AND GENETIC STUDIES OF THE SYNTHESIS AND

DEGRADATION OF RuBP CARBOXYLASE

Ellen Simpson

Biology Department, Washington University
St. Louis, Missouri 63130

The role of ribulose bisphosphate (RuBP) carboxylase in photosynthesis and photorespiration, its structure, and the genetics of
subunit transmission have been well studied (1). Other aspects of
the biology of RuBP carboxylase have received less attention; in
particular, the timing and the genetic regulation of its synthesis
and degradation require further study. Knowledge of the timing and
the effects on it of environmental and developmental cues will contribute to an understanding of whether the amount of RuBP carboxylase
protein is a limiting factor in photosynthesis. Examination of the
large and small subunits separately should clarify whether the small
subunit is necessary to induce synthesis of the large one. The or-
derly progression of plastid development (2) and of leaf senescence
(3) indicates that these are well regulated processes. Nonetheless
the genetic regulation of RuBP carboxylase synthesis and degradation
must be characterized and described before these developmental
processes can be fully understood.

DEGRADATION OF RuBP CARBOXYLASE

Synthesis and degradation of RuBP carboxylase have been studied
in intact seedlings and in excised leaves. Peterson et al. (4) reported that in the first leaf of 6-day-old barley seedlings kept in
continuous light following a $^{14}CO_2$ labeling period (6 hr) the amount
of RuBP carboxylase protein and the radiospecific activity of this
protein remained constant, which indicated that no degradation was
occurring. In continuous darkness there was a loss of RuBP carboxylase protein and a low level of incorporation of $^{14}CO_2$ into it.
These workers suggest that the majority of the RuBP carboxylase is
synthesized in the light and degraded in the dark. After completion

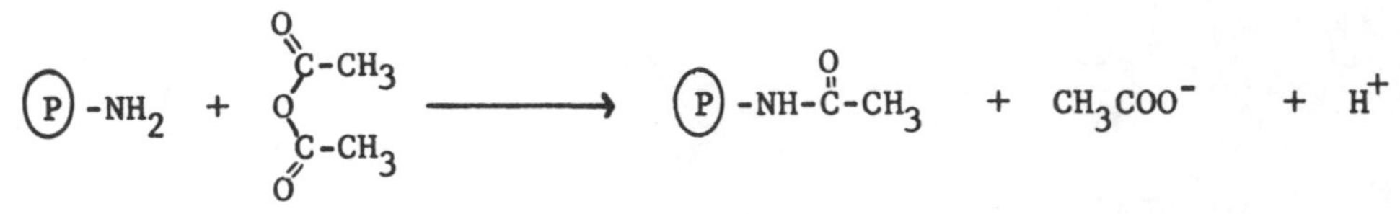

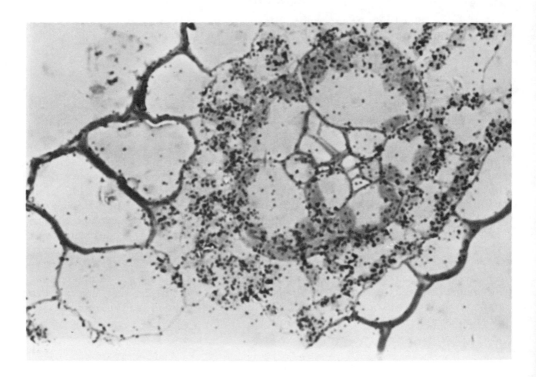

Figure 1.   Acetylation of protein with acetic anhydride.
From Means and Feeney (33).

Figure 2.   Autoradiography of ^{3}H-acetylated corn leaf.   Leaf was
given 40 μCi ^{3}H-acetic anhydride in 2.5 μl benzene, fixed in glutar-
aldehyde and osmium, coated with Kodak NTB emulsion, and exposed
for 5 weeks at 5°C.   Application of label to the upper leaf surface
has resulted in even labeling across the leaf.   Total magnification
is 1280x.

of expansion of <u>Perilla</u> (5) and wheat (6) leaves there is a net loss
of RuBP carboxylase protein, suggesting that degradation exceeds synthesis at this time. It is clear that in senescing leaves RuBP carboxylase protein is lost (7, 8) and synthesis of RuBP carboxylase is
depressed (6, 9). Quantitative determination of the half-time for
degradation of RuBP carboxylase has not been reported for the various
stages of leaf development, however.

       Commonly used methods for <u>in vivo</u> protein labeling involve feed-
ing radioactive or isotopically labeled precursors which are subsequently incorporated into newly synthesized protein (10-12). This
allows determination of protein synthesis and degradation, but requires knowledge of the specific activity of the intracellular precursor pool for calculation of actual rates for comparison of any
two treatment conditions. In addition, reincorporation of labeled
amino acids can result in overestimation of the protein half-
life (13).

       In principle these problems are circumvented by the use of ^{3}H-
acetic anhydride, which labels preexisting protein by acetylating
primary amino groups, such as the N-terminus and the ϵ-amino group
of lysine (Figure 1). This method has the following advantages. (i)
A brief pulse label is attained because the acetic anhydride rapidly
enters the cell and reacts with available molecules, and unreacted
acetic anhydride spontaneously and completely hydrolyzes to acetic acid within minutes of exposure to water. (ii) Incorporation of
labeled amino acids from degraded labeled protein is unlikely since
they are modified. (iii) Labeling is independent of protein synthesis precursor pool size. (iv) Because the labeled compound is applied directly to the leaf suface, turnover can be studied on whole
plants and in field situations (14).

       When 25 μCi of ^{3}H-acetic anhydride (500 μCi/μmole) is applied
to a corn leaf, 1% of the label is recovered as material precipitated
by trichloroacetic acid (TCA) (Table 1a) and an average of one out of
every 200 RuBP carboxylase molecules in the leaf is labeled on one
residue (Table 1b). The ^{3}H-acetic anhydride applied to the upper
leaf surface penetrates across the leaf and labels all cell types
(Figure 2).

       Control experiments were performed to show that acetylation
does not induce turnover, as some abnormal proteins are degraded
more rapidly in bacterial and mammalian cells (15). To determine
the effect of acetylation on enzymatic activity of RuBP carboxylase,
acetylation was performed <u>in vitro</u> in the presence and absence of
its substrate, RuBP. Enzymatic activity was assayed by $^{14}CO_2$ fixation (16). Partially purified RuBP carboxylase, modified in the
presence of 5 mM RuBP (a 15-fold excess of RuBP light subunit),
bound 26 moles acetic anhydride per mole enzyme and showed enzymatic
activity 74% that of the control (unmodified, +RuBP) enzyme (Figure 3). Acetylation in the absence of RuBP resulted in approximately the same extent of acetylation, 22 moles acetic anhydride per
mole enzyme, but enzymatic activity was much lower, 32% that of the
control (unmodified, -RuBP). RuBP apparently protects a "critical"

Table 1

In Vivo Labeling With ^{3}H-Acetic Anhydride

The third leaf of a 3-week-old corn plant was given 25 μCi of ^{3}H-acetic anhydride in 2.5 μl benzene. The leaf was homogenized, and the homogenate was centrifuged at 30,000 g for 15 min. The supernatant was made 15% in TCA by dilution with 30% TCA. The TCA precipitate was collected by centrifugation at 9000 g and washed with 15% TCA and 3:1 ethanol:acetic acid in succession. The pellet was dissolved in 0.2 N NaOH, and the supernatant and pellet were counted separately in Packard scintillation fluid. An aliquot of the high speed supernatant was immunoprecipitated with anti-RuBP carboxylase serum, the immunoprecipitates were counted, and the protein was determined by the Lowry method.

a. Recovery of label applied as ^{3}H-acetic anhydride (500 μCi/ mole):

	cpm	% of total
Total applied	8.76×10^7	100
TCA precipitable	0.12×10^7	1.4
TCA soluble	2.64×10^7	30.4

b. ^{3}H-acetic anhydride labeling of RuBP carboxylase in vivo:

    500 cpm in RuBP carboxylase per μCi applied
    30,000 cpm per mg RuBP carboxylase
    1 labeled RuBP carboxylase molecule per 200 available
        RuBP carboxylase molecules

site, possibly the active site, of the enzyme from acetylation. The protection of RuBP carboxylase enzymatic activity by substrate levels of RuBP during modification by a lysine-reactive reagent was previously shown by Schloss and Hartman (17), using the affinity label 3-bromo-1,4-dihydroxy-2-butanone 1,4-bisphosphate. Since the level of RuBP in the chloroplast in the light is about 5 mM (18), in vivo acetylation may be presumed to be occurring under conditions analogous to the in vitro conditions that result in modified, active RuBP carboxylase. In vivo acetylation results in an average of 1 mole acetic anhydride per modified RuBP carboxylase molecule, that is, those RuBP carboxylase molecules which are modified are acetylated on one residue only, on the average. Only 0.5 to 1.0% of the RuBP carboxylase molecules in the leaf are modified (Table 1b).

To determine whether acetylated enzyme is degraded faster than unmodified enzyme, excised corn leaves were double labeled, first with [14]C- and then with ^{3}H-glutamate after 24 or 72 hr. Half of the set of leaves were painted with unlabeled acetic anhydride 12 hr after ^{14}C-glutamate feeding was begun. Degradation of RuBP carboxylase is measured as loss of ^{14}C radioactive label, and preferential degradation of acetylated RuBP carboxylase would be detected as a higher ^{3}H/^{14}C ratio in RuBP carboxylase from acetylated leaves than in that from unmodified leaves. RuBP carboxylase protein was recovered from extracts with antiserum to spinach holoenzyme. Application of 10.5 μmoles acetic anhydride per leaf results in modification of 33% of the RuBP carboxylase molecules, on the basis of labeling of leaves at several levels of acetic anhydride (data not shown). With the observed standard deviation in the ^{3}H/^{14}C ratio, this level of acetylation is sufficient to reveal differential degradation of RuBP carboxylase if it is occurring. The observation (Table 2) that the ^{3}H/^{14}C ratios of RuBP carboxylase protein from control and acetylated leaves are not significantly different indicates that acetylated RuBP carboxylase is not degraded faster than unmodified enzyme. Taken with the evidence given above that enzyme acetylated in the presence of RuBP is not inactivated, these results show that acetylation does not induce degradation of RuBP carboxylase; therefore, the loss of radioactivity from the enzyme over time may be used as an assay of in vivo physiological degradation rate.

The degradation of labeled RuBP carboxylase was followed in third leaves of 27-day-old (from planting) corn. During the 6 days of the experiment the radioactive label in RuBP carboxylase protein decreased to 10% of the initial value (Figure 4), with an apparent half-time for degradation of 2 to 3 days. RuBP carboxylase protein per gram fresh weight decreased to 60% of the initial value between days 4 and 6 (Table 3).

Table 2

Double Labeling of Excised Leaves With U-^{14}C-Glutamic Acid and 2,3-^{3}H-Glutamic Acid, With and Without Acetic Anhydride (unlabeled)

The ^{3}H/^{14}C ratio of immunoprecipitated RuBP carboxylase is given for control (unacetylated) and treated (acetylated) leaves for each of two experiments. The lag between application of cold acetic anhydride and feeding of ^{3}H-glutamate was 24 hr in experiment a and 72 hr in experiment b. Each figure is the mean of determinations on 5 leaves with the standard deviation.

	Interval	^{3}H/^{14}C	
		Control	Acetylated
a.	24 hr	0.420±0.04	0.445±0.036
b.	72 hr	0.550±0.03	0.510±0.050

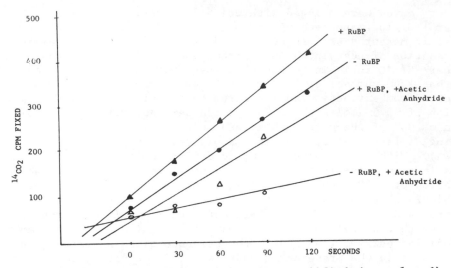

Figure 3.   Enzymatic activity of _in vitro_ modified (acetylated) and unmodified RuBP carboxylase.   Partially purified RuBP carboxylase was reacted with ^{3}H-acetic anhydride (500 μCi/μmole) with 500 moles acetic anhydride present per mole enzyme.   The four treatments were (i) -RuBP: unmodified RuBP carboxylase; (ii) +RuBP: unmodified enzyme incubated in the presence of 5 mM RuBP; (iii) -RuBP, +AcAn: enzyme acetylated in the absence of RuBP; and (iv) +RuBP, +AcAn: enzyme acetylated in the presence of 5 mM RuBP.   Acetylation was performed on ice in Hepes buffer, after Means and Feeney (33).   Enzyme activity was assayed by $^{14}CO_2$ fixation with preincubation for 10 min in 10 mM $NaHCO_3$, 25 mM $MgCl_2$, and 3 mM dithiothreitol in 0.16 M Hepes pH 8.0.   Reaction was initiated by the addition of 10 μCi $NaH^{14}CO_3$ and 5 mmoles RuBP.

It is unlikely that this loss of protein is due to preferential degradation of acetylated protein, since such a loss of RuBP carboxylase has been reported to occur in cucumber and _Perilla_ leaves after the completion of leaf expansion (5).   The corn leaves used in this experiment were fully expanded; the leaf area was constant throughout the experiment (data not shown).   Also, in experiments (reported above, Table 2) in which some leaves received unlabeled acetic anhydride while others did not, there was no observable difference in the loss of modified and unmodified RuBP carboxylase labeled with ^{14}C- and ^{3}H-glutamate.   The data suggest that for the first 4 days RuBP carboxylase protein is being degraded and resynthesized, and that degradation exceeds synthesis during the next period.   Thus in these mature, photosynthesizing leaves, there is concurrent synthesis and degradation.

Table 3

RuBP Carboxylase Protein

Corn plants 27 days old were labeled with 25 μCi each of ^{3}H-acetic
anhydride. Five leaves were harvested on days 0, 2, 4, and 6, and
extracts were prepared as described in the text. The extracts were
immunoprecipitated with anti-RuBP carboxylase serum, and the protein
precipitated was determined by the Lowry method. Each figure is
the mean of determinations on 5 leaves with the standard deviation.

Day	μg protein per g fresh weight
0	2734±615
2	2449±197
4	2679±529
6	2679±529

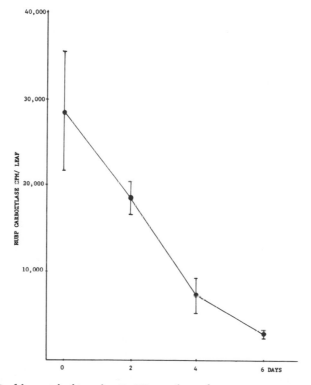

Figure 4. Radioactivity in RuBP carboxylase. Extracts were prepared from leaves labeled with 25 μCi ^{3}H-acetic anhydride by homog-
enization in Hepes buffer, centrifugation at 30,000 g, and Amicon
filtration of the supernatant. Aliquots of the extracts were immunoprecipitated with anti-RuBP carboxylase serum, the immunoprecipitates
were counted in Packard scintillation fluid, and the protein precipitated was determined by the Lowry method. Bars indicate the standard deviation of determinations on 5 leaves.

The demonstration that in vivo ^{3}H-acetic anhydride labeling is
a valid tool for studying the degradation of RuBP carboxylase allows
many interesting experiments with intact seedlings and plants. These
include tests with leaves of different developmental stages, plants
under stress, plants under different fertilization regimes, and
plants kept in continuous light or darkness.

GENETIC REGULATION OF RuBP CARBOXYLASE

The pattern of synthesis of RuBP carboxylase during leaf development and the genetic regulation of its synthesis and degradation are of great interest, as they determine when and how much
carboxylase will be present in the leaf. The genetic regulation of
the development of photosystems I and II has been studied in higher
plants (19, 20) and in Chlamydomonas (21), but data on the developmental genetics of components of the Calvin cycle are sparse.

Although the accumulation of RuBP carboxylase protein and the
development of CO_2 fixation activity are typically correlated with
greening, some etiolated tissues do manufacture the enzyme in the
dark. There is considerable species variability in dark levels of
RuBP carboxylase protein and activity in both monocots and dicots
(22-25). Mohr and co-workers (26) demonstrated that RuBP carboxylase
accumulation is not dependent on attainment of normal plastid organization, using mustard cotyledons grown under light regimes that produce chloroplasts in several stages of ultrastructural differentiation but with equal levels of RuBP carboxylase protein.

Maize leaves offer several advantages for attempts to separate
developmental events such as chlorophyll synthesis and RuBP carboxylase synthesis: their cells have an age gradient along the leaf
with the oldest tissue at the tip, and many well characterized pigment mutants are available. Use of a virescent mutant, with leaves
that are initially yellow but green progressively from the oldest
tissue at the tip toward the younger basal tissue, allows study of
the timing of appearance of chlorophyll and RuBP carboxylase. With
this type of maize mutant, it has been reported that greening is
highly correlated with RuBP carboxylase production (27, 28) as is
also found during greening of etiolated leaves of normal maize.

Screening a large number of such virescent mutants as well as
genotypes genetically incapable of greening, such as yellow (luteus,
1) and whites, should reveal whether chlorophyll production and RuBP
carboxylase synthesis are genetically linked. Both F_2 material of
"unclassified" mutants produced by EMS treatment of pollen with
selfed F_1 progeny to reveal recessives, and well characterized mutants such as iojap and 1*-blandy 4 (Figure 5) were screened for the
·presence of RuBP carboxylase protein by Ouchterlony double diffusion

Table 4

Mutants for Synthesis of RuBP Carboxylase

Plants were grown at 25°C in continuous light. Extracts of leaf tissue were tested for the presence of RuBP carboxylase protein by immunodiffusion in an Ouchterlony plate against antiserum to spinach
holoenzyme. A positive (+) response indicates a normal level of
RuBP carboxylase protein, and a negative (-) response indicates <10%
of the normal level. L and S indicate the long and short arms of
the chromosomes. The luteus phenotype is yellow.

Chlorophyll phenotype	Mutant designation	Genetic location	RuBP carboxylase phenotype at 25°C	Ref.
white	w-21A	5L	-	29
	iojap, white sectors	7L	-	30
luteus	1-129	unknown	-	29
	1-392	10L	+	29
	1*-blandy 4	4S (tentative)	+	31
	chloroplast mutator, yellow sectors	10L	+	32
virescent	w-828, green tips	9S	-	29
	w-828, yellow base		-	
	v-806c, green tips	9L	+	29
	v-806c, yellow base		-	

against antiserum to spinach holoenzyme. Anti-spinach RuBP carboxylase recognizes maize holoenzyme.

All green plants screened were positive and all white plants
and sectors were negative. The identification of luteus mutants
with normal levels of RuBP carboxylase (Table 4) and virescent mutants in which greening precedes RuBP carboxylase synthesis by several days clearly shows that chlorophyll synthesis is not required
for the induction or continuation of RuBP carboxylase synthesis.
It is not clear whether carotenoid synthesis is required for RuBP
carboxylase synthesis. To determine this, the 1*-blandy 4 mutation
could be put into an albino background and the RuBP carboxylase
phenotype tested. Characterization of the red-light and temperature
sensitivity of mutants will more clearly define the environmental
component of the regulation of RuBP carboxylase synthesis.

It is possible that genotypes found to be low in RuBP carboxylase are mutants with greatly increased degradation rather than with

Figure 5. Examples of chlorophyll phenotypes of mutants
screened (left to right): iojap, a green and white sec-
toring mutant; 1*-blandy 4, a luteus mutant; a normal
green seedling; w-21A, a white seedling.

decreased synthesis of RuBP carboxylase. This can now be tested
with the ^{3}H-acetic anhydride method. Albinescent mutants, seedling
lethals which are initially green but lose their pigment as they
age, are a likely class to screen for identification of mutants for
regulation of degradation.

    Acknowledgments: I would like to acknowledge valuable discus-
sions of turnover studies and advice given me by Joseph Varner. I
would also like to thank E. H. Coe, Jr., M. G. Neuffer, and P. Mascia
for gifts of seeds used in these studies and V. Walbot for assistance
with mutant analyses. Research was supported by grants from the
National Science Foundation to J. Varner (PCM 76-14178), V. Walbot
(PMS FS-06818), and M. G. Neuffer (PCM 76-10562).

REFERENCES

1. Kung, S. D., Annu. Rev. Plant Physiol. 28, 401 (1977).
2. Hoober, J. K., in Genetics and Biogenesis of Chloroplasts and Mitochondria, pp. 87-94, T. Bucher et al., Editors, North-Holland, New York, 1976.
3. Choe, H. T. and Thimann, K. V., Planta 135, 101-7 (1977).
4. Peterson, L. W., Kleinkopf, D. E., and Huffaker, R. C., Plant Physiol. 51, 1042 (1973).
5. Kannangara, C. C. and Woolhouse, H. W., New Phytol. 67, 533-42 (1968).
6. Brady, C. J., Scott, N., and Munns, R., Royal Soc. New Zealand Bull. 12, 403 (1974).
7. Peterson, L. W. and Huffaker, R. C., Plant Physiol. 55, 1009-15 (1975).
8. Hall, N., Keys, A., and Merrett, M., J. Exp. Bot. 29, 31-7 (1978).
9. Callow, J. A., New Phytol. 73, 13-20 (1974).
10. Arias, I., Doule, D., and Schimke, R., J. Biol. Chem. 244, 3303 (1969).
11. Hutterman, A. and Wendleberger, G., Methods Cell Biol. 13, Ch. 8, (1976).
12. Zielke, H. R. and Filner, P., J. Biol. Chem. 246, 1772 (1971).
13. Humphrey, T. J. and Davies, D., Biochem. J. 156, 561 (1976).
14. Roberts, R. M. and Yuan, B. O.-Ching, Arch. Biochem. Biophys. 171, 226 (1975).
15. Goldberg, A. L., in Intracellular Protein Catabolism, Vol. 2 pp. 49-66, V. Turk and N. Marks, Editors, Plenum Press, New York, 1977.
16. Andrews, T. J., Badger, M. R., and Lorimer, G. H., Arch. Biochem. Biophys. 171, 93 (1975).
17. Schloss, J. V. and Hartman, F. C., Biochem. Biophys. Res. Commun. 75, 320 (1977).
18. Laber, L. J., Latzko, E., and Gibbs, M., J. Biol. Chem. 249, 3436 (1974).
19. Miles, D., Stadler Genet. Symp. 7, 135-54 (1975).
20. Smillie, R., Nielsen, N., Henigsen, K., and von Wettstein, D., Aust. J. Plant Physiol. 4, 415 (1977).
21. Levine, R. P., Annu. Rev. Genet. 4, 397 (1970).
22. Bradbeer, J. W., in Biosynthesis and Its Control in Plants, p. 279, B. V. Milborrow, Editor, Academic Press, New York, 1973.
23. Feierabend, J. and Pirson, A., Z. Pflanzenphysiol. 55, 235 (1966).
24. Criddle, R. S., Dau, B., Kleinkopf, G. E., and Huffaker, R. C., Biochem. Biophys. Res. Commun. 41, 62 (1970).
25. Fox, S. and Naylor, A., Plant Physiol. Suppl. 57, 4 (abstr. 19) (1976).
26. Frosch, S., Bergfeld, R., and Mohr, H., Planta 133, 53 (1976).
27. Chollet, R. and Paolillo, D. R., Z. Pflanzenphysiol. 68, 30 (1972).
28. Chollet, R. and Ogren, W. L., Z. Pflanzenphysiol. 68, 45 (1972).

29.   Neuffer, M. G., Sheridan, W. F., and Bendbow, E., Maize Genet.
      Coop. Newslett. 52, 84-8 (1978).
30.   Shumway, L. K. and Weier, T. E., Am. J. Bot. 54, 773 (1967).
31.   Mascia, P. and Robertson, D., Maize Genet. Coop. Newslett. 51,
      38 (1977).
32.   Stroup, D., J. Hered. 61, 139 (1970).
33.   Means, G. E. and Feeney, R. E., Chemical Modification of Pro-
      teins, p. 203, Holden-Day, San Francisco, 1971.

DISCUSSION

HUFFAKER: Did you compare the actual rates of carboxylase
breakdown for the acetic anhydride method and for the ^{3}H and ^{14}C
method?

SIMPSON: No. I used double labeling because I had a great
deal of trouble getting reproducible incorporation of tritiated
amino acids into a leaf of that size, the variation being $\pm 15\%$.
With the ratios, the variation was only $\pm 5\%$.

HUFFAKER: It seems like an extremely neat method, but one
worries about the high turnover rate you got.

SIMPSON: I worry about it, too.

HUFFAKER: Did chlorophyll disappear? You said your carboxylase protein was essentially stable for 4 days.

SIMPSON: That's right. Then it started to decrease. Apparently other people have seen that in other leaves. The thing to do
is just to work within the 4 days.

HUFFAKER: What percent does the carboxylase make up of the
total soluble protein in a corn leaf?

SIMPSON: I find 40%.

HUFFAKER: That is quite a bit.

CANVIN: It is of course crucial whether the acetic anhydride
has any effect itself. Have you measured the photosynthesis rate
after acetic anhydride treatment?

SIMPSON: Yes. I measured oxygen evolution and oxygen consumption; but, because we had rather crude equipment, I had to do it
with pieces of leaves submerged in water. I found no difference
after treatment.

CANVIN: You put forward the mechanism as the acetylation of
ε-amino groups. How do you prevent acetylation of all the sulfhydryl
compounds? The method you use is a common technique for acetylating
acidic sulfhydryl groups.

SIMPSON: At pH 7 to 8, modification of protein SH groups is
minimal, and any modification accomplished is highly unstable. In
any case, the double-labeling work indicates no induced degradation
of the acetylated RuBP carboxylase.

CANVIN: The reaction is the most common procedure for making
acetyl CoA; it occurs in seconds.

WILDMAN: Regarding your iojap mutant, in which you did not
find RuBP carboxylase, that is a maternally inherited mutant and
it also makes aberrant chloroplasts.

SIMPSON: It is not surprising that RuBP carboxylase is not present.

WILDMAN: But it is in stark contrast with the situation in tobacco, in which, if you do have a white mutant of a varigated strain, you can separate the white tissue and then isolate RuBP carboxylase and crystallize it.

SIMPSON: Are you saying that the white sectors are positive for RuBP carboxylase?

WILDMAN: In the case of N. tabacum, which, like the iojap, is also maternally inherited.

BURR: The iojap gene is a nuclear gene which, once introduced through the pollen, is subsequently cytoplasmically inherited.

WILDMAN: That's right, but aren't there two classes of chloroplasts?

BURR: Yes, but in different sectors of the leaf.

WILDMAN: The presence of two classes of chloroplast is what makes the difference because, in the case of tobacco, the chloroplasts that lack chlorophyll contain RuBP carboxylase.

COMPARATIVE BIOCHEMISTRY OF RIBULOSE

BISPHOSPHATE CARBOXYLASE IN HIGHER PLANTS

William L. Ogren[*] and Larry D. Hunt

Department of Agronomy, University of Illinois
Urbana, Illinois   61801

The agronomically important aspects of the comparative biochemistry of RuBP carboxylase are locating a natural enzyme, creating a
mutant enzyme, or identifying compounds which differentially alter
the enzyme so as to allow CO_2 to be fixed more efficiently or O_2 to
be fixed less efficiently. Two interrelated approaches can be pursued in trying to increase the agronomic efficiency of the enzyme at
atmospheric CO_2 concentration. (i) The specific activity of the
enzyme could be increased by increasing the maximal velocity of the
enzyme's action with respect to both substrates. This will cause
a proportional increase in both photosynthesis and photorespiration,
but, because the rate of CO_2 uptake by photosynthesis exceeds the
rate of CO_2 loss by photorespiration under natural conditions, net
photosynthesis will increase. (ii) The ratio of oxygenase activity
to carboxylase activity could be decreased, thereby increasing net
photosynthesis by decreasing photorespiratory CO_2 evolution.

SPECIFIC ACTIVITY

Genotypic differences in specific activity of RuBP carboxylase
have been reported for tomato (1), Nicotiana (2), and tall fescue
(3) (Table 1). In tomato, Andersen et al. (1) found three specific
activity classes, the most active enzymes showing a specific activity nearly four times as high as that of the least active enzymes.
The specific activity of the crystalline enzyme from one species
of Nicotiana was 50% higher than that of enzyme crystallized from
six other species, and the factor(s) causing increased activity was

---

[*]Science and Education Administration, U.S. Dept. of Agriculture.

Table 1

Reported Genetic Variation in RuBP Carboxylase Specific Activity

Species	Specific Activity
Tomato (1)	Rel. units
High activity (2 genotypes)	3.0
Medium activity (19 genotypes)	1.3
Low activity (2 genotypes)	0.8
Nicotiana (2)	μmoles CO_2 per mg protein per hr
N. gossei	0.36
Nicotiana (6 species)	0.24
Hybrids	
N. gossei (female parent)	0.29
N. gossei (male parent)	0.24
No N. gossei parents	0.24
Tall fescue (3) (crude extracts)	
Hexaploid cultivars (2)	0.34
Decaploid cultivar	0.44

at least partially carried in the chloroplast genome. Since no measurements of leaf photosynthesis were made in either tomato or
Nicotiana, no determination can be made as to whether the observed
differences represent intrinsic alterations in enzyme activity or
are artefacts of the isolation and assay procedures used.

   The rate of photosynthesis in a leaf, when expressed on an
area basis, is increased by increasing the RuBP carboxylase content.
The leaf RuBP carboxylase content can be altered, for example, by
the amount of illumination received by the plant (4, 5). Randall
et al. (3) demonstrated that ploidy level can also affect the leaf
carboxylase content. Because the RuBP carboxylase activity was assayed in crude extracts, it cannot be determined whether the specific
activity of the enzyme in the decaploid cultivar was higher than that
in two hexaploid cultivars or whether RuBP carboxylase in the decaploid leaf simply constituted a greater percentage of the total soluble protein.

CARBOXYLASE/OXYGENASE RATIO

   The second route to increase the efficiency of net CO_2 fixation
by modifying carboxylase, and the one most intensively pursued, is
alteration of the ratio of RuBP carboxylase to RuBP oxygenase activity. The ratio of the two activities (v_c/v_o) can be expressed in
terms of four kinetic parameters of the enzyme, V_c and V_o, the V_{max}

for carboxylase and oxygenase, respectively, and K_c and K_o, the K_m with respect to the two substrates (6):

$$v_c/v_o = V_c K_o [CO_2]/V_o K_c [O_2] \cdot \qquad (1)$$

Thus to increase the ratio v_c/v_o it will be necessary either to increase V_o or K_o or to decrease V_o or K_c (7-9). These parameters also control the photosynthesis/photorespiration ratio (PS/PR) in leaf photosynthesis (6):

$$PS/PR = V_c K_o [CO_2]/t V_o K_c [O_2] \qquad (2)$$

where t is 0.5, the stoichiometry of O_2 taken up by RuBP oxygenase to CO_2 released by photorespiration. Γ, the CO_2 compensation point (6), is described by

$$\Gamma = t V_o K_c [O_2]/V_c K_o \cdot \qquad (3)$$

TEMPERATURE

Several reports of alterations in the ratio of the two activities have been published, and each in turn has been or can be challenged. Temperature is the best characterized of the parameters influencing v_c/v_o, with higher temperatures increasing photorespiration in leaves and v_o/v_c in the isolated enzyme. The evidence supporting increase of photorespiration at higher temperatures has been reviewed by Jackson and Volk (10) and by Goldsworthy (11). The effect of temperature on the carboxylase/oxygenase ratio, with higher temperatures increasing oxygenase activity, has been demonstrated by simultaneous assay of the two activities in the same reaction vessel (6) and by a differential temperature dependence of K_c and K_o. K_c increases with increasing temperature, whereas K_o is independent of temperature (6, 12). Increased O_2 inhibition of soybean leaf and cell photosynthesis at higher temperatures has also been demonstrated, with the K_c for cell photosynthesis increasing with increasing temperature (13). It has been suggested that higher temperatures increase photorespiration by increasing the V_o/V_c ratio (14), but activation energy measurements made elsewhere (6, 12) indicate that this ratio is independent of temperature. Ku and Edwards (15) measured the effect of temperature on O_2 inhibition of wheat photosynthesis and concluded that K_c and K_o were similarly affected by temperature, and that the increased photorespiration observed at higher temperature was due to a reduced ratio of CO_2 to O_2 in solution. However, the magnitudes of the temperature effect on several parameters of photorespiration, including the O_2 inhibition of the quantum yield of photosynthesis (16), the CO_2 compensation point (17), and the K_c/K_o ratio in leaves, cells, and carboxylase, are all about 8 kcal/mole, whereas the temperature dependence of the CO_2/O_2 solubility ratio is only 1.8 kcal/mole (13).

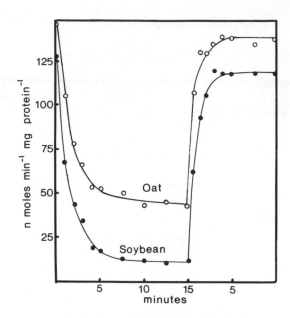

Figure 1.   The deactivation and reactivation of RuBP carboxylase
from soybean and oat.   RuBP carboxylase was partially purified (30)
and 2.5 mg enzyme was activated prior to assay in 0.2 ml 100 mM Tris-
Cl, pH 8.2, 10 mM dithiothreitol, 10 mM $MgCl_2$, and 10 mM $NaHCO_3$.   After dilution to 5.0 ml in the same medium minus added $NaHCO_3$, 0.05-ml
aliquots were withdrawn at the indicated times and assayed under N_2
for carboxylase activity in the reaction medium described previously
(30) containing 1.24 mM $NaHCO_3$ at pH 8.2 and 25°C.   Reactions were
terminated after 30 sec.   The enzyme was reactivated by the addition
of $NaHCO_3$ to a final concentration of 10 mM, and aliquots were withdrawn and assayed under the conditions above.

ASSORTED FACTORS

Several other factors which are reported to alter the carboxyl-
ase/oxygenase ratio have not been so well characterized.   These include the developmental stage of tomato fruits (18), the method of
enzyme preparation (19), the enzyme age (20), and the effect of exogenous chemicals such as glycidate (21) and sugar phosphate (22).   In
experiments on all these factors, the two activities were assayed in
substantially different reaction media.   The primary difference was
that the CO_2 concentration was saturating during the carboxylase assay and very low during the oxygenase assay.   The pH and the concentrations of Mg^{2+}, sulfhydryl reagent, and enzyme were also different.

Table 2

Deactivation Characteristics of RuBP Carboxylase Isolated
From Several Species

(Enzymes were isolated and assays conducted as described
in legend to Figure 1. A is given in units of nmoles CO_2
per mg protein per min.)

Species	$A_{Init.}$	$A_{Deact.}$	$A_{Restor.}$	Deactivation $t_{1/2}$ (min)	$A_{Deact.}/A_{Init.}$
Soybean	129	12	120	1.2	0.09
Oat	148	42	138	1.5	0.28
Sunflower	94	10	89	1.0	0.11
Wheat	108	37	98	1.2	0.34
Pea	88	6	81	1.6	0.07

In view of the complicated kinetics of RuBP carboxylase with respect
to CO_2, Mg^{2+}, and pH (23-28) and the rapid rate of enzyme deactivation in low CO_2 concentrations (28) (Figure 1, Table 2), the interpretation given to these experiments is highly questionable.   In
the glycidate experiment (21), for example, the presence of CO_2
during the carboxylase assay maintained the enzyme in a fully activated state, while the absence of CO_2 from the oxygenase assay led
to deactivation of the enzyme and an apparent differential effect
of glycidate on the two activities (G. F. Wildner, personal communication).   Separate assays misled Ryan and Tolbert (22) into conclud-
ing that sugar phosphates differentially altered the v_c/v_o ratio
(cf. 22, 23).
        It is evident that firm conclusions cannot be drawn about the
v_c/v_o ratio of enzymes when the two activities are assayed in different states of enzyme activation.   A precipitous loss of enzyme
activity occurs upon dilution into CO_2-free buffer, the recommended
procedure for oxygenase assay (28, 29) (Figure 1).   The half-time of
decay of enzyme activity was about 1.0 to 1.6 min for partially
purified enzyme from five crop species (Figure 1, Table 2).   Since
most of the initial activity was restored on the addition of 10 mM
bicarbonate, the rapid decline in activity was not due to denaturation.   Also, the extent of deactivation, as a proportion of initial
activity, was not the same for all species even though the isolation
and assay conditions were identical in all experiments (Table 2).
        Activation of RuBP carboxylase is a function of pH, Mg^{2+}, CO_2,
and certain sugar phosphates.   Although we have not yet examined the
kinetics of enzyme deactivation with respect to these factors, we
expect to find that they influence the rate of deactivation as greatly
as that of activation.   Standard assays (28, 29) of oxygenase

activity are suitable for locating activity, but at present they do
not give good estimates of actual oxygenase activity and are of little value in determining the v_c/v_o ratio. This ratio can be accurately determined only by assaying both activities under identical
conditions, preferably simultaneously, in the same reaction vessel
(6, 23, 25, 26, 31).

pH

    One parameter which has received considerable study with re-
spect to the relationship between photorespiration and photosynthesis at the cellular and subcellular level is pH (32-36). Previous
work with purified RuBP carboxylase (32, 36), chloroplasts (32),
and isolated leaf cells (36) indicated that increasing pH decreased
the $K_m(CO_2)$ of carboxylase and thereby decreased the rate of photorespiration relative to photosynthesis. In these experiments, bicarbonate was added to the reaction mixtures, and the CO_2 concentration

---

Table 3

pH Dependence of Soybean Cell Photosynthesis

Soybean cells were isolated and assayed as described
earlier (36). The data for 10 µM CO_2 were taken from
ref. 36, where 10 µM CO_2 was calculated from the amount
of added bicarbonate by the Henderson-Hasselbalch equation. For the experiment at 345 µl/l CO_2, vials containing assay medium were connected in series to a pump
and infrared gas analyzer. $NaH^{14}CO_3$ was injected into
a vial in the system containing H_3PO_4 until 345 µl/l
CO_2 was reached. After 15 min equilibration, the individual reaction vials were clamped off, and soybean
cells were injected into the vials in the light. The
reactions were terminated after 15 min.

	10 µM CO_2 (calculated from HCO_3^-)			345 µl/l CO_2 (IRGA)		
	µmole CO_2 fixed per mg Chl per hr			µmole CO_2 fixed per mg Chl per hr		
pH	2% O_2	50% O_2	% O_2 inhibition	2% O_2	50% O_2	% O_2 inhibition
7.2	19	7	63	14.5	7.2	63
7.8	34	18	47	16.4	8.2	64
8.8	57	36	37	13.8	7.2	69

Table 4

pH Dependence of O_2 Inhibition of RuBP Carboxylase

The data for 10 μM CO_2 were taken from ref. 36, where 10 μM CO_2 was calculated from the amount of added bicarbonate by the Henderson-Hasselbalch equation. The system contains 345 μl/l CO_2 in the gas phase of the reaction vessels as described for Table 3. After the vials were equilibrated and clamped off, 50 μl enzyme in 10 mM $NaHCO_3$ was injected into them. The assays were terminated after 30 sec. Percentage inhibition was calculated from reactions run under N_2 abd 50% O_2.

	% O_2 inhibition (50% O_2)						
	10 μM CO_2 (calculated from HCO_3^-)			345 μl/l CO_2 (IRGA)			
Species pH:	7.7	7.8	8.8	7.2	7.8	8.2	8.8
Soybean	44	38	29	47	44	47	63
Oat				49	57	60	67
Sunflower				26	46	40	46
Wheat				43	58	64	72
Triticales				56	61	66	73
Mean	44	38	29	44	53	55	64

was calculated from the Henderson-Hasselbalch equation. Recent experiments with enzyme and cells, in which the CO_2 concentration was determined in the gas phase over the reaction mixtures with an infrared gas analyzer (IRGA), do not support this conclusion. Photosynthesis in isolated cells at 345 μl CO_2 per liter and 50% O_2 was inhibited by about 65% compared with that at 2% O_2 at all three pH values examined, pH 7.2, 7.8, and 8.8 (Table 3). This result is contrast to those of previous experiments, which showed O_2 inhibition ranging from 63% at pH 7.2 to 37% at pH 8.8. The CO_2 concentration in this experiment (36), as calculated from the amount of added bicarbonate, was 10 μM. When RuBP carboxylase, partially purified from five species, was assayed under a constant CO_2 concentration in the gas phase, O_2 inhibition tended to increase with increasing pH (Table 4). This is in contrast to previous enzyme experiments, in which O_2 inhibition decreased with increasing pH (32, 36).

The reason for the discrepancy between the previous (32, 36) and the present results (Tables 3, 4) is that the Henderson-Hasselbalch equation used to calculate CO_2 concentration from added bi-

Table 5

pH Dependence of CO_2 Distribution in 22.5-ml Reaction Vessels

(Calculations were based on a solubility of 11.36 μM CO_2 at 345 μl/l CO_2 and an ionization constant of 6.82 x 10-7 at 25°C.)

	Percentage distribution					
	Calculated from Henderson-Hasselbalch equation			Calculation including air volume in reaction vessel		
CO_2 species pH:	7.2	7.8	8.8	7.2	7.8	8.8
CO_2 in solution	8.47	2.27	0.23	2.38	1.35	0.22
CO_2 in gas phase	nil	nil	nil	69.47	39.28	6.32
Bicarbonate	91.53	97.73	99.77	28.14	59.37	93.46

carbonate does not account for the partitioning of CO_2 between the aqueous and gas phases of the reaction vial. In the 22.5-ml vials used in our experiments, the CO_2 concentration in solution at pH 7.2 is only 28% of that calculated from the Henderson-Hasselbalch equation (Table 5). At pH 7.8 the CO_2 concentration is 59% of that calculated, and the actual CO_2 concentration rises to 96% of the calculated value at pH 8.8.

The partitioning of CO_2 between the gas and aqueous phases does not appear to have been generally considered in the calculations of previous assays of CO_2 fixation by carboxylase, chloroplasts, or cells. Consideration of this partitioning will affect some conclusions which have been drawn about pH. For example, it has been reported that those sugar phosphates which stimulate RuBP carboxylase at low concentration are more effective at low pH than at high pH (23). Because the magnitude of sugar phosphate stimulation is inversely related to the CO_2 concentration (23, 24), it is probable that the pH dependence observed (23) is, in fact, a CO_2 dependence. At pH 7.8, a commonly used pH for photosynthesis and RuBP carboxylase studies, the actual CO_2 concentration is only 59% of that calculated by the Henderson-Hasselbalch equation in 22.5-ml reaction vessels (Table 5). The error is increased with larger reaction vessels, and decreased with smaller ones. Thus most $K_m(CO_2)$ values calculated for carboxylating systems at this pH have been overestimates.

GENETIC CHANGES

Although the demonstration of altered ratios of carboxylase/oxygenase activity in vitro provides encouragement that permanent

in vivo alterations might be accomplished, it is clearly most desirable to induce enzyme changes in the leaf itself. Not only is
this where alterations must occur if we are to realize gains in
photosynthetic productivity, but this is the only place where such
changes can be unequivocally demonstrated to be more than experimental artefacts.

      If the ratio of carboxylase to oxygenase activities determines
the relative rates of photosynthesis and photorespiration, as has
been suggested (7-9, 36), then any modification of this enzyme in
vivo must be reflected in the gas exchange characteristics of the
intact leaf. The leaf parameters most sensitive to changes in the
photosynthesis/photorespiration ratio are the CO_2 compensation point
(Γ) and O_2 inhibition of photosynthesis. If photorespiration is
truly reduced in any leaf, then Γ and O_2 inhibition must also be
reduced. We consider these parameters to be definitive in C_3 plants
and the standard against which all reported in vivo modifications of
the enzyme should be evaluated. Reports of genetic variability
based solely on enzyme measurements will and should be met with considerable skepticism because of the demonstrated lack of naturally
occurring variability in the photosynthesis/photorespiration ratio
and the considerable possibility of isolation and assay artefacts
due to the complicated biochemistry of the enzyme. The Γ of C_3
plants, when measured in a closed system where the processes of
photosynthesis and photorespiration can reach equilibrium, falls
within a very narrow range. In our laboratory, at 25°C and 21% O_2,
Γ was 41 ± 3 $\mu l/l$ CO_2 for all the many C_3 species and hundreds of
soybean and oat cultivars we have examined over the past several
years. A similar narrow range of Γ values has been reported by
Moss and co-workers (37, 38). In those species where reduced photorespiration has been convincingly demonstrated, for example, Panicum
milioides (30, 39), Γ and O_2 inhibition were also reduced. It should
be emphasized that Γ values of about 40 $\mu l/l$ CO_2 in C_3 plants are
for healthy, mature leaves. The Γ in developing and senescing leaves,
where high rates of dark respiration occur, are higher and not definitive. The magnitude of O_2 inhibition is similar for C_3 species (39,
40). But to be definitive, the CO_2 concentrations used in determining inhibition must be the internal, not external, concentrations.

      To date two reports of altered carboxylase within species have
appeared (41, 42). One is supported by leaf measurements (41), although the leaf measurements are not as definitive as those mentioned
above. Garrett (41) reported that the $K_m(CO_2)$ of ryegrass was a
function of ploidy level. The $K_m(CO_2)$ was about 50 μM CO_2 for four
diploid cultivars and 22 μM CO_2 for four tetraploid cultivars. For
both groups of plants the $K_m(O_2)$ was about 510 μM O_2. From Eq. (2),
which describes the photorespiration/photosynthesis ratio, the $K_m(CO_2)$
differences predict a doubled rate of photorespiration in the diploid
cultivars. Photorespiration, measured as the post-illumination CO_2
burst, was 19 mg CO_2 dm^{-2} hr^{-1} in the diploids and 10 mg CO_2 dm^{-2}
hr^{-1} in the tetraploids. Thus it appears that ploidy level affects
photorespiration rate in ryegrass and that this is correlated to the

kinetic properties of RuBP carboxylase. However, measurements of
absolute photorespiration rate are not as definitive as the assays
mentioned above (Γ, O_2 inhibition) because the rate of photorespiration in C_3 plants is directly proportional to photosynthesis rate
(43, 44).

CONCLUSION

Photosynthetic carbon dioxide fixation by higher plants might
be increased by genetic, developmental, or chemical modification of
RuBP carboxylase. The desired modification will either increase the
overall specific activity of the enzyme or alter the carboxylation/
oxygenation ratio in favor of carboxylation. Several reports of
genetic alteration in specific activity, and genetic, ontogenic,
chemical, and environmental alteration of RuBP carboxylase/oxygenase
activity can be found in the literature. Except for temperature,
the significance of these reports with respect to photosynthesis
in vivo remains to be determined. Because of the complicated interactions of the enzyme with substrates, cofactors, and pH, current
standard assay procedures do not permit definitive conclusions to be
drawn about the effects of various parameters on the ratio of the
two activities. It is necessary that both activities be assayed under identical conditions.
Genetic alterations of RuBP carboxylase activities are most desirable, because plant variants having altered photorespiration can
be used in breeding programs to evolve their agronomic potential.
Genetic changes are also most easily determined, because they can be
directly assayed by simple measurements: Γ and O_2 inhibition of photosynthesis in intact leaves. Failure to demonstrate changes in
these physiological characteristics when enzyme changes are found
indicates that the observed enzyme changes are most likely due to
artefacts of the isolation or assay system used.

REFERENCES

1. Andersen, W. R., Wildner, G. F., and Criddle, R. S., Arch.
    Biochem. Biophys. 137, 84-90 (1970).
2. Singh, S. and Wildman, S. G., Plant Cell Physiol. 15, 373-9
    (1974).
3. Randall, D. D., Nelson, C. J., and Asay, K. H., Plant Physiol.
    59, 38-41 (1977).
4. Bowes, G., Ogren, W. L., and Hageman, R. H., Crop Sc. 12, 77-9
    (1972).
5. Björkman, O., Physiol. Plant. 21, 1-10 (1968).
6. Laing, W. A., Ogren, W. L., and Hageman, R. H., Plant Physiol.
    54, 678-85 (1974).
7. Ogren, W. L., in Environmental and Biological Control of Photo-
    synthesis, pp. 45-52, R. Marcelle, Editor, Junk, The Hague, 1975.

8.   Ogren, W. L., in CO$_2$ Metabolism and Plant Productivity, pp. 19-29, R. H. Burris and C. C. Black, Editors, University Park Press, Baltimore, 1976.

9.   Ogren, W. L., in Photosynthesis '77, (Proc. 4th Int. Congr. on Photosynthesis), pp. 721-33, D. O. Hall et al., Editors, Biochem. Soc., London, 1978.

10. Jackson, W. A. and Volk, R. J., Annu. Rev. Plant Physiol. 21, 385-432 (1970).

11. Goldsworthy, A., Bot. Rev. 36, 321-40 (1970).

12. Badger, M. R. and Collatz, G. J., Carnegie Inst. Washington Yearb. 76, 355-61 (1977).

13. Servaites, J. C. and Ogren, W. L., Plant Physiol 61, 62-7 (1978).

14. Badger, M. R. and Andrews, T. J., Biochem. Biophys. Res. Commun. 60, 204-10 (1974).

15. Ku, S. B. and Edwards, G. E., Plant Physiol. 59, 986-90 (1977).

16. Ehleringer, J. and Björkman, O., Plant Physiol. 59, 86-90 (1977).

17. Björkman, O., Gauhl, E., and Nobs, M. A., Carnegie Inst. Washington Yearb. 68, 620-33 (1969).

18. Bravdo, B-A., Palgi, A., Lurie, S., and Frenkel, C., Plant Physiol. 60, 309-12 (1977).

19. Harris, G. C. and Stern, A. I., Plant Physiol. 60, 697-702 (1977).

20. Andrews, T. J., Lorimer, G. H., and Tolbert, N. E., Biochemistry 12, 11-18 (1973).

21. Wildner, G. F. and Henkel, J., Biochem. Biophys. Res. Commun. 69, 268-75 (1976).

22. Ryan, F. J. and Tolbert, N. E., J. Biol. Chem. 250, 4234-8 (1975).

23. Chollet, R. and Anderson, L. L., Arch. Biochem. Biophys. 176, 344-51 (1976).

24. Chu, D. K. and Bassham, J. A., Plant Physiol. 55, 720-6 (1975).

25. Laing, W. A., Ogren, W. L., and Hageman, R. H., Biochemistry 14, 2269-75 (1975).

26. Laing, W. A. and Christeller, J. T., Biochem. J. 159, 563-70 (1976).

27. Lorimer, G. H., Badger, M. R., and Andrews, T. J., Biochemistry 15, 529-36 (1976).

28. Lorimer, G. H., Badger, M. R., and Andrews, T. J., Anal. Biochem. 78, 66-75 (1977).

29. Kung, S. D., Chollet, R., and Marsho, T. V., Methods Enzymol., in press (1978).

30. Keck, R. W. and Ogren, W. L., Plant Physiol. 58, 552-5 (1976).

31. Laing, W. A., Ph.D. Dissertation, U. of Illinois, Urbana, 1974.

32. Bowes, G., Ogren, W. L., and Hageman, R. H., Plant Physiol. 56, 630-3 (1975).

33. Dodd, W. A. and Bidwell, R. G. S., Plant Physiol. 47, 779-83 (1971).

34. Orth, G. M., Tolbert, N. E., and Jiminez, E., Plant Physiol. 41, 143-7 (1966).

35. Robinson, J. M., Gibbs, M., and Cotler, D. N., Plant Physiol. 59, 530-4 (1977).

36.   Servaites, J. C. and Ogren, W. L., Plant Physiol. 60, 693-6
      (1977).
37.   Krenzer, E. G. Jr., Moss, D. N., and Crookston, R. K., Plant
      Physiol. 56, 194-206 (1975).
38.   Moss, D. N., Krenzer, E. G. Jr., and Brun, W. A., Science
      164, 187-8 (1969).
39.   Brown, R. H. and Brown, W. V., Crop Sci. 15, 681-5 (1975).
40.   Hesketh, J. D., Planta 371-4 (1967).
41.   Garrett, M. K., in 4th Int. Congr. on Photosynthesis Abstracts,
      p. 123, J. Coombs, Compiler, U. K. Science Committee, London,
      1977.
42.   Kung, S. D. and Marsho, T. V., Nature London 259, 352-6 (1976).
43.   Nelson, C. J., Asay, K. H., and Patton, L. D., Crop Sci. 15,
      629-33 (1975).
44.   Ogren, W. L. and Bowes, G., Nature New Biol. 230, 159-60 (1971).

DISCUSSION

LORIMER: The concentration of RuBP carboxylase active sites
within the chloroplast is 3 to 40 mM, while the concentration of
CO_2 is ~10 μM. Your application of Michaelis-Menten kinetics to
in vivo photosynthesis appears to yield the correct answers, but
it is not, strictly speaking, kinetically valid. Might you be
correct for the wrong reasons?

OGREN: Yes. On the other hand, if a considerable percentage
of the enzyme is inactive, it is, in a sense, not part of the
catalyst cycle.

LORIMER: We are talking about an almost 500-fold difference
in concentration. You are not suggesting that only 0.2% of the
enzyme is active?

OGREN: No.

BLACK: You made a very strong statement that, to prove the
oxygenase:carboxylase ratio had changed, leaf data on photosynthesis
should be available. Panicum milioides leaf data are available; do
they indicate that the carboxylase:oxygenase ratio has changed?

OGREN: The proof should go the other way. If you show a
change in the enzyme, you must also show a similar change in leaf
photosynthesis. P. milioides is actually a different case from a
plant with an altered carboxylase.

BLACK: But it meets all of the strict criteria you laid down
at the end of your talk.

OGREN: Except that the Γ is not a linear function of oxygen
concentration, whereas for C_3 plant it is linear. Any alteration
in the carboxylase would also be linear and produce a change in
slope. P. milioides has a break in the curve which indicates re-
cycling of CO_2. That does not indicate altered carboxylase.

BLACK: That indicates phosphoenol pyruvate carboxylase.

OGREN: Yes, PEP carboxylase refixation.

REUTILIZATION OF RIBULOSE BISPHOSPHATE CARBOXYLASE

R. C. Huffaker and B. L. Miller

Plant Growth Laboratory, Department of Agronomy & Range Science
University of California, Davis, California 95616

INTRODUCTION

Ribulose bisphosphate (RuBP) carboxylase is truly a multi-functional protein. Not only does it exhibit the well-known carboxylase and oxygenase activities, but also its high concentration and turnover characteristics in the leaf fit the classification of a storage protein. RuBP carboxylase varies in concentration by species but can be up to 65% of the total soluble protein of grass or alfalfa leaves. It is assembled and sequestered in a discrete organelle, the chloroplast, wherein it is protected from the proteolytic enzymes in the cytoplasm. In fact, it appears that carboxylase degradation is a cytoplasmically driven process. After its synthesis and assembly, little turnover is detected until plants require its remobilization. RuBP carboxylase can then be mobilized during senescence or when the plant requires its reserves because of deficits of either nitrogen or carbohydrates. As such, the RuBP carboxylase concentration in the leaf is very responsive to environmental stresses.

Perturbing the environment allows an approach to many of the inducible processes in the plant. By growing plants in darkness and then placing them in the light, induction of several enzymes can be followed. In dark-grown barley plants, about half the maximal concentration of RuBP carboxylase is synthesized. When the plants are placed in light, synthesis continues until a final quite constant concentration is attained. During the greening period, radioactive amino acids can be incorporated into the protein to follow its synthesis and degradation.

TURNOVER DURING INITIAL SYNTHESIS

To determine whether turnover of RuBP carboxylase occurs during its initial synthesis, leaves were detached from dark-grown barley plants and placed base down in light and in the presence of ^{14}C-amino acids (1). After several hours, the labeled amino acids were removed and chased with cold amino acids. No loss of specific antibody-precipitable label was detected during the greening period, which showed the absence of degradation while the enzyme was undergoing synthesis.

TURNOVER AFTER STEADY CONCENTRATION IS ATTAINED

After the carboxylase protein reached a steady maximum concentration, no turnover was detected in leaves from intact barley plants (2). This was shown by placing dark-grown plants into light in ^{14}C-CO$_2$ for 6 hr, then removing the labeled CO$_2$ and chasing with stable CO$_2$ and following the specific radioactivity of the carboxylase protein with time. The concentration of carboxylase protein remained essentially constant as did the specific radioactivity, which indicated that little turnover occurred. In contrast, soluble protein other than RuBP carboxylase turned over rapidly. Since the observation of turnover depended on changes in the specific radioactivity of previously highly labeled carboxylase protein, it is possible that a small rate of synthesis would be undetected. It could well be that a normal but slow turnover of chloroplasts occurs, the rate of which may be species dependent. This concept will be amplified below.

Other evidence for lack of turnover of carboxylase in barley leaves is that, after the carboxylase reached its maximum concentration, no messenger activity was detected in isolated polyribosomes in an _in vitro_ heterologous protein-synthesizing system (3). During the period of rapid synthesis of the carboxylase, isolated polyribosomes in a heterologous system synthesized the large subunit of the carboxylase. Patterson and Smillie (4) showed that, in wheat leaves, RNA synthesis decreased and incorporation of labeled amino acids into RuBP carboxylase greatly decreased after the concentration of RuBP carboxylase reached a constant level. Zucker (5) further observed the stability of RuBP carboxylase in green Xanthium leaf discs. Little radioactivity was detected in Fraction I protein (F-I-p) even though significant amounts were incorporated into phenylalanine ammonia lyase. Likewise, when Kannangara and Woolhouse (6) fed ^{14}CO$_2$ to fully expanded _Perilla_ leaves, only a small percentage of the activity was in F-I-p.

RuBP CARBOXYLASE AND SENESCENCE

The processes of either synthesis or degradation of RuBP carboxylase may be sequentially separated. During its synthesis and while it is at a steady concentration, little turnover may occur. In contrast, when the leaf senesces because of age or limiting environmental factors, RuBP carboxylase protein seems to be the major protein degraded. Earlier workers (7-10) showed that F-I-p was the main protein constituent lost in senescing tobacco and Perilla leaves (6).

It is well known that light retards the onset of senescence and reduces the rate of senescence once under way. Seasonal variation in irradiance can influence the rate of leaf senescence. The shading of lower leaves by the upper leaves in a crop canopy can cause a constant senescing of lower leaves. Therefore, we have induced or simulated senescence in barley plant leaves by detaching the leaves (11) or placing intact barley plants in darkness (2), and have followed the resulting changes in protein. The loss of RuBP carboxylase showed similar kinetics in both detached and intact leaves of barley, accounting for 95% of the protein lost over the several-day time course. Chlorophyll was similarly lost in the senescing leaves. These results suggest that the chloroplast is the major site of degradation early in senescence. When the barley plants are placed in darkness, the time course of RuBP carboxylase and carbohydrate loss indicates that RuBP carboxylase is being reutilized as a source of both carbon and nitrogen after carbohydrates have been exhausted.

The environmentally induced loss of RuBP carboxylase is totally reversible up to 72 hr in darkness. When the intact plants are returned to light, the leaves again synthesize the full complement of RuBP carboxylase (2). Wittenbach (12) reported recovery of photosynthesis, chlorophyll, and total protein when wheat leaves were returned to light after 2 days of darkness, but after 4 days in dark, recovery did not occur.

SIMULTANEOUS SYNTHESIS OF RuBP CARBOXYLASE DURING DEGRADATION

To determine whether a small amount of synthesis could be detected at a time when RuBP carboxylase was undergoing rapid net degradation, detached barley leaves were placed into tritiated amino acids (11). A low level of amino acids was incorporated into RuBP carboxylase protein while it was being rapidly degraded. In comparison, high levels of labeled amino acids were incorporated into the total soluble protein while its concentration remained quite constant, which showed a high level of turnover.

LOSS OF RuBP CARBOXYLASE AND PROTEOLYTIC ACTIVITY

In detached leaves of barley, the loss of RuBP carboxylase protein was negatively correlated with an increase in endopeptidase activity (measured against azocasein) while carboxypeptidase activity against N-carbobenzoxy-L-tyrosine-p-nitrophenyl ester (CTN) remained constant during the experiment (11). On the other hand, no increase in endopeptidase activity occurred in leaves of intact plants while RuBP carboxylase was being degraded (Figure 1). Exo- and endopeptidase activities were detected in leaves of intact barley plants under normal conditions before the onset of senescence and apparently function in turnover of cytoplasmic proteins (13).

The amount of proteinase activity found in leaves of different species depends on the physiological age of the plant and the environmental treatments it has undergone. Soong et al. (14) showed that when corn leaves were detached before the onset of senescence, carboxypeptidase, aminopeptidase, and endopeptidase activities remained constant during the following 48 hr of darkness. When the experiment was run with leaves detached after the onset of senescence, aminopeptidase and carboxypeptidase activities decreased while endopeptidase activity against casein increased. Feller et al. (15) determined proteolytic activities as foliar senescence symptoms developed in corn leaves in the field. Increasing senescence was paralleled by decreases in exopeptidase activities and increases in endopeptidase activities. Decreases in chlorophyll and protein accompanied these observations. During later stages of grain development, amino- and carboxypeptidase activities decreased while endopeptidase activity increased. The highest caseolytic activity was observed when protein loss from leaves was greatest. Dalling et al. (16) showed a high correlation between protein loss from wheat leaves and acid proteinase activity.

It is not yet known whether the increase in proteolytic activity observed in vitro is required for an increased rate of protein loss in vivo. Several reports show that protein loss can be just as rapid even though endopeptidase activity remains constant during senescence. Beevers (17) showed that caseolytic activity was not correlated with protein loss in leaf discs of Nasturtium. Soong et al. (14) showed that a high level of endopeptidase activity is not a prerequisite for rapid protein degradation in detached corn leaves. Similar results were obtained with detached oat leaves (18). Endopeptidase activity did not increase in intact barley leaves in darkness, but protein was lost almost as rapidly as in detached leaves, which showed greatly increased endopeptidase activity (Figure 1). As described below, cycloheximide (CHI) prevented the loss of RuBP carboxylase and chlorophyll in detached barley leaves while simultaneously preventing the increase in proteolytic activity. CHI stopped the

formation of a new proteolytic activity which, in the control
leaves, was seen as a second band appearing after isoelectric
focusing (Figure 5). The relationship of the appearance of the
second band of proteolytic activity to RuBP carboxylase degradation is unknown.

PROTEOLYTIC SUBSTRATES

A partially purified system from barley leaves shows high activity against RuBP carboxylase, casein, methylated casein, azocasein,
and hemoglobin, and less against serum albumin, and high activity
against artificial substrates such as CTN, α-naphthyl acetate (ANA),
and others (Figure 2) (11).
The proteinases have been somewhat difficult to purify because
they appear to hydrolyze during the purification. Hence, we have
attempted to minimize the time required and have compared the
electrophoretic and isoelectrophoretic band patterns of endopeptidase activity at the different steps of purification. Disc gel

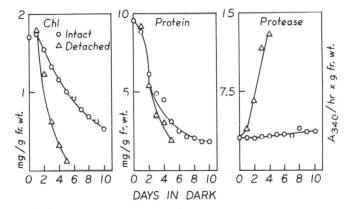

Figure 1.   Time course for chlorophyll and protein concentrations and proteolytic activity in leaves from intact and detached
leaves of barley. Plants were grown in a chamber at 28°C and 60%
relative humidity, at 540 μeinstein·sec^{-1}·m^{-2} to 7 days of age and
then in darkness. Leaves were harvested and ground in 0.1 M K
phosphate buffer, pH 6, and 1 mM dithiothreitol (DTT), the homogenate was filtered through cheesecloth, and the filtrate was centrifuged at 30,000 x g for 20 min. Aliquots of the supernatant were
assayed for total protein by the Lowry method (32) and for proteolytic activity against azocasein (11.) Chlorophyll was extracted
from the pellet with 80% acetone and determined according to
Arnon (31).

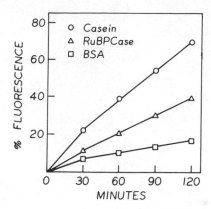

Figure 2. Proteolytic activity by a partially purified preparation from barley leaves. Detached barley leaves grown as described for Figure 1 were placed in darkness for 3 days and then extracted as for Figure 1. The supernatant from the centrifugation was brought to 35% ammonium sulfate, and the precipitate was discarded. The supernatant was then brought to 70% ammonium sulfate, and the pellet was resuspended in extraction medium and dialyzed for 17 hr against extraction medium. The reaction mixture contained 200 μg of partially purified proteinase, 2.5 mg of each protein substrate, 25 mM phosphate buffer, pH 6.0, and 1 mM DTT. Assay temperature was 40°C.

electrophoresis showed initially single bands of endopeptidase activity from leaves of senescing intact plants and detached leaves (Figure 3). A preparation containing the proteolytic activity from senescing detached leaves was stored for 2½ weeks at -20°C as the ammonium sulfate precipitate. Multiple bands of activity appeared when the preparation was again subjected to electrophoresis, which shows the problem of degradation to other active components during storage (Figure 3).

Isoelectric focusing of cell-free extracts showed one band of proteolytic activity from senescing leaves of intact barley plants and two bands from senescing detached leaves, but CHI or kinetin prevented the formation of the second band (Figure 4); chloramphenicol had no effect.

Drivdahl and Thimann (19) purified two proteases from their senescing oat leaf system. An acid and a neutral protease were purified about 500-fold by a combination of ammonium sulfate precipitation, affinity chromatography on hemoglobin-Sepharose, and

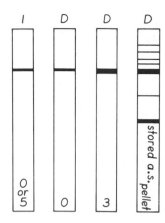

Figure 3.   Acrylamide electrophoresis of a cell-free extract
from intact (I) and detached (D) leaves and a partially purified
preparation from detached leaves of barley.   Electrophoresis was
done according to Davis (33).   See Figure 2 for extraction and
purification procedures.   100 μl of cell-free extract and 400 μg
of partially purified proteinases were added to the respective gels.
Proteolytic bands were detected by slicing the gels longitudinally,
placing them on a thin layer of agar-gelatin, and incubating them
at 40°C for 2 hr.   The disc gels were removed, and the zymograms
were developed by pouring a solution of acidified $HgCl_2$ over the
thin layer.   Cleared areas are a result of proteolytic digestion.
The numbers at the bottom denote days of dark treatment.

ion-exchange chromatography.   The molecular weight of each was
estimated at 76,000.   The neutral protease was apparently a sulfhydryl enzyme.   The corn and barley leaf endopeptidases are also
apparently sulfhydryl enzymes (15, 35).

INTRACELLULAR LOCATION OF RuBP CARBOXYLASE DEGRADATION

Several lines of evidence indicate that the proteinases responsible for RuBP carboxylase degradation are located in the cytoplasm.   In detached barley leaves, cycloheximide totally prevented
the loss of RuBP carboxylase protein and chlorophyll and prevented
the formation of a new proteolytic activity, whereas chloramphenicol
had no effect on either (Figure 4).   Hence, it appears that the new
proteolytic activity was developed on cytoplasmic 80 S polyribosomes.
An actual loss of proteolytic activity occurred in the cycloheximide
treatment, indicating that some of the proteinases themselves may be
turned over.   Also, we have found no evidence for proteolytic activity in intact chloroplasts or chloroplast membrane fractions either

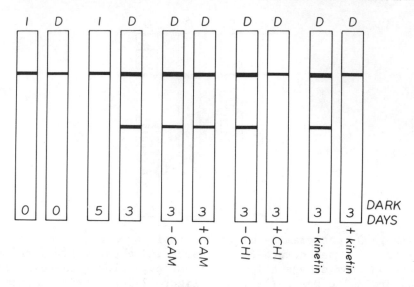

Figure 4.   Isoelectrically focused proteolytic activities from cell-free extracts from intact and detached barley leaves. Extraction was as described for Figure 1.   Two g of leaves was ground in 3 ml of 0.1 M phosphate buffer, pH 6.0, and 4 mM DTT. The mixture was centrifuged at 30,000 x g.   Then 1.5 ml of the extract was added to a slurry of ampholytes (pH 3 to 10) and G-75 Sephandex.   After preparation, a thin strip of Brinkman print paper was rolled over the thin layer and allowed to absorb some of the protein.   The paper was then laid over the agar-gelatin plate, and zymograms were developed as described for Figure 3.

treated or untreated with detergents.   On the other hand, this might indicate that the chloroplasts do not all degrade simultaneously (20) and that during the chloroplast isolation those senescing may be lost while the healthy ones would show no proteolytic activity.   Thomas (21) showed that CHI prevented the loss of chlorophyll in forage grass, and Martin and Thimann (23) showed that chloramphenicol did not prevent the loss of chlorophyll in oat leaves.   Choe and Thimann (24) showed that isolated chloroplasts lost chlorophyll and protein much more slowly than did oat leaves. This further indicates that the senescence process is probably initiated and driven by cytoplasmic reactions.

Instances are reported that show simultaneous loss of chlorophyll and RuBP carboxylase protein (e.g., from detached leaves of barley) (11) or of chlorophyll and total protein (22).   Thomas (20) recently showed, however, that these events can be separated. Using a mutant of _Festuca pratensis_ Huds.   That loses very little of its chlorophyll throughout senescence, he showed that RuBP carboxylase protein was lost at the same rate in both the mutant and the normal type. The normal type loses all its chlorophyll during the same time period that the carboxylase protein is lost. Electron microscopy of mutant chloroplasts indicated that the stroma matrix was destroyed but that thylakoid membranes persisted in a loose, unstacked condition. The activities of photosystems I and II declined at similar rates in both mutant and normal leaves. Polypeptides of chloroplast membranes were separated into about 30 components. Of five major components, two declined similarly in both mutant and normal, whereas the other three were more stable in the mutant. One of the components was tentatively identified as the apoprotein of the light-harvesting chlorophyll-protein complex.

The separation of chlorophyll degradation and some of the membrane proteins from the normal senescing pattern shows another degree of complexity. Indications are that significant manipulation of the events occurring during senescence or crop maturation may be possible. Identification of soybean variants showing a "green leaf character" further into the maturation period is indicative of this.

CHEMICAL REGULANTS

Besides inhibitors of protein synthesis such as cycloheximide, kinetin greatly retarded the loss of both RuBP carboxylase and chlorophyll as well as inhibiting newly formed proteolytic activity in detached barley leaves (Figure 4) (11). Kinetin is well known for its retardation of senescence in many plants.

A possible interpretation of results thus far is that RuBP carboxylase may well be protected from degradation by being discretely packaged in the chloroplast. Since the loss of RuBP carboxylase protein and of chlorophyll often show identical kinetic patterns, the turnover of carboxylase protein may not occur until the entire chloroplast is turned over. This might mean that the chloroplast membrane is first attacked by hydrolases, which would allow exchange of both RuBP carboxylase and proteinases. Proteolytic activity is present in the cytoplasm at all times, and cytoplasmic proteins show high rates of turnover whereas turnover of carboxylase appears negligible.

RELATIONSHIP OF RuBP CARBOXYLASE TO YIELD AND QUALITY

Maintenance of RuBP carboxylase through the growing season is critical for biomass production. Photosynthate furnishes the sink requirements of the developing seed for reduced carbon compounds. It also furnishes energy for both N_2 fixation and nitrate assimilation to meet nitrogen sink requirements of the developing seed.

When the supply of reduced N does not meet the seed's sink requirement, the leaf protein is degraded to furnish the N requirement.
The available information shows that RuBP carboxylase is among the
first leaf proteins to be lost at that time. As RuBP carboxylase
protein is degraded, photosynthetic CO_2 fixation capacity can be
lost. This has been termed a "self-destruct" phenomenon in soybeans
(25).

      Thus, maintaining a flow of reduced N through the plant is
critical to protecting RuBP carboxylase from being degraded. Crop
varieties showing the characteristic of maintaining photosynthetic
competence further into seed maturation have a potential for higher
yields. Frye (36) reported the selection of an oat variety which
retained chlorophyll longer during seed fill and subsequently had
higher yields. In crop plants utilizing nitrate as the principal
source of N, this means selecting for those plants which can assimilate nitrate more efficiently in terms of acquiring it and trans-
locating the reduced products to the seed. Wheat varieties vary
considerably in the ability both to assimilate nitrate and to trans-
locate the reduced products from the straw to the developing seeds
(26).

      Sinclair and de Wit (25, 27) analyzed the photosynthate and N
requirements of various seed crops and found that soybean was
unique among 24 species studied. Since soybean has a high content
of both protein and lipid, it must assimilate large quantities of
N to supply the needs of the seeds, yet it produces biomass at one
of the lowest rates. As the seed develops, its sink capacity for
N may become so great that the nodules cannot supply it; then reduced N could be translocated from the vegetative parts. The resulting loss in photosynthetic CO_2 fixation capacity would not
sustain both the seed and the nodule requirements for photosynthate.
This self-destruct characteristic could limit the length of the
seed development period and could potentially limit total seed production. In a field study at UCD, the presence of RuBP carboxylase
protein in soybean leaves was correlated with total protein, nitrogenase activity, and H_2 production over a season (Figure 5).
Mondal et al. (30) recently showed that when the seed sink of soybeans was removed, the concentration of both RuBP carboxylase and
total protein remained higher in the leaves. Increased N input
during rapid seed development could be achieved, according to Hardy
et al. (28), by extending the exponential phase of N_2 fixation or,
according to Sinclair and de Wit (25), by increasing the proportion
of photosynthate allocated to support N_2 fixation. These objectives
might be achieved by selecting for variants that maintain more
photosynthetically active leaves throughout the reproductive growth.

      To test the above hypothesis, several thousand plants in a
soybean nursery at UCD were examined for variation in senescence
characteristics (29). From among these, five plants were chosen
that had completely green leaves with brown mature pods. Seeds
from the plant showing the highest N_2 fixation produced 14 F_3
plants in a growth chamber. Of these, five plants showed delayed

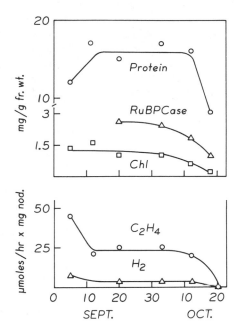

Figure 5. Time course of total protein, RuBP carboxylase protein, and chlorophyll concentrations in field-grown soybean leaves and corresponding acetylene reduction and H_2 evolution. Three leaves from each of five different plants were ground in 0.2 M Tris buffer, pH 8.0, and the mixture was centrifuged at 30,000 x g for 20 min. An aliquot of the supernatant was assayed for total protein by the Lowry method (32). Another aliquot was subjected to acrylamide electrophoresis (33), and the bands were stained with Coomassie Blue. The RuBP carboxylase band was identified by comparison with authentic purified RuBP carboxylase, and its concentration was determined by scanning the gel in a spectrophotometer and determining the area under the peak in relation to a standard curve. Chlorophyll was determined according to Arnon (31). N_2 fixation was estimated by acetylene-dependent ethylene production (34), and H_2 evolution by gas chromatography.

senescence. These in turn had much higher levels of chlorophyll, total leaf protein, RuBP carboxylase protein and activity, and nitrogenase activity than did the normally senescing control plants. F_4 plants produced from those displaying delayed and normal senescence yielded comparable data. These results suggest that maintenance of chlorophyll and RuBP carboxylase is a heritable quality which can be exploited to allow continued N_2 fixation during seed

production. It seems that the variant plants showing delayed leaf
senescence had an increased capacity to partition photosynthate
to maintain proportionally the requirements of both the nodule and
the seed sinks. Indications are that much genetic variability
is already available in nature regarding the regulation of senescence. The process of photosynthesis and nitrogen assimilation
seems tightly correlated with the regulation of senescence.

REFERENCES

1. Smith, M. A., Criddle, R. S., Peterson, L. W., and Huffaker,
    R. C., Arch. Biochem. Biophys. 165, 494-504 (1974).
2. Peterson, L. W., Kleinkopf, G. E., and Huffaker, R. C., Plant
    Physiol. 51, 1042-5 (1973).
3. Alscher, R., Smith, M. A., Peterson, L. W., Huffaker, R. C.,
    and Criddle, R. S., Arch. Biochem. Biophys. 174, 216-25 (1976).
4. Patterson, B. D. and Smillie, R. M., Plant Physiol. 47, 196-8
    (1971).
5. Zucker, M., Annu. Rev. Plant Physiol. 23, 133-56 (1972).
6. Kannangara, C. G. and Woolhouse, H. W., New Phytol. 67, 533-42
    (1968).
7. Dorner, R. W., Kahn, A., and Wildman, S. G., J. Biol. Chem.
    229, 945-52 (1957).
8. Kawashima, N., Imai, A., and Tamaki, E., Plant Cell Physiol.
    8, 447-58 (1967).
9. Kawashima, N. and Mitake, T., Agric. Biol. Chem. 33, 539-43
    (1969).
10. Kawashima, N. and Wildman, S. G., Annu. Rev. Plant Physiol. 21,
    325-58 (1970).
11. Peterson, L. W. and Huffaker, R. C., Plant Physiol. 55, 1009-15
    (1975).
12. Wittenbach, V. A., Plant Physiol. 59, 1039-42 (1977).
13. Huffaker, R. C. and Peterson, L. W., Annu. Rev. Plant Physiol.
    25, 363-92 (1974).
14. Soong, T-S. J., Feller, U. K., and Hageman, R. H., Plant
    Physiol. Suppl. 59, 112 (1977).
15. Feller, U. K., Soong, T-S. J., and Hageman, R. H., Plant
    Physiol. 59, 290-4 (1977).
16. Dalling, M. J., Boland, G., and Wilson, J. H., Aust. J. Plant
    Physiol. 3, 721-30 (1976).
17. Beevers, L., in Biochemistry and Physiology of Plant Growth
    Substances, pp. 1417-35, F. Wightman and G. Setterfield,
    Editors, Runge, Ottawa, 1968
18. van Loon, L. C. and Haverkort, A. J., Plant Physiol. Suppl. 59,
    113 (1977).
19. Drivdahl, R. H. and Thimann, K. V., Plant Physiol. 59, 1059-63
    (1977).
20. Thomas, H., Planta 137, 53-60 (1977).

21.   Thomas, H., Rep. Welsh Pl. Breed. Stn. for 1975, pp. 133-8.
22.   Thomas, H. and Stoddart, J. L., Plant Physiol. 56, 438-41 (1975).
23.   Martin, C. and Thimann, K. V., Plant Physiol. 49, 64-71 (1972).
24.   Choe, H. J. and Thimann, K. V., Plant Physiol. 55, 828-34 (1975).
25.   Sinclair, T. R., and de Wit, C. T., Agron. J. 68, 319-24 (1976).
26.   Rao, K. P., Rains, D. W., Qualset, C. O., and Huffaker, R. C.,
      Crop Sci. 17, 283-6 (1977).
27.   Sinclair, T. R. and de Wit, C. T., Science 189, 565 (1975).
28.   Hardy, R. W. F., Burns, R. C., Hebert, R. R., Holsten, R. D.,
      and Jackson, E. K., in Biological Nitrogen Fixation in Natural
      and Agricultural Habitats, p. 561, T. A. Lie and E. G. Mulder,
      Editors, Nijoff, The Hague, 1971.
29.   Abu-Shakra, S. S., Phillips, D. A., and Huffaker, R. C.,
      Science 199, 973-5 (1978).
30.   Mondal, M. H., Brun, W. A., and Brenner, M. L., Plant Physiol.
      61, 394-7 (1978).
31.   Arnon, D. I., Plant Physiol. 24, 1-15 (1949).
32.   Lowry, O. H., Rosebrough, N. J., Farr, A. L., and Randall,
      R. J., J. Biol. Chem. 193, 265-75 (1951).
33.   Davis, B. J., Ann. N. Y. Acad. Sci. 121, 404-27 (1964).
34.   Bethlenfalvay, G. J. and Phillipps, D. A., Plant Physiol. 60,
      868-71 (1977).
35.   Huffaker, R. C. and Miller, B. L., Unpublished, 1978.
36.   Frye, K., presented at Conf. on Capture, Conversion, and Con-
      servation of Energy in Plants, Michigan State U., 1977.

DISCUSSION

WAGNER: Is it possible to determine whether turnover of RuBP
carboxylase during dark-induced senescence is due to increased
leakiness of the chloroplast to the enzyme or to total degeneration
of the whole chloroplast? Recovery or resynthesis of RuBP carboxylase after light is restored seems to imply that chloroplasts do not
totally degenerate. If one assumes the absence of protease in the
chloroplast, one might expect the turnover to be due to leakage of
the enzyme into the cytosol, where it could be degraded by cytosol
protease.

HUFFAKER: We hypothesize that the loss of RuBP carboxylase
may occur after a lytic attack produces a "nick" in the chloroplast
membrane. Cytoplasmic proteinases may then enter and cause ultimate
total destruction of the chloroplast. The evidence indicates that
loss of RuBP carboxylase and loss of chlorophyll occur nearly at
the same time during the onset of senescence. We think that recovery of RuBP carboxylase and chlorophyll after removal of the environmental stress may be due to formation of new chloroplasts, as
indicated by the time course of recovery. Your question is a good
one, and the answer has not yet been worked out.

WAGNER: Are there data suggesting that some of the Calvin cycle enzymes are more rapidly turned over than the carboxylase appears to be in your experiments?   If there are no proteases in the chloroplasts, how do you explain this?

HUFFAKER: I do not know of any turnover figures for Calvin cycle enzymes.   Activity is not a very good indicator because it can be changed greatly without a change in enzyme concentration just by effector molecules.   For definitive results one must obtain data on the enzyme protein itself.

ZELITCH: Many years ago Chibnall showed that excising leaves caused a large decrease in protein, but allowing them to root stopped or reversed this trend.   He assumed there was a rooting hormone, which I assumed was kinetin.   How does this relate to your work?

HUFFAKER: I would predict, from his data, that until the leaf rooted it would use some protein as a source of carbon and nitrogen to produce the new root, but, as soon as it rooted, the new plant would be capable of synthesizing its normal components.

ZELITCH: So the proteases must be turned off at some point?

HUFFAKER: I do not think the proteases are ever turned off. I think the plant may develop new chloroplasts which protect the new complement of carboxylase again.

A MUTATIONAL APPROACH TO THE STUDY OF PHOTORESPIRATION

Mary B. Berlyn

Department of Biochemistry
Connecticut Agricultural Experiment Station
New Haven, Connecticut   06504

The ideal situation for applying a somatic cell genetics approach to a pathway present in cells of higher plants is the study of a pathway well defined in terms of the biochemistry of synthesis and regulation. This condition has been most nearly approached where bacterial and fungal systems have provided a model, as in selections for overproduction of aspartate-derived amino acids in rice (1) [also proposed for corn (2)] and for tryptophan pathway studies (3). However, the area of photosynthesis and photorespiration presents special problems (and opportunities) for a microbial genetics approach. The genetically best known microorganisms do not photosynthesize, and photosynthetic bacteria or even algae do not provide a high-fidelity model for the biochemical and genetic study of photorespiration in plants. In plants, we have not reached an understanding of some of the fundamental aspects of the pathway of photorespiration. Under these circumstances we are dependent on the plant mutants which we can obtain, rather than microbial mutants, to elucidate the pathway as well as eventually to provide a new means of regulation.

We are using cell and callus cultures of Nicotiana tabacum for selections. This system allows regeneration of plants and verification of the genetic or nongenetic nature of the variations induced in culture. We are working on photorespiration and its effect on net photosynthesis rather than on the carboxylase activity itself. However, the approach may prove applicable to more direct studies of CO_2 fixation by RuBP carboxylase. Although great variation of carboxylase is not observed in nature, possibilities for experimentally selected variation are not necessarily limited by observations in natural populations.

SURVEY OF SELECTIONS

Since it is regulatory phenomena that I would most like to ma-
nipulate genetically, my first consideration concerned the factors
known to regulate photorespiration and photosynthesis. One of the
oldest observations, of course, was the effect of oxygen on net
photosynthesis, and subsequently correlation between these effects
and photorespiration rates was observed (4). One of the approaches
we will discuss is concerned with oxygen levels in the atmosphere.
Although our work in this area is at very early stages, I will describe it because I hope that eventually this approach will bear
directly on the question of RuBP oxygenase function. The regulation
of carbon fixation by intermediates in the pentose phosphate carbon-
reduction cycle, although well studied, did not suggest immediately
feasible means of selecting for specific and negotiable effects on
cell metabolism, because of the complexity of interactions and fundamental nature of carbon and phosphate metabolism involved. Blocking the glycolate pathway by chemical inhibitors was shown by
Zelitch to increase net photosynthesis in tobacco leaf discs (5, 6).
These inhibitors include glycidate (5) and α-hydroxysulfonates (6-
8). Isonicotinic acid hydrazide (INH), which inhibits the conversion of glycine to serine and causes excretion of glycolate and accumulation of glycine in algae (9, 10), also blocks glycolate synthesis, photorespiration, and serine formation in tobacco (8, 11).
Selections with such inhibitors might provide us with a means of
probing the pathway and learning about its regulation. Resistance
to INH will be discussed in detail later.

In addition we made some speculations about what kinds of regulatory control might exist in such a pathway and attempted selections based on these speculations. For example, if the syntheses of
glycine and of serine are important functions of the pathway, regulation of glycolate synthesis might be related to biosynthesis and
utilization of these amino acids. Fortunately, David Oliver has
recently discovered that some common metabolites actually do decrease glycolate synthesis and photorespiration and increase net
photosynthesis in tobacco leaf discs. The most effective metabolites are glyoxylate, the product of glycolate oxidation (12), and
aspartate and glutamate (13), which decrease photorespiration by
~60% and increase net photosynthesis by 15 to 100%. These may in
the long run provide the most important tools for devising and analyzing effective selection procedures. One of the selections we will
mention, resistance to α-methylaspartate, is based on these findings.

Oliver has recently extended his studies to a C_4 plant. His
results suggest that glutamate and aspartate levels may play a role
in regulation of net photosynthesis in C_4 plants (14), but that they
do not appear to have a direct effect on the isolated RuBP carboxylase enxyme. Similarly, glyoxylate and INH do not affect isolated
enzyme activity (D. Oliver, personal communication).

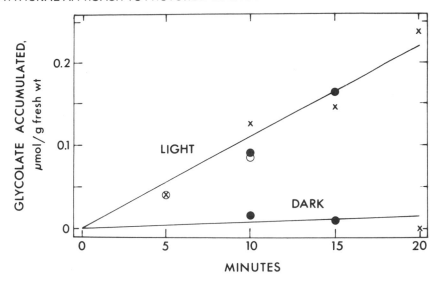

Figure 1. Effect of time and light on the accumulation of glycolate in the presence of sulfonate in autotrophically grown callus
cells. About 330 mg of cells were placed on Miracloth moistened
with 5 mM α-hydroxy-2-pyridinemethanesulfonic acid in the light
or dark in air for the times shown. The average rate of glycolate
accumulation in the light for the three experiments was about 0.6
μmol per g fresh wt per hr. From Berlyn et al. (16).

GROWTH OF AUTOTROPHIC CULTURES AND SELECTION ON OXYGEN

Selection for survival in the presence of high levels of oxygen
was made on autotrophic cultures of tobacco. We have maintained
tobacco callus cultures through many passages in the light with an
elevated CO_2 atmosphere as the only carbon source (15). Although
growth is slow relative to growth on sucrose-containing media, this
photosynthesis-dependent growth can continue indefinitely. Thus
photosynthesis is necessary for survival of the cells under these
conditions; similarly, I hope we can develop a system of glycolate-
utilizing cells which will make the glycolate pathway requisite for
survival. We have demonstrated glycolate synthesis and an oxygen
(Warburg) effect on photosynthesis in autotrophic cultures (16)
(Figure 1 and Table 1). In addition we observed a deleterious effect of high oxygen levels during 3-week incubations of autotrophic
cultures but not of heterotrophic cultures (Table 2). This suggests
that the toxicity of high oxygen may be related to the inhibition
of photosynthetic ability in these cultures, although it certainly
does not establish a direct and specific effect on photosynthesis.

When autotrophic cultures were shifted from 1 or 3% CO_2, to
70% O_2, the cultures bleached and decreased in weight. Viability

Table 1

Effect of O_2 Concentration on Photosynthetic $^{14}CO_2$ Uptake
by Tobacco Callus Cells

Cells were gassed with air or 100% O_2 and kept in the light
at 30°C for 10 min before $^{14}CO_2$ (final concentration 0.18%)
was released in a closed system. Fixation time was 30 min.
Fresh wt of tissue was about 800 mg in Expt. 1 and 400 mg
in Expt. 2 and 3. The values are the means of duplicate
flasks. From Berlyn et al. (16).

Expt. No.	Subclone number	Photosynthesis rate in 0.18% CO_2 (μmol CO_2 per g fresh wt per hr)		% Inhibition by 100% O_2
		Cells in 21% O_2	Cells in 100% O_2	
1	14B	1.72	0.86	50
2	13-1	1.68	0.98	42
3	13-1	1.58	0.96	38

was tested by transferring all cells on a plate to standard
(heterotrophic) medium. Most cells were inviable, but a few (15)
small areas of growth were observed on a total of 22 plates after
several weeks of incubation. In retests of autotrophic cultures
of eight of these isolates, four selected cultures remained viable
in high oxygen atmospheres longer than did unselected cultures.
These preliminary results are not sufficient to establish these
isolates as stable variants, but do suggest that such selections
may be feasible. Oxygen effects in short-term photosynthetic assays
are being compared in selected oxygen lines and unselected lines.

In order to perform these selections more efficiently, we are
trying to develop continuous autotrophic suspension (rather than
callus) cultures of tobacco. Bergmann (17) first reported short-
term autotrophic growth in tobacco suspension cultures with an
elevated CO_2 atmosphere. Chandler et al. (18) documented growth of
autotrophic suspensions of specific clones of tobacco during three
2 to 2½-week passages. Hüsemann and Barz (19) recently reported
growth of a _Chenopodium rubrum_ suspension culture maintained in
a two-tiered vessel (20) with a carbonate-bicarbonate buffer solution in the lower flask providing an enriched CO_2 atmosphere (21).
A 165% increase in fresh weight was observed during the tenth 18-
day passage. Using these methods with suspensions of tobacco, we
have maintained viable green cultures, but to date have observed
negligible weight increases during 2- or 4-week incubations.

Table 2

Autotrophic and Heterotrophic Growth of JWB Tobacco Callus in
1% CO_2 in Air Compared With That in 60% O_2 and 0.03% CO_2

The cells were grown in Petri plates in the light, in trans-
parent plastic cabinets continuously gassed with the atmospheres shown.   The growth was normalized to a standard 21-
day passage. Data for growth on autotrophic medium in air
are for cells of JWB grown outside the cabinets in other
experiments.   From Berlyn et al. (16).

			Relative increase in dry weight (Ratio of harvest/inoculum)					
Medium	Atmosphere	Passage No: 1	2	3	4	5	Av.
Auto* ** Hetero	1% CO_2-air	2.6 4.8	3.7 5.4	3.0 6.5	2.7 3.9	6.0 8.8	3.6 5.9
Auto Auto	60% O_2-0.03% CO_2	2.0 1.5	0.9 1.5	0.6 1.0	0.6 0.7	- -	
Hetero Hetero	60% O_2-0.03% CO_2	6.0 3.3	3.5 4.8	8.0 8.9	5.9 4.8	7.1 -	6.1 5.5
Auto Auto Auto	Air	- - 1.8	2.0 2.2 1.4	1.8 1.5 1.4	 1.5		

*Autotrophic growth was carried out in medium with no sucrose.
**Heterotrophic growth was carried out in the same medium ex-
cept that 2% sucrose was present.

ISONICOTINIC ACID HYDRAZIDE RESISTANCE

In Chlorella, treatment with INH in the presence of $^{14}CO_2$ in
the light resulted in increased excretion of ^{14}C-glycolate, accumulation of ^{14}C-glycine, and failure to convert glycine to serine (9,
10).   In tobacco leaf discs or segments, increases in glycine, decreases in serine, and decreases in glycolate synthesis and photorespiration are also observed after INH treatment (8, 11, 22).
Since serine hydroxymethyltransferase, the enzyme complex catalyzing
the glycine to serine conversion, requires pyridoxal phosphate, and
INH can act as an inhibitor of pyridoxal phosphate-requiring enzymes
(23-25), this is the mechanism assumed in plants (22, 26).   However,

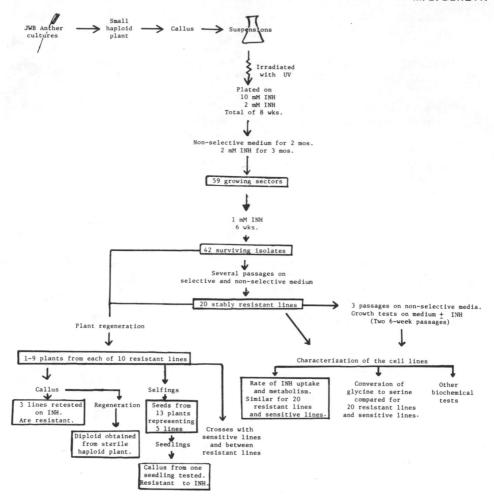

Figure 2.   A diagram of the selection and testing of INH-
resistant cell lines.

this is not the only mechanism proposed for INH toxicity and resistance in mycobacteria.   It has been proposed that toxicity is due
to the formation of large quantities of isonicotinic acid, which
acts as a nicotinamide analog, and resistance results from a low
rate of conversion of INH to isonicotinic acid due to low levels of
peroxidase activity (27-29).   This also results in restriction of
uptake of INH.

Thus the relationship of INH resistance to RuBP carboxylase is
at best an indirect one, involving the metabolism of glycine as a
possible regulator of photorespiration, which in turn may be a function of RuBP oxygenase.   The study of the INH-resistant lines is,
however, the furthest advanced and serves best to illustrate the

process and techniques we are using. A summary of the process is
shown in Figure 2. Small haploid tobacco plants arising from anther
cultures of variety John Williams Broadleaf (JWB su) were used to
obtain callus cultures and then heterotrophic suspension cultures.
Suspensions were irradiated with UV and incubated in the presence
of 10 mM INH, then 1 or 2 mM INH. After transfer and further incubation, small areas of growth were observed on media containing
2 mM INH. Of 59 such isolates transferred to media containing 2 mM
INH, 42 survived 6-week incubation on the inhibitor. After several
cycles of growth on selective and nonselective media, 20 isolates
appeared to be stably resistant to 1 mM INH. Plants were regenerated
from a number of these isolates. A quantitative growth test was made
to confirm the previous qualitative observations of resistance, and
Figure 3 illustrates the results, showing data for 12 of the cell
lines in comparison with data for unmutagenized, sensitive lines of
unselected cultures, JWB su. Each passage was 6 weeks long. This
prolonged period was used with the intent that the numbers would reflect total growth capacity rather than growth rate. It can be seen
in Figure 3 that many of the resistant lines grow better than the
sensitive lines on nonselective as well as selective medium. In all
the lines shown, and in 17 of the 20 lines tested, growth in the
presence of INH is clearly greater than that of sensitive cultures;

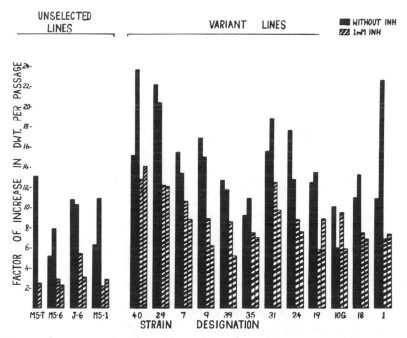

Figure 3. Growth of callus from 4 unselected cultures and
12 INH-resistant lines during two 6-week passages on media
with (striped bars) and without (black bars) 1 mM INH.

in several of the lines, growth on INH is >70% of the growth of the
same line on standard nonselective medium.

I. Zelitch and I are examining these lines for biochemical
differences that may confer INH resistance. In no case do we observe (unpublished results) deficient uptake of INH or failure to
convert INH to isonicotinic acid, the reported basis for resistance
in <u>Mycobacterium</u> <u>tuberculosis</u> (28, 29). We are examining in detail
intact callus and enzymes extracts of several lines that take up
and metabolize radioactive glycine faster than do the sensitive
(wild-type) lines.

The other aspect of this study, which is of the greatest importance, is the establishment of whether these variant lines are
in fact true mutants. This is the point at which the availability
of regenerated plants becomes critical. A total 0f 80 plants have
been regenerated from 10 of the INH[R] lines, and we are attempting
regeneration from the remaining lines. Callus cultures from explants
of 40 of these plants are being cultivated for tests of resistance
to INH. To date, three cultures have been tested, and all were resistant to 1 mM INH. Cultured explants from plants regenerated
from unselected cell lines were sensitive to this level of INH. An
example of such a comparison is given in Figure 4.

Since these cultures were initiated from haploid plants, it is
not surprising that many of the regenerated plants are sterile. A

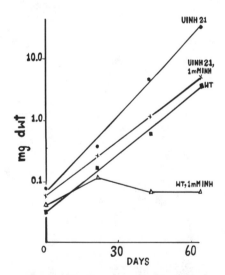

Figure 4. Growth of callus from plants derived from an
unselected cell line (WT) and an INH-resistant cell line
(UINH21) on medium lacking INH (● and ■) and containing
1 mM INH (X and Δ). Total growth during each passage is
calculated on a dry weight basis, from data on samples
of each culture.

few, however, are fertile, presumably as a result of spontaneous diploidization. Seeds have been obtained from 13 plants, representing 5 different lines. In one case, callus from seedling has been tested and appears to be resistant. Thus, in this case, the resistant trait has been passed through a sexual generation and can be considered a true mutation.

Combinations of cutting back of the plants, colchicine treatment, and midrib culture are being used in attempts to obtain diploids from sterile plants. In the case of one sterile UINH21 plant, microspectrophotometric evidence indicated that it was haploid. A number of plants were regenerated from resistant callus derived from midrib or pith cultures of the original plant. One of these is bearing seed and is presumably diploid.

We have deferred tests of photorespiration and of glycolate pathway metabolism in the whole plant because we were reluctant to match culture-derived sensitive and resistant plants in view of differences in their growth rates and morphology. We hoped that seedlings would provide uniform control and mutant plants for testing. However, the two types of INH-resistant seedlings so far examined have been smaller and less vigorous than wild-type (also callus-derived) seedlings of the same age.

RESISTANCE TO AMINO ACID ANALOGS

Because of our interest in glycine or serine as a possible regulator of the glycolate pathway, we used procedures similar to those described above (Figure 2) to isolate 27 variant lines resistant to 2 mM glycine hydroxamate. After many cycles of growth and retesting, seven vigorously growing, stably resistant lines have been retained for analysis. Arthur Lawyer has demonstrated that glycine hydroxamate is a competitive inhibitor of glycine decarboxylase activity in particulate fraction extracts of tobacco callus. He observed no differences in activity or inhibition of this enzyme in extracts of variant and wild-type cultures. He has demonstrated that none of the variant lines fails to take up glycine hydroxamate or exhibits unusual glycine hydroxamate metabolism. He found, when examining glycine and serine levels in callus, that some of the variant lines grown on standard heterotrophic medium have higher levels of glycine than does wild-type callus (Lawyer, unpublished data).

There is no large body of evidence concerning regulation of and by glycine or serine levels (30). The existence of three possible pathways of glycine synthesis added a great deal of complexity to considerations of glycine effects (31). Oliver (13) found that exogenous glycine did not affect photorespiration and net photosynthesis in tobacco leaf discs, but he did not monitor the intracellular levels of glycine. Vallee et al (32) report that proline administration to leaf discs leads to changes in the glycolate pathway and to increases in endogenous glycine levels. It will be interesting to see whether amino acids so directly synthesized from photosynthate and utilized in biosynthesis of a wide variety of other cellular components will turn out to be subject to direct and strict kinds of regulation.

Finally, I will briefly mention the α-methylaspartate-resistant lines which we isolated in response to the finding that aspartate and glutamate levels may be important factors in regulation of photorespiration (13). Sixty-eight lines were isolated from dip-loid, heterotrophic suspension cultures of Wisconsin-38 which were UV-irradiated and incubated in and subsequently plated on media containing 5 mM α-methylaspartate. Oliver determined aspartate and glutamate levels in callus cultures of 46 of these lines and found 4 lines that had levels higher than wild-type callus (Table 3) (33). Plants have been regenerated from 10 lines.

SUMMARY

A number of quite different types of variant cell lines of to-bacco have been isolated. This multiplicity of approaches has been in part dictated by the uncertainities of the photorespiratory path-way. We are in the process of determining the relationship between the observed alterations and the glycolate pathway. The INH-resis-

---

Table 3

Aspartate and Glutamate Concentration in Wild-Type Tobacco Callus and in Variants Resistant to α-Methylaspartate

The callus tissue was grown for 3 weeks in a standard medium containing ^{14}C-sucrose. A sample of tissue was then killed and homogenized, and the ^{14}C-aspartate and ^{14}C-glutamate concentrations were determined by Dowex-50 fractionation and Dowex-1 acetate chromatography. From Zelitch et al. (33).

Cell line	$\mu mol/g$ fresh wt	
	Aspartate	Glutamate
Wild type	2.3	2.5
A 3	5.4	4.4
A 30	4.5	5.0
A 40	5.5	4.6
A 45	5.8	3.2
Mean of all 46 variant lines	3.5	3.2

tant trait has been observed in callus derived from plants regenerated from the original resistant callus and in cone case from seedlings of the subsequent generation. It is hoped that these cultures
and the plants derived from them will further our understanding of
photorespiration and will eventually lead to the ability to regulate
photorespiration and net photosynthesis in higher plants.

Acknowledgements: I am indebted to Arthur Lawyer, David Oliver,
and Israel Zelitch for allowing me to cite unpublished data from
their work as well as collaborative work and for helpful discussions.
Technical assistance in various aspects of our studies was ably provided by Pamela Beaudette, Katherine Clark, Kenneth Lattimore,
Marilyn Newman, and Cyntha Spoor.

REFERENCES

1. Chaleff, R. and Carlson, P., in Modification of the Information
    Content of Plant Cells, pp. 197-214, R. Markham et al., Editors,
    American Elsevier, New York, 1975.
2. Green, C. and Phillips, R., Crop Sci. 14, 827-30 (1974).
3. Widholm, J., Crop Sci. 17, 597-600 (1977).
4. Zelitch, I., Photosynthesis, Photorespiration, and Plant Pro-
    ductivity, Academic Press, New York, 1971.
5. Zelitch, I., Arch. Biochem. Biophys. 163, 367-77 (1974).
6. Zelitch, I., Plant Physiol. 41, 1623-31 (1966).
7. Zelitch, I., Plant Physiol. 43, 1829-37 (1968).
8. Zelitch, I., Plant Physiol. 51, 299-305 (1973).
9. Pritchard, G., Griffin, W., and Whittingham, C., J. Exp. Bot.
    38, 176-84 (1962).
10. Pritchard, G., Whittingham, C., and Griffin, W., J. Exp. Bot.
    41, 281-9 (1963).
11. Zelitch, I., Plant Physiol. 50, 109-13 (1972).
12. Oliver, D. and Zelitch, I., Science 196, 1450-1 (1977).
13. Oliver, D. and Zelitch, I., Plant Physiol. 59, 688-94 (1977).
14. Oliver, D., Plant Physiol. 62, in press (1978).
15. Berlyn, M. and Zelitch, I., Plant Physiol. 56, 752-6 (1975).
16. Berlyn, M., Zelitch, I., and Beaudette, P., Plant Physiol. 61,
    606-10 (1978).
17. Bergmann, L., Planta 74, 243-9 (1967).
18. Chandler, M., de Marsac, N., and deKouchkovsky, Y., Can. J.
    Bot. 50, 2265-70 (1972).
19. Hüsemann, W. and Barz, W., Physiol. Plant. 40, 77-81 (1977).
20. Street, H. E., in Plant Tissue and Cell Culture, pp. 64-5,
    H. E. Street, Editor, U. of California Press, Berkeley, 1973.
21. Warburg, O. and Krippahl, G., Z. Naturforsch. 15b, 364-7 (1960).
22. Asada, K., Saito, K., Kitoh, S., and Kasai, Z., Plant Cell
    Physiol. 6, 47-59 (1965).
23. Yoneda, M., Kato, N., and Okajima, M., Nature London 170, 803
    (1952).

24. Youatt, J., Biochem. J. <u>68</u>, 193-7 (1955).
25. Davison, A., Biochim. Biophys. Acta. <u>19</u>, 131-40 (1956).
26. Gore, M., Hill, H., Evans, R., and Rogers, L., Phytochemsitry <u>13</u>, 1657-65 (1974).
27. Krieger-Thiemer, E., Ber. Borstel <u>4</u>, 299 (1957). (Cited in ref. 28.)
28. Seydel, J., Schaper, K. J., Wempe, E., and Cordes, H., J. Med. Chem. <u>19</u>, 484-91 (1976).
29. Devi, B., Shaila, M., Ramakrishnan, T., and Gopinathan, K., Biochem. J. <u>149</u>, 187-97 (1975).
30. Miflin, B., in <u>Biosynthesis and Its Control in Plants</u>, pp. 49-68, B. V. Milborrow, Editor, Academic Press, New York, 1973.
31. Miflin, B. and Lea, P., Annu. Rev. Plant Physiol. <u>28</u>, 299-329 (1977).
32. Vallee, J., Vansuyt, G., and Prevost, J., Physiol. Plant. <u>40</u>, 269-74 (1977).
33. Zelitch, I., Berlyn, M., Oliver, D., and Lawyer, A., in <u>Proc. 2nd Latin Am. Bot. Congr.</u>, <u>Brasilia, 1978</u>, M. A. A. Hermans, Editor.

THE OPPORTUNITY FOR AND SIGNIFICANCE OF ALTERATION

OF RIBULOSE 1,5-BISPHOSPHATE CARBOXYLASE ACTIVITIES

IN CROP PRODUCTION

R. W. F. Hardy, U. D. Havelka, and B. Quebedeaux

Central Research & Development Department, Experimental Station
E. I. du Pont de Nemours & Company, Inc., Wilmington, Delaware 19898

SUMMARY

Increased population and the dietary changes accompanying increased affluence are creating a need for a suggested doubling of world cereal grain production (a 3% per year compounding rate) and quadrupling of grain legume production (a 6% per year compounding rate) during this quarter century (1). CO_2 enrichment of field-grown crops has demonstrated the possibility of enhancing RuBP carboxylase activity to achieve improved crop production; it increases the production of grain legumes by 50 to 100% and that of cereal grains, for which the studies are less complete, by perhaps 10 to 50%. Results of O_2 alteration of growth-room legumes and cereal grains are consistent with the results of CO_2 enrichment except for a second role of O_2 in assimilate partitioning. It may be necessary to include other components of the system, e.g., additional soil fertility, especially for non-N_2-fixing plants, to enable an improved RuBP carboxylase to increase production. No practical method--chemical, genetic, or physical--of improving RuBP carboxylase activity has been reported.

INTRODUCTION

Many significant advances in the study of photosynthetic carbon assimilation have been or may be relevant to increased crop production (Table 1) (2). These advances extend from the biochemical to the agronomic areas; only the chemical area is poorly represented, showing a single selected example of the fixation of CO_2 to transition metals (3), which may be important in chemical fixation of CO_2 but is probably not relevant to biological fixation. Carbon dioxide

Table 1

Exploratory Advances in Photosynthetic Carbon Assimilation (2)

	Chemistry	Biochemistry	Biology	Agronomy
		Coupled mitochondrial oxidation of glycine	Mutants with altered carboxylase activity	
1975	$M(CO_2)$	CO_2-Mg^{2+} activation of RuBP carboxylase	O_2 concentration regulates reproductive growth	
1970		Oxygenase activity of RuBP carboxylase	O_2 concentration regulates growth of C_3 plants	CO_2 enrichment of wheat, rice, soybeans, oats
		Peroxisomes and serine-glycine pathway		
1965		C_4 photosynthetic pathway	Crop growth modeling - LAI	
1960		Glycolate as photorespiratory substrate		Planting Density
1955		Post-illumination burst and photorespiration		Lodging Resistance
		C_3 photosynthetic cycle		Photoperiodic insensitivity
1950				
1930		O_2 inhibition of photosynthesis		

enrichment of enclosure-produced crops, altered planting density, photoperiod-insensitive crops, and lodging-resistant dwarfed cereals have contributed major new technologies for increased photosynthetic carbon assimilation and crop productivity (Table 2).

Other exploratory advances suggest the opportunity for further improvements in photosynthetic carbon assimilation. The highly significant series of discoveries extending from the recognition of the oxygen inhibition of photosynthesis to the proposal that the dual oxygenase-carboxylase activity of RuBP carboxylase is the major primary reaction leading to photorespiration has provided a fundamental

Table 2

Photosynthetic Carbon Assimilation Technologies
for World Crop Production (4)

Generation	Technology	Period of initial impact
1	CO_2 enrichment of enclosure-produced crops	20th Century, 2nd quarter
2	Planting density	20th Century 3rd quarter
3	Lodging-resistant crops: rice, wheat	20th Century 3rd quarter
4	Photosynthetically efficient legumes and cereals	?
5	Elimination of nonessential dark respiration	?
6	Deregulated photosynthetic CO_2-fixing system	?
7	Delayed senescence	?
8	Maximized partitioning of assimilate into economic (harvested) yield	?

basis from which to develop photosynthetically efficient legumes and cereals. This goal is one of the best possibilities for increasing production of most crops, as documented below. Other suggested future means of increasing photosynthetic carbon assimilation include elimination of nonessential dark respiration, deregulation of the photosynthetic CO_2-fixing system, regulation of senescence, and maximized partitioning of assimilate into the economic and harvestable yield (Table 2) (4).

This paper considers specifically the opportunity for and the significance of the alteration of RuBP carboxylase activities in crop production. The activities are considered in the broadest terms. First the existing biochemical, physiological, genetic, and agronomic knowledge of the activities is summarized from the viewpoint of what's wrong with RuBP carboxylase for maximum crop production. Then the effects of CO_2 enrichment on yield of field-grown crops are summarized with emphasis on legumes, for which more normal field-condition results are available. These results indicate broad opportunities for practical alteration of RuBP carboxylase activities.

Besides identification of limitations and quantitative assessment of their significance at the field level, several other types of information are required for developing a new technology to increase crop productivity, but these will not be discussed here (5). For example, specific but rapid and simple screening methods are

necessary for selecting genetic, chemical, or other means of over-
coming the limitations, and this area has not been covered at this
Symposium. The screening methods must be put into operation, and
the potential means must be evaluated on the basis of economic,
safety, and possibly even nutritional factors; and the practical
means that satisfy these criteria must be implemented for increased
crop production to become a reality.

THE OPPORTUNITY, OR, WHAT'S WRONG WITH RuBP CARBOXYLASE?

Biochemistry

Almost all the biochemical characteristics of RuBP carboxylase
with the possible exception of its in vivo stability (6) are inappropriate for maximum crop production. However, many of these inappropriate characteristics are probably almost impossible to alter
and others probably do not comprise important limitations. Fortunately, a few that may be susceptible to alteration appear to represent major limitations. Expansion of our basic knowledge will
undoubtedly identify other significant inappropriate characteristics
and provide further opportunities for practical improvement.

RuBP carboxylase from higher plants has an unacceptably large
molecular weight of 560,000 so that at most only a fraction of a
mole is present per acre of crop. The complexity of the enzyme increases from bacteria, in which each enzyme molecule has only one
component, the large subunit, to higher plants, in which each molecule
contains eight small subunits and eight large ones (7,8). This
multicomponent composition requires a specific ratio of one large
subunit to one small subunit, the small one being synthesized in the
cytoplasm and the large one in the chloroplast. The unknown but
undoubtedly metabolically costly system for orchestrating the
synthesis and transfers to provide equivalent amounts of subunits
must be highly sophisticated and probably energy consumptive (9, 10).
Is the ratio of small to large subunits, in fact, always 1:1 in
the chloroplast, or might altered ratios occur (11)? Do altered
ratios influence specific activity, as the altered ratios of nitrogenase have been shown to do (12)? The occurrence of eight active
sites per molecule appears to be not an advantage but rather a disadvantage, since complexation of substrates with one active site
appears to have no beneficial influence on complexation at the
remaining active sites. The behavior of the enzyme appears to be
allosteric, with the small subunit probably having a role in activation of the large one. The enzyme also has specific activation requirements such as a certain pH range, Mg^{2+}, CO_2, and light (13).
Its substrate is CO_2, but HCO_3^- may provide advantages, as noted below.

RuBP carboxylase under ambient substrate conditions has a low
turnover number of about 200 moles of CO_2 fixed per mole of enzyme
per minute, which is almost as low as the turnover number of 100 for
nitrogenase. At saturating conditions, the turnover number may be

1000 or more. Up to 50% of soluble leaf protein is RuBP carboxylase, which appears to represent an excessive and inappropriate amount of this enzyme. Is only part of the enzyme functional while the remainder is nonfunctional and serves as an expensive storage or structural material? The enzyme is localized in a single organelle, the chloroplast, whereas a more diverse distribution might be advantageous. These characteristics appear to limit the CO_2-fixing capability of RuBP carboxylase, but it is highly unlikely that many of them can be altered in the foreseeable future in a practical way.

     There are other limitations, some of which it should be possible to alter. One of the major things "wrong" with the enzyme is its substrate promiscuity. This enzyme functions as both a carboxylase and an oxygenase, interacting with both CO_2 and O_2 (14). This, as well as several other characteristics, makes it similar to nitrogenase, whose substrate promiscuity is even broader. However, the alternative substrates of nitrogenase (N_2O, N_3^-, RCCH, RCN, RNC) are not highly concentrated in the ambient environment of the enzyme as is the single known alterative substrate (O_2) of RuBP carboxylase. Moreover, the high $K_m(CO_2)$ of RuBP carboxylase ($\sim$10 μM) vs. the low ambient CO_2 concentration and the tendency of the $K_m(CO_2)$ to increase with increased temperature, coupled with its low $K_m(O_2)$ ($\sim$200 μM) vs. the high ambient O_2 concentration and the tendency of the $K_m(O_2)$ to decrease with increased temperature, produces about 25% or more of wasteful oxygenase reactions (15). Furthermore, the plant cannot directly utilize the product of the oxygenase reaction, phosphoglycolate, but must transfer it through a variety of organelles. The transfers are presumably energy consuming in order to recoup part of the carbon of phosphoglycolate. So far, no chemical reagent that alters the carboxylase/oxygenase ratio has been reported, and the chemical structure of the active site is only beginning to be elucidated (16). The differential solubility of oxygen and carbon dioxide with increasing temperature tends to favor the oxygenase vs. the carboxylase reaction and may partly explain the higher percentage of photorespiration at higher temperatures (17). The investigators themselves may be providing a limitation in this area via the long-standing disagreement about the importance of the oxygenase reaction as a source for the photorespiratory substrate glycolate (18), although recent $^{18}O_2$ experiments strongly support oxygenase being the major source (19).

     It is difficult to do work with the enzyme in vitro that is relevant to the in vivo situation because of the kinetic differences between the in vitro and in vivo conditions. The identification of the activation of the enzyme by CO_2 and Mg^{2+} in vitro has resolved some of this difficulty. The form of the substrate, CO_2^- vs. HCO_3^-, can be viewed as a major limitation. At ambient CO_2 concentrations phosphenolpyruvate carboxylase, which utilizes HCO_3^- as a substrate, is highly efficient, but RuBP carboxylase, which uses CO_2, is highly inefficient. The concentrations of RuBP, ATP, and reductant may also be limiting.

The mechanism of regulation of the enzyme is not yet resolved, but phosphate, sugar phosphates, and starch have been indicated as regulatory molecules (20-22). The unusually long stability of the enzyme appears to be a positive attribute, but it precludes the opportunity for anything except gross regulation at the gene expression level.

Physiology

Several physiological characteristics can be considered as part of "what's wrong" with RuBP carboxylase. The photosynthesis rate in C_3 plants is generally substantially lower than in C_4 plants, the differential being greater at higher temperatures (23). Plants with C_3 photosynthesis have a high compensation point (40 to 60 ppm CO_2), whereas plants with C_4 photosynthesis have a compensation point approaching zero, which increases the inferiority of C_3 plants at subambient CO_2 concentrations that may occur in a dense crop canopy. The post-illumination burst of CO_2 presumably represents the loss of CO_2 from a photorespiratory substrate accumulated during illumination.

The common port for CO_2 uptake and water loss may create limitations under conditions of water restriction because stomata close to conserve water and thereby restrict the availability of CO_2 to RuBP carboxylase. Separate systems for CO_2 uptake and water loss would have obvious beneficial effects, especially considering that normal plants lose several hundred molecules of water for each molecule of CO_2 fixed (24). Stomata also tend to close in response to elevated CO_2 concentrations, but this may not decrease carbon assimilation because probably the elevated concentrations will make adequate CO_2 available in spite of the reduced stomatal aperture. There are various intracellular resistances to the transfer of CO_2 to the active site of RuBP carboxylase, and no CO_2 transport system has been developed in C_3 plants to facilitate CO_2 diffusion the way leghemoglobin facilitates rapid diffusion of low concentrations of oxygen in the legume nodule. The CO_2 concentration in the environs of the active site of RuBP carboxylase with CO_2 fixation occurring must be substantially lower than in the surrounding ambient air. C_4 plants appear to have overcome the physiological limitation of low CO_2 concentration with their phosphenolpyruvate carboxylase-initiated CO_2-pumping system, but this system has a substantial energy cost.

C_3 plants make specialized organelles, peroxisomes, to enable them to utilize products of the oxygenase reaction of RuBP carboxylase. The rate of photosynthetic carbon assimilation adapts slowly in intact plants exposed to increases in light intensity. Light quality may affect photosynthetic carbon assimilation activity. Its variation from the top to the lower parts of the canopy probably alters carboxylase activity. What advantage does such alteration provide? Since photosynthetic carbon assimilation is inactive in the dark, 50% of the soluble enzyme of the leaf is of no benefit during 50% of the time. Active proteolytic destruction of the

carboxylase appears to be an early event in senescence (6). Do the
activities of photosynthetic carbon assimilation change with age?
Some studies suggest occurrence of both C_4 and C_3 characteristics
in the same plant (25). Photosynthesis appears to have a single
main export product, sucrose. Is this the most desirable export
product for maximum productivity of various crops?

The sink has effects on photosynthesis, which may include
altered O_2 concentrations around reproductive structures (26), pod
growth, and sink removal (27). What is the molecular structure of
the signal that correlates sink needs with source activities? In
at least one case, sink removal is reported to alter leaf angle
for decreased light interception and to cause stomata to close
partially (28). The translocation rate is greater in C_4 than in
C_3 plants. Why?

Present understanding of physiology of RuBP carboxylase and
associated activities has not enabled us to design ideal assay
methods. For example, many assays are run under reduced CO_2 partial
pressure, but the validity of their results is questionable in relation to effects in ambient CO_2. Cultured cells, in general, have
low photosynthesis rates and thus are of limited value in studying
RuBP carboxylase and its associated activities (29).

Genetics

Studies on genetics, especially at the molecular level, began
relatively recently. It is hoped that expanded studies may provide
information about regulation at the gene and/or enzyme level, as
similar studies have done for the nitrogenase system (30). The
most obvious genetic limitation of RuBP carboxylase is the segrega-
tion of the genes between the nucleus and the chloroplast. This is
a major limitation in conventional plant breeding efforts to improve RuBP carboxylase activity because the large subunit, which
appears to contain the active site, has its genetic information
located in the chloroplast. There is little information on repres-
sors and/or inducers for either the nuclear or chloroplast genes
involved in RuBP carboxylase synthesis. The number of genes involved is not well defined. The gene dose for the large subunit,
which occurs in the chloroplast, is severalfold that of the gene
dose for the small subunit. Is the gene dose adequate or excessive
for the formation of each of the components of RuBP carboxylase?
Might it be desirable to decrease or increase the gene dose?

Little genetic diversity of RuBP carboxylase has been found,
but this may simply reflect the limited search. Some suggest that
the C_3-C_4 intermediate species, Panicum milioides, has a carboxyl-
ase/oxygenase with altered kinetic parameters (31); others attribute
the altered CO_2 compensation point to a suboptimal content of enzymes associated with the CO_2 pump of C_4 plants (32), and further
support for this view is provided by a poster presentation at this
Symposium (Chollet, personal communication). Improved but undiscussed

genetic diversity may already exist in natural environments such
as the tropics, where there are selection pressures for an improved
carboxylase. The chance finding only a few years ago of a natural
Rhizobium nonlegume association, the Rhizobium-Trema N_2-fixing
symbiosis, exemplifies the opportunities existing in nature (33).
The preliminary report of possible improved carboxylase activity
in certain tetraploid rye grasses is of great interest (34). So
far, conventional breeding and/or screening of Atriplex, tobacco,
soybeans, and wheat--as well as extensive mutational program on
soybeans at Illinois--have failed to produce an improved carboxylase.
Will unconventional genetic approaches, such as recombinant DNA,
somatic fusion, organelle transplants, etc., enable the construction
of crop plants with improved carboxylase activities (35, 36)? The
developing activities in molecular genetics provide opportunities
to understand and maybe to improve carboxylase (37-39).

Agronomy

     Several agronomic shortcomings of RuBP carboxylase have been
identified. The low ambient CO_2 concentration may suffer even
further depletion by photosynthesis within the dense canopy of a
crop at midday, worsening the already less than saturating concentrations. However, the high diffusion rate of CO_2 even in canopies
precludes practical CO_2 enrichment of nonenclosed agronomic crops.
All crops have a shockingly low conversion of incident solar energy
to harvestable component: 0.1 to 0.2%. The growth rate of C_3
crops (wheat, soybeans) is in general less than that of C_4 crops
(corn, sugar cane), the difference being influenced by temperature.
     Photosynthetic carbon assimilation is inadequate to meet the
simultaneous needs of biological N_2 fixation and reproductive
growth in high-protein crops such as soybeans (40); this ability
is the basis of the self-destructive hypothesis (41). The rate
of photosynthetic carbon assimilation per leaf area for various
soybean cultivars is correlated with leaf thickness but not with
yield, and is therefore useless as the basis for a screening technique (42). In general, the harvest index (ratio of mass of economic component to total mass) of highly domesticated C_3 crops is
greater than that of C_4 crops. This difference probably represents
selection by breeders rather than an inherent metabolic difference
between C_3 and C_4 crops with respect to assimilate partitioning.
Certain leaves provide most of the photosynthate to the developing
seed, e.g., the flag leaf of wheat and the leaf subtending each pod
in soybeans. Clearly nature provides the means to enable RuBP
carboxylase activities to support adequate seed production for survival in the adverse natural environment, but these means often
appear as "wrong" characteristics in the protected agricultural
environment, where maximum space-time yield of the economic component is the grower's objective.

SIGNIFICANCE OF THE ALTERATION OF RuBP CARBOXYLASE ACTIVITIES

We have utilized CO_2 enrichment of field-grown grain legumes and some cereal grains to assess the quantitative significance of the alteration of RuBP carboxylase. The technique has many useful attributes, and we think it should be more extensively applied to all major field crops in their major geographic areas (43). Perturbation of the plant by CO_2 enrichment is minimal compared with that by other techniques such as shading and leaf removal. Multiplot experimentation and scale-up are possible. The CO_2 concentration within the canopy can be easily monitored, and it can be altered as desired. The equipment is sufficiently simple and inexpensive to be used in the major crop production areas throughout the world, consisting of open-top, side-enclosed chambers through which air or CO_2-enriched air is forced from a perforated pipe lying on the soil surface (42). Thus, the temperature and light within the chamber are the same as in the field, and only the CO_2 concentration is different. The data reported below for grain legumes were obtained with such chambers, but most of that for cereal grains and other crops was obtained with chambers in which light and/or temperature were also altered.

Increases in seed yield (Table 3) (43, 44) of 50 to 100% were produced in three grain legumes (soybeans, peas, and beans) by CO_2 enrichment to 1500 ppm in the canopy from anthesis to senescence. The increase was only 21% for peanuts, but interference with pegging and an inadequate growth period prevented maximum expression of yield. Note that the yields of the air-grown controls are excellent, so that the yield of almost 100 bu/acre for CO_2-enriched soybeans is three to four times the national average. The increased yield is due to at least three factors, (i) increased photosynthetic carbon assimilation, (ii) increased biological N_2 fixation, and (iii) delayed senescence, the first being the primary factor. We attribute the increased photosynthetic carbon assimilation to a decrease in photorespiration caused by the elevated CO_2/O_2 ratio and possibly to a higher degree of saturation of RuBP carboxylase with CO_2. A decrease in O_2 concentration at constant CO_2 content produces an increase in photosynthetic carbon assimilation similar to that produced by an increase in CO_2 concentration at constant O_2 content (26); this supports the central role of the CO_2/O_2 ratio and carboxylase/oxygenase activity in photosynthetic carbon assimilation.

CO_2 enrichment has also produced increases in yields of cereal grains and cotton but smaller than those for N_2-fixing legumes (43). For wheat, rice, and barley the yield was increased 34 to 50% by CO_2 enrichment before anthesis and 14 to 28%, after anthesis. For a given cereal, the increase was always greater for pre-anthesis than post-anthesis enrichment, and CO_2 enrichment did not delay senescence.

Why is the response of cereal grains to CO_2 enrichment inferior to that of grain legumes? Recent data comparing nodulating (N_2-fixing) and nonnodulating (non-N_2-fixing) isolines of soybeans may

Table 3

Increase in Seed Yield of Various Field-Grown
Grain Legumes in Response to CO_2 Enrichment
of Canopy From Anthesis to Senescence (43, 44)

	Seed yield (kg/ha)		
	Air	1500 ppm CO_2	% Increase
Soybeans	3003	5946	98
Peas	3446	5262	53
Dry beans	2930	4660	59
Peanuts[*]	3472	4205	21

[*]Erect growth habit in chambers prevents normal pegging, and
senescence was frost initiated.

Table 4

Role of Fertility in Seed Yield Response of Nodulating and
Nonnodulating Isolines of Field-Grown Soybeans to
CO_2 Enrichment of Canopy From Anthesis to Senescence (45)

	Seed yield (kg/ha)		
	Air	1500 ppm CO_2	% Increase
Nodulating isoline			
None	4597	6544	42
Soil N	4007	6659	66
Nonnodulating iso-line			
None	3039	3723	22
Soil N	2959	4732	60

provide the answer. CO_2 enrichment alone produced as high a yield
of nodulating soybeans as did CO_2 enrichment plus nitrogen fertilization (Table 4) (45), but the combination was required for maximum
yield of nonnodulating soybeans. Additional soil fertility may be
required for an improved RuBP carboxylase to express its activity
in terms of photosynthetic carbon assimilation. Support for this
hypothesis is provided by some recent greenhouse cotton experiments
(Table 5), in which the increases in yield of blooms and bolls

Table 5

Influence of Nutrient Concentration on Yield Components of
Greenhouse-Grown Cotton in Response to CO_2 Enrichment (46)

	Air	630 ppm CO_2	% Increase
Normal Nutrient			
Blooms/plant (No.)	67	86	28
Bolls/plant (No.)	39	59	51
2X nutrient			
Blooms/plant (No.)	69	108	57
Bolls/plant (No.)	40	78	95
Lint yield (g/plant)	61*	170*	179
Seed Yield (g/plant)	114	274	140

*Calculated as 1740 kg/ha (3.1 bales/acre for air and 3250 kg/ha
(5.8 bales/acre) for CO_2 enriched (+88%) assuming LAI of 5 in
both cases.

per plant in response to CO_2 enrichment were substantially greater
with a doubled nutrient solution (46).

In conclusion, CO_2 enrichment experiments demonstrate the
remarkably broad opportunities that await a practical solution for
some of the things "wrong" with RuBP carboxylase.

REFERENCES

1. Hardy, R. W. F., in Report of the Public Meeting on Genetic
   Engineering for Nitrogen Fixation, pp. 77-106, U. S. Government
   Printing Office, Washington, DC, 1977.
2. Hardy, R. W. F. and Havelka, U. D., in Protein Resources and
   Technology: Status and Research Needs, pp. 204-35, M. Milner
   et al., Editors, AVI Publ. Co., Westport, CT, 1978.
3. Herskovitz, T., J. Am. Chem. Soc. 99, 2391 (1977).
4. Hardy, R. W. F., in Energy, Food, Population and World Interde-
   pendence, pp. 8-18, Am. Chem. Soc., Washington DC, 1978.
5. Hardy, R. W. F., in Genetic Engineering for Nitrogen Fixation,
   pp. 369-99, A. Hollaender et al., Editors, Plenum, New York, 1977.
6. Huffaker, R. C. and Miller, B. L., See paper in this Symposium.
7. McFadden, B. A. and Purohit, K., See paper in this Symposium.
8. Akazawa, T. et al., See paper in this Symposium.
9. Kung, S. D. and Rhodes, P. R., See paper in this Symposium.
10. Chua, N.-H. and Schmidt, G. W., See paper in this Symposium.
11. Roy, H., Costa, K. A., and Adari, H., Plant Sci. Lett. 11, 159-
    68 (1978).

12. Davis, L. C., Shah, V. K., and Brill, W. J., Biochim. Biophys. Acta 403, 67-78 (1975).
13. Lorimer, G. et al., See paper in this Symposium.
14. Ogren, W. L. and Hunt, L. D., See paper in this Symposium.
15. Servaites, J. C. and Ogren, W. L., Plant Physiol. 61, 62-7 (1978).
16. Paech, C. et al., See paper in this Symposium.
17. Ku, S.-B. and Edwards, G. E., Plant Physiol. 59, 986-90; 991-9 (1977).
18. Chollet, R. C., TIBS 2, 155-9 (1977).
19. Lorimer, G., Woo, K. C., Berry, J. A., and Osmund, C. B., in Proc. 4th Int. Congr. Photosynthesis 1977, pp. 311-22, D. O. Hall et al., Editors, Biochem. Soc., London, 1978.
20. Walker, D. A. and Robinson, S., See paper in this Symposium.
21. Bassham, J. A. et al., See paper in this Symposium.
22. Jensen, R. G. et al., See paper in this Symposium.
23. Keys, A. J., Sampaio, E. V. S. B., Cornelius, M. J., and Bird, I. F., J. Exp. Bot. 28, 525-33 (1977).
24. Raschke, K., Philos. Trans. R. Soc. London Ser. B 273, 551-60 (1976).
25. Raghavendra, A. S., Rajendrudu, G., and Das, V. S. R., Nature London 273, 143-4 (1978).
26. Quebedeaux, B. and Hardy, R. W. F., in CO_2 Metabolism and Plant Productivity, pp. 185-204, R. H. Burris and C. C. Black, Editors, University Park Press, Baltimore, MD, 1976.
27. Thorne, J. H. and Koller, H. R., Plant Physiol. 54, 201-7 (1974).
28. Koller, H. R. and Thorne, J. H., Crop Sci. 18, 305-10 (1978).
29. Berlyn, M. B., See paper in this Symposium.
30. Shanmugam, K. T., O'Gara, F., Andersen, K., Morandi, C., and Valentine, R. C., in Recent Developments in Nitrogen Fixation, pp. 321-30, W. Newton et al., Editors, Academic Press, London, 1977.
31. Kech, R. W. and Ogren, W. L., Plant Physiol. 58, 552-5 (1976).
32. Goldstein, L. D., Ray, T. B., Kestler, D. P., Mayne, B. C., Brown, R. H., and Black, C. C., Plant Sci. Lett. 6, 85-90 (1976).
33. Trinick, M. J., Nature London 244, 459-60 (1973).
34. Keating, J. D. H. and Garrett, M. K., in 4th Int. Congr. Photosynthesis Abstracts, pp. 191-2, J. Coombs, Editor, UKISES, London, 1977.
35. Day, P. R., Science 197, 1334-9 (1977).
36. Kleinhofs, A. and Behki, R., Annu. Rev. Genet. 11, 79-101 (1977).
37. Link, G. L. et al., See paper in this Symposium.
38. Howell, S. H. and Gelvin, S., See paper in this Symposium.
39. Andersen, K. et al., See paper in this Symposium.
40. Hardy, R. W. F. and Havelka, U. D., in Symbiotic Nitrogen Fixation in Plants, pp. 421-39, P. S. Nutman, Editor, Cambridge U. Press, 1976.
41. Sinclair, T. R. and de Wit, C. T., Agron. J. 68, 319-24 (1976).
42. Shibles, R. M., Anderson, I. C., and Gibson, A. H., in Crop Physiology: Some Case Histories, pp. 151-89, L. T. Evans, Editor, Cambridge U. Press, 1975.

43.   Hardy, R. W. F. and Havelka, U. D., in Biological Solar Energy
      Conversion, pp. 299-322, A. Mitsui et al., Editors, Academic
      Press, New York, 1977.
44.   Havelka, U. D. and Hardy, R. W. F., Agron. Abstr., p. 72 (1976).
45.   Havelka, U. D. and Hardy, R. W. F., Agron. Abstr., p. 86 (1977).
46.   Mauney, J. R., Fry, K. E., and Guinn, G., Crop Sci. 18, 259-63
      (1978).

DISCUSSION

BASSHAM: I cannot accept that it will not be practical, at
some time in the foreseeable future, to use CO_2 enrichment for
agronomic crops. I believe your estimates are far too pessimistic.
You have lots of data on legumes, but alfalfa has not been studied.
One of the limiting factors for many legumes is that they go through
a cycle and senesce. For a plant that could be harvested continuously year round, the possibilities for enhancing productivity with
CO_2 enrichment are far greater. Regarding the cost of delivering
CO_2, obviously bottled CO_2 is not practical, but huge amounts of
CO_2 are being produced from fossil fuel power generation, and these
will increase. This CO_2 could be used, although there may be some
problems with removing SO_2, etc. Another benefit could result from
using CO_2: much of our country suffers severe water shortages, and
the conservation of water by using covered agriculture could be very
important in terms of land cost and water utilization. As you have
shown, nitrogen-fixing capability is greatly increased by CO_2 enrichment. All these factors together make covered agriculture possible in the future, although we may not be ready for it yet.
    HARDY: I urge you to do a CO_2 enrichment experiment on alfalfa
to see what the actual yield improvement is, and then to do a base-
case economic evaluation to see how far your method is from being
economically viable. I did this for nitrogen fixation about a year
ago. I calculated the economic tradeoff of putting nitrogen-fixing
nodules on corn vs. the saving in fertilizer nitrogen. The ratio
was not good. It was a shock to me and my nitrogen-fixing colleagues, including Ray Valentine, yet they found nothing wrong with
the calculations; that is why I tend to have a negative attitude.
For alfalfa, you should measure, under normal crop production conditions, how much CO_2 is needed to get a certain increase in yield
over the entire growing season of this perennial crop. Where might
you get the CO_2? You mentioned generating stations. CO_2 wells have
been considered but never made practical. You will be dealing with
small areas, but any major agronomic crop, including alfalfa, involves huge acreages, on which I cannot visualize how point sources
could be effectively used. Regarding increased efficiency of water
utilization, certainly anything that would improve the ratio of CO_2
fixed per water lost, which is one to several hundred, would be bene-

ficial. C_4 plants have a better water efficiency than C_3, as Clanton Black pointed out many years ago, but I am sure (although we have not measured it) that our CO_2 enriched C_3 soybean plants have a better water utilization than do C_4 plants.

CHEMOSYNTHETIC, PHOTOSYNTHETIC, AND CYANOBACTERIAL

RIBULOSE BISPHOSPHATE CARBOXYLASE

Bruce A. McFadden and Kris Purohit

Program in Biochemistry and Biophysics, Department of Chemistry
Washington State University, Pullman, Washington 99164

INTRODUCTION

A.   The Evolutionary Context

There are compelling reasons to believe that the initial atmosphere of the earth after its formation about 4.7×10^9 years ago
was a reducing one consisting chiefly of methane, ammonia, water,
and hydrogen (1).   In the last two decades considerable research
has been described in which numerous organic precursors of biopolymers have been synthesized under conditions simulating the primitive
earth and its atmosphere (for a review see ref. 2).   Formaldehyde
and hydrogen cyanide are known to be formed so readily under a
variety of these conditions that they are considered not only as
products of ancient chemical evolution but as likely precursors
of a variety of important components of the prebiotic soup, including sugars, amino acids, and purine bases (2).
  It is reasonable to assume that the first organisms appeared
in waters enriched in organic compounds and that they had limited
biosynthetic capacity, as first suggested by Oparin in 1924 (for a
discussion in English see ref. 3) and later independently by Haldane
(4).   Moreover, the organisms should have been anaerobic and had
complex nutritional requirements, as do many fermentative bacteria
even today (5, 6).   Presumably ancient organisms conducted substrate-
level phosphorylations to synthesize ATP, and oxidation-reduction
reactions were mediated by pyridine nucleotides.   Thus ATP and NADH
(or structurally similar primitive counterparts) were available to
couple with newly acquired reductive biosynthetic steps which enabled the synthesis of an essential nutrient.   Horowitz was the
first to emphasize the profound selective advantage accruing to
organisms which acquired the capacity to synthesize essential

179

nutrients that had been depleted from the environment. Logically, this led to the postulate that the evolution of biosynthetic sequences had occurred from their termini (as we know them today) towards earlier steps (7)--a process often referred to as retro-evolution.

In analyzing contemporary biology it is invariably helpful to examine the origins of present-day species. With adequate circumspection, biochemical relationships that existed in ancient organisms may become evident, and present-day relationships can be rationalized in terms of molecular evolution. In the context of the present Symposium, we may ask about the ancient origin of the utilization of CO_2 via the reductive pentose phosphate (Calvin) cycle. In terms of carbon metabolism this presumably devolves to the acquisition of two enzymes, D-ribulose 1,5-bisphosphate (RuBP) carboxylase and 5-phospho-D-ribulokinase. Elsewhere (5), we have proposed that the latter enzymic activity was recruited from pre-existing phosphofructokinase and that the acquisition of RuBP carboxylase would have triggered the appearance of CO_2-fixing species. We have chosen to study the properties of microbial RuBP carboxylases because of our interest of many years in the evolution of autotrophism (5, 6, 8-11). In this Symposium we will describe salient features of this enzyme, largely from prokaryotes.

B.   Size and Quaternary Structure of RuBP Carboxylase

It is now evident that RuBP carboxylase is most easily classified in terms of quaternary structure. Two types of subunits have been found in nature--large (MW ca. 55,000) and small (MW ca. 15,000). The nomenclature L for large and S for small subunits has been suggested (5). Only enzymes from ostensibly more highly evolved autotrophic species contain both subunit types in a combination of 8 and 8; i.e., an L_8S_8 unit is isolable. This dual subunit type of structure has been designated T (12). These large T enzymes (MW ca. 525,000) have been isolated from <u>Chromatium</u> D (<u>vinosum</u>) (13, 14) and <u>Ectothiorhodospira</u> <u>halophilia</u> (13), both members of the Chromatiaceae. Although the RuBP carboxylase from <u>E. halophilia</u> contains an unusually large S subunit (MW 18,000) and is therefore the largest RuBP carboxylase known (MW 601,000), it is closely similar to the counterpart from the nonhalophilic <u>Chro</u>-<u>matium</u> D in other properties including amino acid composition (13).

Among the chemolithotrophic bacteria, only four--<u>Hydrogenomonas</u> <u>eutropha</u> (12, 15, 16), <u>Thiobacillus</u> A2 (17, 18), <u>T</u>. <u>novellus</u> (19), and <u>Paracoccus</u> <u>denitrificans</u> (20)--contain the enzyme of T structure. These L_8S_8 enzymes have also been isolated and characterized from four blue-green algae: the filamentous <u>Plectonema</u> <u>boryanum</u> and <u>Anabaena</u> <u>variabilis</u> (21) and the unicellular forms <u>Aphanocapsa</u> (22) and <u>Microcystis</u> <u>aeruginosa</u> (23).

Ribulose biphosphate carboxylase from all eukaryotic sources examined to date also has a T (L_8S_8) structure. These include the following green algae: <u>Chlorella</u> <u>ellipsoidea</u> (24), <u>Chlorella</u> <u>fusca</u> (25),

Chlamydomonas reinhardi (26), Euglena gracilis (27), and a marine
form, Halimeda cylindracea (28). The same T structure has been
inferred for well characterized RuBP carboxylases from the following
higher plants: spinach (29), wheat (30), spinach beet (31), five
species of tobacco (32-34), curly dock or Rumex (35), French bean
(36), pea (37), barley (38), and evening primrose or Oenothera (39).
Recent studies of the enzyme from nine plant species including mono-
and dicotyledons (40) and from several other higher plants (41)
also revealed the T type of quaternary structure.

Observations have been presented which suggest that the higher
plant enzymes, especially those from tobacco, consist of heterogeneous large and heterogeneous small subunits (see, e.g., ref. 41).
Heterogeneity was detected after isoelectric focusing of the isolated enzymes in the presence of 8 M urea; it is crucial to stress,
however, that post-cell-rupture proteolysis may explain this.
For example, Gray and Kekwick (36) found that unless the enzyme
from Phaseolus was isolated in the presence of 1 mM diisopropyl-
fluorophosphate, final preparations not only rapidly lost RuBP
carboxylase activity but exhibited a number of components after
isoelectric focusing. Moreoever, sodium dodecyl sulfate (SDS)
electrophoresis showed degradation of the large subunits. It has
been our experience with two other eukaryotic organisms (both
nematodes) that extensive proteolysis occurs after cell rupture
and that it results in major modification of several enzymes (42).

Proteolysis in plant extracts is probably variable but should
always be considered as a source of subunit variability. Recently
Gray et al. suggested that the variability of large subunits in
higher plants is the result of modification of a single gene product but that the two small subunit polypeptides of tobacco RuBP
carboxylase are separate gene products (43). In the prokaryotes,
heterogeneity of large subunits has been observed with the H.
eutropha enzyme isolated under conditions that would virtually
eliminate proteolysis in extracts, and studies have suggested a
$L_5 L_3' S_8$ structure (15). Data establishing the heterogeneity of
small subunits for the enzyme from Thiocapsa roseopersicina will
be presented in this paper.

In contrast to the T enzymes described, we discovered another
class of enzymes, a one subunit type (designated O). To date, these
enzymes, consisting of the 55,000-dalton L polypeptides but lacking
the smaller S polypeptides, have been isolated from Rhodospirillum
rubrum (44) and from Chlorobium limicola f. thiosulfatophilum (45)
in stable dimeric and hexameric states, respectively, and from
Thiobacillus intermedius (46), Agmenellum quadruplicatum (47), and
Anabaena cylindrica (48) in an octameric state. The last two organisms are unicellular and filamentous blue-green algae, respectively, and it should be stressed that the O enzyme from them differs from the previously mentioned T enzymes isolated from four
other blue-green algae. Perhaps this variability of structure
within these organisms is not surprising when one considers the

well-known morphological and physiological diversity among the
blue-green algae. On the other hand, other research on the enzyme
from A. cylindrica has suggested that the enzyme has an L8S8 structure and that in some cases the purification procedure (especially
an acid precipitation step) may remove small subunits (49), as
appears to be the case for the Aphanocapsa enzyme (22). In the
face of these conflicting results, it should be mentioned that
RuBP carboxylase has been isolated free of small subunits under
extremely mild conditions from T. intermedius (46) and C. limicola
f. thiosulfatophilum (45).

     Of interest is the recent finding of both T and O enzymes in
each of two rhodopseudomonads, Rps. capsulata (50) and sphaer-
oides (51). In the latter organism these enzymes do not cross-
react immunologically (51, 52).

C.   Composition of RuBP Carboxylase

     Quantitative analysis (53) of the amino acid compositions
of large subunits suggests that these polypeptides may be homologous from R. rubrum through higher plants (6, 54). The evolutionary conservation of this structure is in harmony with the
fact that the large subunits harbor the catalytic potential in
T enzymes (5, 55-59).

     On the other hand, the small subunits from Halimeda, Chlo-
rella, spinach, and Hydrogenomonas eutropha are very dissimilar
in composition (6, 12, 54). Even among different genera of
higher plants, the small subunits are quite variable (54).

D.   Immunological Comparisons

     It is of interest that antibodies of RuBP carboxylase from
R. rubrum fail to yield precipitates with the enzyme from any
source tested, including spinach, green algae, blue-green algae,
chemolithotrophic bacteria, and green and purple sulfur bacteria, nor do they inhibit the enzyme from H. eutropha (44).
In some cases the failure to cross-react may involve masking of
antigenic determinants on the large subunits by the small subunit. Lord et al. (60) reported that antiserum to the enzyme
from E. gracilis inhibited RuBP carboxylase from Chlorella
fusca and four species of blue-green algae, in addition to forming precipitates with enzymes in extracts from all these sources.
Akazawa and Osmond (28) reported that RuBP carboxylase from the
marine green alga Halimeda cylindracea shows immunological
cross-reactivity with the spinach enzyme. Other immunological
comparisons of this enzyme from divergent sources have been
summarized (5).

     It will be important to quantify immunological relationships
between large (L) subunits and between small (S) subunits if antibodies to each native structure can be prepared. In this regard,
immunological comparisons of each subunit type prepared by gel

chromatography of SDS-dissociated native enzyme from French beans
and subsequent removal of SDS by anion exchange chromatography in
the presence of urea have been made by eliciting antibodies to
urea-free subunits. Anti-S sera showed some precipitation of
native enzyme but no inhibition of enzyme activity, whereas anti-L
sera precipitated L but not S subunits and inhibited the carboxylase
(61). These results dealing with inhibition of the carboxylase
activity have been confirmed with anti-L and anti-S sera obtained
by using p-hydroxymercuribenzoate-dissociated subunits from the
spinach enzyme as antigens. After dissociation, subunits were
separated by gel chromatography and freed of mercurial by β-
mercaptoethanol. Of interest was the observation that 13% of the
original carboxylase activity could be recovered in oligomeric L
but no activity could be found in S (58). These results imply that
the suggested quantitative immunological comparisons between L and
S subunits are feasible with a technique such as micro-comple-
ment fixation.

E.   Function of the Small Subunits

The presence of small subunits seems not to confer unusual
catalytic or regulatory properties upon RuBP carboxylase. For
example, the enzyme from R. rubrum, which lacks small subunits,
has a turnover number of 1.7 (moles of CO_2 utilized per second per
mole of catalytic site) if one assumes one catalytic site per subunit of the dimer, whereas the spinach enzyme, which is composed of
eight large (catalytic) and eight small subunits, has a turnover
number of 1.5. Both enzymes have similar Michaelis constants for
CO_2 and RuBP (62). Neither enzyme displays striking cooperativity
in kinetic response to substrates. The only major difference is
that RuBP carboxylase from R. rubrum is not inhibited by 1 mM 6-
phosphogluconate, whereas the spinach enzyme is quite sensitive to
this ligand. Sensitivity to 6-phosphogluconate, a competitive
inhibitor with respect to RuBP, is a general feature of all large
RuBP carboxylases (63), including the recently characterized octamer lacking small subunits from T. intermedius (46). The turnover
number of this enzyme is 2.6.

It is well to emphasize that the precise function of the small
subunits in RuBP carboxylase remains an enigma. Perhaps these
relatively nonconserved polypeptides play a hitherto unrecognized
regulatory role or serve as membrane attachment sites to anchor the
catalytically functional large subunits.

F.   Isolation of RuBP Carboxylase

Goldthwaite and Bogorad (35) used the technique of sedimentation of leaf extracts into a sucrose density gradient to isolate
large RuBP carboxylases from several higher plants. We have established that an analogous procedure can be utilized for extensive
purification of the large enzyme from several microbial sources (64).

The large enzyme from <u>Euglena gracilis</u> may be sedimented from
0.1 M Tris buffer at 160,000 x g for 2 hr after previous sedimentation of ribosomes (6). The RuBP carboxylase recovered is at
least 80% pure by gel electrophoretic criteria, and crystalline
enzyme is easily obtained from the fraction sedimented. An analogous pelleting procedure has been used in purifying the enzyme
from <u>T. intermedius</u> (46). We have also established that RuBP
carboxylase is easily crystallized from <u>T. intermedius</u>, <u>H. eutropha</u>,
and <u>Pseudomonas oxalaticus</u> (in preparation). These results afford
an alternative to the approaches used by other groups to crystallize the tobacco enzyme (65-71).

G.   Catalytic Pathway for RuBP Carboxylase and RuBP Oxygenase

The RuBP carboxylase-catalyzed reaction leading from RuBP and
CO_2--the known substrate (72)--to 3-phosphoglycerate is known to
proceed via the 2,3-enediol of RuBP, which is carboxylated to a
relatively unstable 6-carbon intermediate (5, 73, 74). Magnesium
or manganese ions are required for overall activity. Electron
paramagnetic and nuclear magnetic resonance studies of the spinach
enzyme suggest the formation of a quaternary complex containing
enzyme, Mn^{2+}, RuBP, and CO_2, the latter two compounds being in the
inner and second sphere, respectively, of enzyme-bound Mn^{2+} (75).
Whether this complex is catalytically productive remains to be seen,
in view of recent information about the activation of RuBP carboxylase (to be discussed).

It has become evident that the 2,3-enediol may also be cleaved
by oxygen in a reaction catalyzed by highly purified spinach RuBP
oxygenase (76, 77). Oxygen and water add to the double bond of
the 2,3-enediol intermediate in such a way that ^{18}O from molecular
oxygen appears in phosphoglycolate whereas that from water appears
in phosphoglycerate. These results, coupled with previous observations that O_2 was a competitive inhibitor with respect to CO_2
(78), clearly established that RuBP carboxylase of higher plants
catalyzes oxygenolysis of RuBP.

H.   RuBP Oxygenase and Photorespiration

Production of phosphoglycolate during photorespiration in
plants has been attributed to the oxygenase activity of RuBP carboxylase (77). Alternatives to account for photorespiration by
higher plants have been discussed (79). In considering the total
body of evidence it seems likely that the oxygenase activity of
RuBP carboxylase accounts for or makes a major contribution to
photorespiration (80-89), a process that opposes photosynthesis
and therefore lowers plant growth rates.

Prokaryotic models for photorespiration in plants are of
interest. The excretion of glycolate by <u>Chromatium</u> (90) and
<u>R. rubrum</u> (91) during photosynthesis in the presence of exogenous
oxygen correlates with the finding that highly purified RuBP

carboxylases from these sources have RuBP oxygenase activity (27, 92, 93), and that glycolate excreted by Chromatium contains 1 atom of ^{18}O in the carboxyl group after incubation under $^{18}O_2$ (94). Codd et al. (95) have reported that glycolate is excreted as a function of oxygen partial pressure by a suspension of H. eutropha, a finding that correlates with the oxygenase activity detected in RuBP carboxylase from this organism (12, 16).

In spite of the potentially profound significance of the oxygenase activity of RuBP carboxylase, relatively little is known about this activity. Of interest are observations that antispinach-L serum inhibits the carboxylase and oxygenase activities of native enzymes from both Chromatium and spinach (56, 59), but that anti-spinach-S serum shifts the pH-optimum of the spinach oxygenase (56). Akazawa and colleagues have suggested a regulatory function for the S subunits but the evidence for this needs buttressing.

I.   The Kinetic Mechanism of RuBP Carboxylase/Oxygenase

With regard to the kinetic mechanism of catalysis, nothing is known. The overall reaction appears to take place by initial combination of CO_2 and Mg^{2+} with RuBP carboxylase followed by addition of RuBP to the complex, on the basis of research on the bacterial enzymes from H. eutropha and R. rubrum (96, 97) and on the plant enzyme from Tetragonia (98). However, the implied sequence of reactant additions may be illusory in view of recent work by Lorimer et al. (99) establishing that the spinach carboxylase must be activated prior to catalysis in a reaction first with CO_2 and then with Mg^{2+}, and by Laing and Christeller (100) suggesting that Mg^{2+} is not required for catalysis. Because the CO_2 participating in the activation is not necessarily the same as that condensing with RuBP, the sequence of reactant additions to RuBP carboxylase is still an open question.

J.   Activation of RuBP Carboxylase/Oxygenase

In 1974 Bahr and Jensen found that a form of RuBP carboxylase with a low $K_m(CO_2)$ is released upon transfer of spinach or tobacco chloroplasts to hypotonic medium (101) and described properties later found to be similar to those of RuBP oxygenase activity from both freshly ruptured spinach (102) and corn chloroplasts (103). Laing et al. (104, 105) found the enzyme from soybean to be unstable in the absence of high bicarbonate concentrations, the accelerated decline in activity at lower concentrations leading to a high apparent $K_m(CO_2)$. They found, as did Badger and Andrews (106) independently working with spinach, that the high-affinity form of RuBP carboxylase has RuBP oxygenase activity with a higher activation energy than that of the carboxylase activity. O_2 was competitive with respect to CO_2 and vice versa for the carboxylase and oxygenase activities, respectively. These results are consistent with the well-known inhibition of photosynthesis by oxygen and of

photorespiration by CO_2; moreover, increasing temperature favors
photorespiration (107, 108), a finding in accord with the relative
activation energies.

It would be important to examine Recently Lorimer et al. (99) have thoroughly characterized
the sequential activation of RuBP carboxylase by CO_2 and Mg^{2+}, as
mentioned earlier, and have established that the enzyme is rapidly
inactivated upon removal of CO_2 and Mg^{2+}. They found the activity
of the ternary complex after preincubation to equilibrium at constant concentrations of CO_2 and Mg^{2+} to increase as the preincubation pH was raised, which indicated that CO_2 reacted with an
enzymic group whose pK was distinctly alkaline. They proposed that
the activation of RuBP carboxylase involves the formation of a
carbamate. They have discussed their results in light of a number
of findings: (a) apparently high but artifactitious K_m values for
CO_2, (b) the effects of increasing pH in reducing the K_m for CO_2
and for Mg^{2+}, (c) the effect of increasing $[Mg^{2+}]$ in decreasing
the apparent $K_m(CO_2)$, (d) incorrectly sharp pH optima, and (e) the
incorrectly inferred alkaline pH optimum (9.3) for the oxygenase
reaction--all of which can be explained in terms of the activation
process. In connection with the last item, it is of interest that
the spinach oxygenase activity requires identical sequential activation by CO_2 and Mg^{2+} and decays with the same kinetics as the carboxylase activity when these activators are removed (109). In addition, Lorimer et al. (99) discuss the effect on RuBP carboxylase
of a number of metabolites such as sugar phosphates or related
compounds, all of which appear to be "secondary" because of the
requirement for the presence of CO_2 and Mg^{2+}. They suggest that
these metabolites act by modifying the basic activation process
brought about by CO_2 and Mg^{2+}. Finally, they suggest that activation of RuBP carboxylase in vivo is caused by the reaction of CO_2
with specific uncharged amino group(s) of the enzyme formed by the
alkalization of the stroma upon illumination. The reverse reactions
would lead to inactivation in the dark. In this connection, it will
be important to examine the effect of 6-phosphogluconate on the inactivation process. This ligand rapidly accumulates in the dark in
Chlorella pyrenoidosa, with apparent inactivation of RuBP carboxylase (110), and it inhibits the T enzymes found in green algae and
higher plants (63, 111).

It would be hazardous to extrapolate the conclusions of Lorimer
et al. (99) to all RuBP carboxylase/oxygenases, especially in light
of the occurrence of O and T types of enzymes. Nevertheless, in
all future kinetic experiments in which CO_2 concentrations below
that required for activation are used, it will be essential to make
instantaneous dilutions of CO_2 (furnished as HCO_3^-) after activation
and to measure initial rates. Moreover, the requirement of CO_2 for
activation of the oxygenase but the opposing competitive inhibition
during catalysis will also necessitate careful experimental design
and standardization of assay conditions.

Ribulose bisphosphate carboxylase remains one of the most
sluggish catalysts in nature. Indeed the natural occurrence of

unusually large amounts of this enzyme may compensate for the low
turnover numbers observed (5), especially when it is recognized
that the enzyme is probably not saturated by substrate(s) in vivo
(5, 6). These considerations, coupled with the findings of Lorimer's
group regarding the in vitro activation of RuBP carboxylase by CO_2,
raise serious questions about the fraction of this enzyme that is
catalytically active in vivo. The central question as to why there
is so much of this enzyme in nature is as yet unanswered.

K.   Native Environment of RuBP Carboxylase

In most, if not all, higher plants and green algae, RuBP carboxylase is apparently located within chloroplasts, as are other
enzymes functioning in the Calvin cycle. In general, the enzyme is
easily solubilized, which suggests that it is a peripheral com-
ponent--in the terminology of Singer and Nicholson (112)--if attached to chloroplast membranes.

Of unusual significance is the research on various strains of
tobacco plants by Wildman and co-workers which suggests that the
small and large subunits are encoded by nuclear and chloroplast
DNA, respectively (113, 114). That the small and large subunits
in eukaryotes are made on cytoplasmic and chloroplast ribosomes,
respectively, has been indicated (115-119). In this connection
the synthesis of a 20,000-dalton precursor of the small 16,500-
dalton subunit has recently been detected in experiments using
cytoplasmic ribosomes and mRNA from Chlamydomonas reinhardi
to supplement a wheat-germ amino acid-incorporating system. This
immunologically cross-reacting precursor could be converted to
small subunits by an endoprotease from C. reinhardi (120). Precisely when the precursor is processed with respect to the transfer
across chloroplast membranes remains to be seen.

Among the prokaryota, polyhedral inclusion bodies have been
observed in blue-green algae and in the following chemosynthetic
bacteria: T. neapolitanus, T. ferrooxidans, a marine Thiobacillus,
Nitrobacteri agilis, and possibly N. winogradskyi (for a review
see ref. 5). Studies of these bodies from T. neapolitanus (121)
and N. agilis (122) have shown that they are greatly enriched in
RuBP carboxylase. In the former organism, RuBP carboxylase is organized into a paracrystalline molecular array within well preserved
inclusion bodies (123). Similar arrays are evident in polyhedral
inclusion bodies in the facultative autotroph, T. intermedius. It
is of special significance that the bodies per cell roughly correlate with the RuBP carboxylase content in this organism and that
there are no bodies and no detectable enzyme after heterotrophic
growth (124). These observations suggest (but do not establish)
that the bodies will prove to be functional in CO_2 assimilation.
The enzyme composition and organization within similar bodies from
other autotrophic prokaryotes is of obvious interest. Recently Codd
and Stewart (125) have isolated these bodies from the blue-green alga

Anabaena cylindrica and have established that they contain about
20% of the RuBP carboxylase.

In many other prokaryotic autotrophs, including photolitho-
trophic bacteria and numerous chemolithotrophic bacteria, polyhedral
inclusion bodies are not evident. In general, RuBP carboxylase from
these forms is easily recovered in a soluble form after gentle cell
rupture. Nevertheless, it will be important to find out whether
this enzyme is a peripheral component of cytoplasmic membranes and
whether other enzymes of the Calvin cycle in these organisms are
similarly attached.

L.   Molecular Evolution of RuBP Carboxylase

In analyzing the evolution of RuBP carboxylase, trends in size,
quaternary structure, and active-site architecture are worthy of
comment. The enzymes of small (MW 114,000) and intermediate size
(MW <350,000) are not inhibited by 6-phosphogluconate, whereas the
large enzymes (MW >450,000) are. Although the data are not conclusive, it is likely that 6-phosphogluconate inhibition occurs
through competition with RuBP for catalytic sites. Thus the in-
sensitivity of smaller enzymes suggests that their catalytic-site
topography differs from that of large enzymes. It will be extremely
important to develop a good active-site reagent for RuBP carboxylase.
Hartman and colleagues have synthesized 3-bromo-2-butanone 1,4-bis-
phosphate, a compound that shows some specificity in covalently
modifying the active site of the spinach (126) and R. rubrum enzymes (127). These studies should be extended to the enzyme from
other sources, and other active-site-directed reagents should be
developed. Recently, Whitman and Tabita (128) have established
that RuBP carboxylase from R. rubrum (an O enzyme) can be inactivated by incubation with pyridoxal phosphate but that RuBP protects
against the inactivation. Presumably pyridoxal phosphate forms a
Schiff base with enzymic amino groups, because the adducts could be
reduced by $NaBH_4$ and the resultant product could not be reactivated.
This research is being extended in Tabita's laboratory to RuBP
carboxylases from other sources. In our laboratory, borate-dependent
inactivation of the enzymes from Pseudomonas oxalaticus and from
barley by 2,3-butadione and protection by 3-phosphoglycerate has
recently placed arginine at the active site (129). In the long
run, results with these and other reagents may assist in crudely
mapping and comparing active-site structures if different functional
groups can be modified.

Although the quaternary structure of only two smaller enzymes
has been examined, the results establish that the enzymes from
R. rubrum and C. limicola f. thiosulfatophilum are composed of two
and six large subunits (MW ~55,000), respectively. A stable octameric state of O enzymes has been found for RuBP carboxylases from
two blue-green algae and from T. intermedius. In marked contrast,
the largest enzymes are composed of large (MW ~55,000) and small
(MW ~15,000) subunits, in a ratio of 8 to 8. Electron microscopy

Table 1

Purification of RuBP Carboxylase/Oxygenase From T. roseopersicina

(The 330 g of wet-packed cells had been grown photoorgano-
trophically on 20 mM D,L-malate. Activity is given in
μmoles RuBP-dependent $^{14}CO_2$ fixed per minute.)

Procedure	Protein (mg)	Activity	Sp. act. (units/mg)	Recovery (%)
105,000 x g (3 hr) super- natant after gel filtra- tion on Sephadex G-25	9870	307	0.03	(100)
Alkaline $(NH_4)_2SO_4$ (68 to 98% saturation) precipitate	654	188	0.29	61
Pooled Sephadex G-100 fractions	332	156	0.47	51
DEAE-cellulose fractions, pooled and concentrated	84	110	1.31	36
Glycerol (14 to 34% v/v) gradient	11	27	2.45	9

has suggested that both L_8 and L_8S_8 enzymes have 4:2:2 symmetry
(16, 27, 46, 130). Thus the existence of a regular cubical structure for both O and T enzymes strongly suggests that the small subunits are peripherally disposed in T enzymes, perhaps in two 4-
membered structures sandwiching the catalytic cubical core. This is
also compatible with dissociation pathways briefly summarized by
Takabe and Akazawa (59). It is interesting that the green algal T
enzyme is frequently seen in configurations of eightfold symmetry
with dimensions clearly that of a staggered cube (27), whereas the
O enzyme from T. intermedius is rarely seen in this configuration
under identical conditions (46). These observations suggest that
the presence of small subunits weakens the interaction between L_4
assemblies in half-cubes in at least some T enzymes. A tentative evolutionary scheme will be presented below in the Discussion section.

METHODS AND RESULTS

     In this section recent unpublished research on RuBP carboxylase/
oxygenase from the photosynthetic bacterium Thiocapsa roseopersicina,
a member of the Chromatiaceae, is presented. Growth of this organism
in the light on malate leads to full derepression of RuBP carboxylase
in comparison with growth on bicarbonate/thiosulfate. The steps in
purification of the enzyme from 330 g of wet-packed malate-grown

cells are summarized in Table 1. The final 82-fold purified product, obtained in 9% yield, was unstable in the absence of glycerol but stable when stored at -80°C at 2 mg/ml in 10% glycerol (v/v), and had a specific activity of 2.45, which is at the high end of the range for this enzyme.

Figure 1 shows data from sedimentation velocity studies of the enzyme in the presence of 4% glycerol. The enzyme sediments as a single symmetrical peak with a $S_{20,w}$ value of 19.0 S, within the normal range for large (MW ~550,000) RuBP carboxylases (5). However, in the presence of 10% glycerol, a significantly lower value, 16.0, suggests dissociation of the enzyme. Studies of sedimentation equilibrium also suggest that the enzyme undergoes glycerol-dependent dissociation. Unfortunately, it is impossible to calculate correct sedimentation coefficients and molecular weights because of the unknown influence of glycerol upon $\bar{v}$, the partial specific volume. Nevertheless, the sedimentation data suggest that the enzyme is homogeneous but dissociates in the presence of higher concentrations of glycerol.

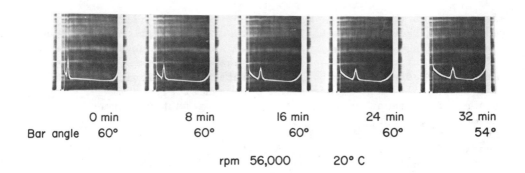

	0 min	8 min	16 min	24 min	32 min
Bar angle	60°	60°	60°	60°	54°

rpm 56,000   20° C

Figure 1. The sedimentation pattern of T. roseopersicina RuBP carboxylase that had been equilibrated with 20 mM Tris containing 1 mM EDTA, 10 mM $MgCl_2$, 50 mM $NaHCO_3$, 0.1 mM DTT, and 10% glycerol adjusted to pH 8.0. The protein concentration was 2 mg/ml.

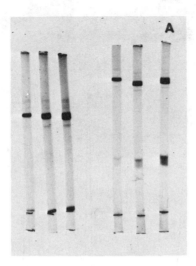

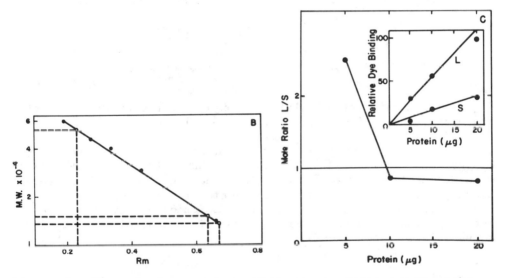

Figure 2. The subunit structure of T. roseopersicina RuBP carboxylase determined by SDS-polyacrylamide gel electrophoresis. A:
Electrophoretograms obtained with gels polymerized from 11% (left
3) and 14% (right 3) acrylamide containing 0.1% SDS, for 5, 10, and
20 μg of protein (from left to right in each triad). B: Molecular
weights obtained by SDS-PAGE in gels polymerized from 14% acrylamide. The standards were 1, catalase; 2, ovalbumin; 3, aldolase;
4, carbonic anhydrase; 5, ribonuclease. C: Mole ratios of the large
(L) to the mixed small (S) subunits (part B). The gels were scanned
at 656 nm after destaining. The inset shows the proportionality of
Coomassie Brilliant Blue R binding in relation to protein applied.

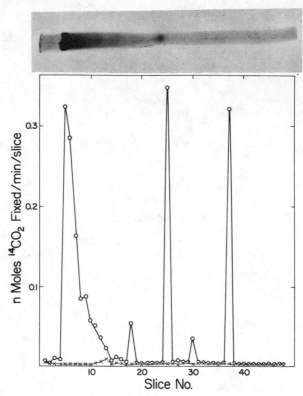

Figure 3. Localization of RuBP carboxylase activity in different
states of aggregation by polyacrylamide gel electrophoresis. Enzyme
(33 µg/gel) was run at 7°C on gels photopolymerized from 7.5% acrylamide in the presence of 10% glycerol (131). After electrophoresis
the gel was longitudinally sliced. One half of the gel was stained
with Coomassie Brilliant Blue R (top) and the other half was ver-
tically sliced into 1.46-mm cross-sectional segments. Each segment
was assayed for enzyme activity at 30°C for 65 min (o—o). The
activity profile in the absence of RuBP in a control gel sliced as
above is also shown (x—x).

        A further indication of homogeneity is the finding that this
enzyme consists of large and small subunits (Figure 2) and is therefore a T enzyme like those found in other members of the Chromatiaceae. However, it should be stressed that the ratio of mixed
small to large subunits deviates considerably from 1.0 at a low protein concentration (Figure 2).
        When RuBP carboxylase from T. roseopersicina was examined elec-
trophoretically at pH 8 in gels polymerized from 7.5% acrylamide in
the presence of glycerol and with riboflavin as a catalyst, the

Table 2

Properties of RuBP Carboxylase From T. roseopersicina

Kinetic parameters:

Ligand	$K_m(\mu M)$ at pH, temp. ($^\circ C$)		
RuBP	110	8.1,	30
CO_2	180	8.1,	30
CO_2	67	8.1,	10
Mg^{2+}	5540	8.1,	30
Mg^{2+}	8400	8.6,	25

Relative activities with 20 mM M^{2+} at $30^\circ C$, pH 8.1:

Mg^{2+}	Mn^{2+}	Co^{2+}	Ni^{2+}	Ca^{2+}	Cd^{2+}
1.00	0.33	0	0	0	0

Table 3

Inactivation of T. roseopersicina RuBP Carboxylase by 2,3-Butadione

RuBP carboxylase (75 μg) in 150 μl of buffer containing 50 mM Tris-Cl, 1 mM EDTA, 10 mM $MgCl_2$, 50 mM $NaHCO_3$, 10% glycerol, and 0.1 mM DTT (pH 8.1) was mixed with BO_3^{3-} (50 mM) and phosphoglycerate (10 mM) (where applicable) and allowed to warm for 15 min at $30^\circ C$. Then 2,3-butadione was added to a concentration of 20 mM. Zero-time activity was determined by assaying enzyme + BO_3^{3-} and was assumed to be 100% throughout the time course for enzyme + BO_3^{3-} + 2,3-butadione. Zero-time activity for enzyme + phosphoglycerate + BO_3^{3-} was determined, and this was considered to be 100% for enzyme + phosphoglycerate + BO_3^{3-} + 2,3-butadione. The final volume was 500 μl, and 50-μl samples were removed at the specified times for assay.

Reaction components	Percent activity after mixing			
	0 min	4 min	13 min	32 min
Enzyme + phosphoglycerate + BO_3^{3-}	100	102	93	103
Enzyme + phosphoglycerate + BO_3^{3-} + 2,3-butadione	-	106	96	72
Enzyme + BO_3^{3-}	100	93	84	80
Enzyme + BO_3^{3-} + 2,3-butadione	-	85	51	16

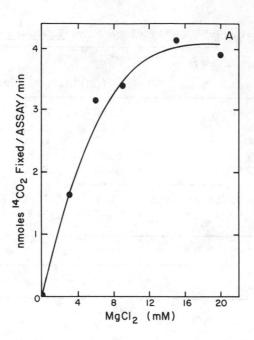

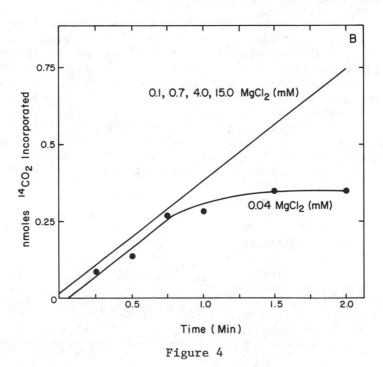

Figure 4

← Figure 4.   Divergence of T. roseopersicina RuBP carboxylase in the requirement of Mg^{2+} for activation and catalysis.   A: For the determination of overall Mg^{2+} dependence, the enzyme was dialyzed for 4 hr at 4°C against 2500 volumes of 100 mM Tris containing 10 mM $NaHCO_3$, 1 mM EDTA, and 10% glycerol and adjusted to pH 8.6. After redialysis as above in the buffer without EDTA the enzyme was activated in the presence of 0, 3, 6, 9, 12, 15, or 20 mM Mg^{2+}.   The $^{14}CO_2$ incorporation was measured in reaction mixtures containing an identical amount of Mg^{2+} for a period of 1 min. The K_m for Mg^{2+} for RuBP-dependent $^{14}CO_2$ incorporation was 8.4 mM. B: The enzyme (0.62 mg/ml) was dialyzed as described above and activated at 25°C in the presence of 15 mM $MgCl_2$.   A 10-μl portion was rapidly diluted 375-fold by adding it to 3740 μl of reaction mixture (pH 8.6) containing 100 mM Tris buffer, 20 mM $NaH^{14}CO_3$ supplemented with 10% glycerol, and 0.8 mM RuBP containing 0 to 15 mM Mg^{2+} prewarmed at 25°C.   One 600-μl aliquot was removed at each specified time, added to 0.4 volume of 60% trichloroacetic acid, prepared for counting, and counted as previously described (62).   The plot showing RuBP-dependent CO_2 incorporation for 0.1, 0.7, 4.0, and 15.0 mM Mg^{2+}, obtained by a linear least-squares regression analysis using all data points from 0.25 to 2 min for two parallel experiments, had a slope of 0.36 with a standard deviation of 0.02.   The corresponding intercept of 0.014 had S.D. 0.027.   Similar data treatment for 0.04 mM Mg^{2+} through 0.75 min yielded a slope of 0.38 (S.D. 0.07) and intercept of 0.04 (S.D. 0.056).

pattern shown in Figure 3 was consistently obtained.   The same general activity and protein profiles were also obtained in the pH interval up to 10.   The top picture shows the longitudinal gel slice stained for protein, and the bottom chart shows the activity distribution for assays in the presence and absence of RuBP.   Clearly RuBP carboxylase was found in three major peaks.   Fraction 30 was where the 68,000-dalton monomer of bovine serum albumin ran in parallel experiments.   We conclude that the Thiocapsa enzyme undergoes dissociative equilibrium and that the fastest activity peak may be the large subunit 53,000-dalton monomer.   Of considerable interest is the high catalytic activity of this fast component, which does not correspond to a Coomassie Blue stained region and accordingly contained <0.05 μg of protein.   The specific activity of this species was therefore a minimum of 6.6 and may have been much higher.

In Table 2 various molecular properties of RuBP carboxylase from Thiocapsa are summarized.   The K_m for CO_2 was measured at pH 8.1 after activation by 7.5 mM HCO_3^- with initial rates determined over a 1-min period by measurements of CO_2 fixed in 15-sec intervals. The resultant K_m of 67 (10°C) and 180 μM (30°C) are considerably higher than the value of 21 measured at 25°C and pH 8.1 under comparable conditions for the tobacco enzyme (132).   The K_m for RuBP (at 30°C, pH 8.1) of 110 μM is higher than that (20 μM) of the purified

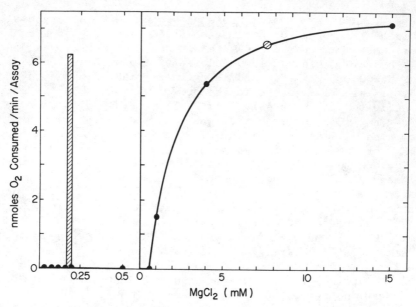

Figure 5. Mg^{2+} requirement for RuBP oxygenase of <u>T. roseopersicina</u>.
The enzyme was depleted of Mg^{2+} by dialysis as described (Figure 4)
and then incubated in the presence of 0.04 to 15 mM $MgCl_2$ at
25°C for 10 min.   The Mg^{2+}-treated enzyme (50 µl) was used to initiate the reaction at 25°C in a final volume of 2.0 ml which contained 100 mM CO_2-free Tris (pH 8.6), 0.8 mM RuBP, 10% v/v glycerol,
and Mg^{2+} matching the concentration used during the enzyme pretreatment.   The RuBP-dependent rate of oxygen consumption was calculated
from the initial linear portion of the oxygen consumption curve obtained with a Gilson Oxygraph.   The bar graph (left) shows oxygen
consumption at 0.1875 mM Mg^{2+} when the enzyme (50 µl) which had
been preincubated at 7.5 mM was used to initiate the reaction in
the buffer without Mg^{2+}.   The shaded circle (right) designates the
theoretical rate of oxygen consumption at 7.5 mM Mg^{2+}.

spinach enzyme at 25°C and a pH of 8.2 (132).   At pH 8.1 and 30°C,
Mn^{2+} (present at 20 mM in both activation and assay) was 33% as
effective as 20 mM Mg^{2+}, and no activity was seen with other metal
ions tested.

    Table 3 presents data establishing that the <u>Thiocapsa</u> enzyme
undergoes borate-dependent inactivation by 2,3-butadione, and that
the presence of the product 3-phosphoglycerate at 10 mM confers protection.   Thus this enzyme behaves like those from barley and <u>P.
oxalaticus</u> (129) and has one or more essential arginines, presumably
at the active site.

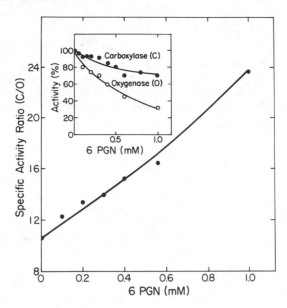

Figure 6. Differential 6-phosphogluconate (6PGN) sensitivity of
T. roseopersicina RuBP carboxylase and oxygenase. Both enzyme activities in the presence of varying concentrations of 6PGN were assayed at pH 8.6 and 25°C in 100 mM Tris buffer containing 20 mM
$MgCl_2$ and 10% v/v glycerol. The enzyme was predialyzed against
the assay buffer containing 10 mM $NaHCO_3$. The carboxylase was assayed in the presence of 20 mM $NaH^{14}CO_3$ and 0.8 mM RuBP whereas the
oxygenase assay mixture contained 0.3 mM $NaHCO_3$ and 0.2 mM RuBP.
In both cases the reactions were initiated by addition of activated
enzyme to the reaction mixtures containing 6PGN. Specific activi
ties were calculated from the constant rates of RuBP-dependent
$^{14}CO_2$ incorporation or of oxygen consumption. The concentration
of protein for oxygenase assay was 4.56 times that used for carboxylase assays.

From the data in Figure 4, the K_m for Mg^{2+} (pH 8.6, 25°C) can
be estimated for RuBP carboxylase, and the value of 8.4 mM is that
for activation because 375-fold dilution of Mg^{2+} had no effect on
the initial rate. Magnesium ion may not be required for catalysis. Similar conclusions are reached for the oxygenase activity of
the same enzyme from Thiocapsa (Figure 5) although the K_m for activation by Mg^{2+} may be lower.

In Figure 6, the sensitivities of RuBP carboxylase and RuBP
oxygenase of the pure enzyme from T. roseopersicina to 6-phospho-
D-gluconate (6PGN) are shown to be different. The decreased sensi-

tivity of the oxygenase probably cannot be attributed to a mixture
of activation by 6PGN (at the low HCO_3^- concentration present during
catalysis) and inhibition by 6PGN, because the initial rates were
unaffected by the presence of 6PGN (data not shown). In this context, it should be mentioned again that with the spinach enzyme
(99, 109) the carboxylase and oxygenase activities are activated
similarly, i.e., by sequential addition of CO_2 (as HCO_3^-) and Mg^{2+}.

DISCUSSION

The work reported here establishes that RuBP carboxylase from
T. roseopersicina is unstable in the absence of glycerol. Although
the enzyme appears to consist of large (53,000-dalton) and mixed
small (15,000- and 13,500-dalton) subunits, the stoichiometry of
combination of L and S subunits is unsatisfactory, once again
raising the question about the role of small subunits. When 10%
glycerol is present, the enzyme dissociates. Two small dissociation
products contain enzyme activity and one, a putative monomer lacking
the small subunit, probably has the highest specific activity
ever observed.

Some of the properties of RuBP carboxylases are summarized in
Table 4. Presumably the ancestral gene encoding the large, catalytic subunits was established first and expressed in anaerobic bacteria. In this connection, it may be significant that T. denitri-
ficans, which contains RuBP carboxylase of intermediate size, can
be cultured anaerobically with nitrate as an electron acceptor. In
contrast, T. novellus and Thiobacillus A2 are aerobic and contain a
larger enzyme, probably of the T type. We postulate (Figure 7) that
the gene specifying the structure of the small subunit was established prior to the appearance of eukaryotes (5). Because of the
ease of its dissociation into subunits (at pH 9), RuBP carboxylase
from the Chromatiaceae (Chromatium D and E. halophilia, for a discussion see ref. 6) may be an archetype of the large T enzyme. The
work described here showed that another member of this family,
T. roseopersicina, also has a large T enzyme which apparently dissociates into a catalytically active large subunit. Since the small
subunit is of unknown function, it is idle to speculate about the
selective advantage that may have accrued to autotrophic organisms
acquiring RuBP carboxylase having this polypeptide. Curiously, as
emphasized earlier, the enzymes from R. rubrum and T. intermedius,
which lack small subunits, are as good catalysts as the spinach
enzyme. For these reasons it is important to stress that the small
subunits may be artifacts of isolation of RuBP carboxylase. For
example, these polypeptides may have a high affinity for oligomeric
large subunits and combine stoichiometrically after cell disruption.

We have established that one T type prokaryotic RuBP carboxylase (from H. eutropha) consists of small subunits and two kinds
of large subunits that differ subtly in molecular weight. The

Table 4

Comparisons of RuBP Carboxylases Isolated From Bacteria,
Cyanobacteria, Green Algae, and Higher Plants

L = large (MW 50,000 to 58,000); S = small (MW 12,000 to 18,000).
A dash indicates not determined.

Enzyme source (MW or $S_{20,w}$)	Quaternary structure	Oxygenase	Inhibition by 1 mM 6PGN
PHOTOLITHOTROPHIC BACTERIA			
(CLASSICAL)			
T. roseopersicina			
(ca. 55,000 & higher)	nL, nS	yes	yes
R. rubrum (112,000)	2L	yes	no
C. thiosulfatophilum			
(360,000)	6L	--	yes
Chromatium D (550,000)	8L, 8S	yes	yes
E. halophila (600,000)	8L, 8S	--	yes
Rps. sphaeroides (360,000)	6L	--	no
(550,000)	8L, 8S	--	yes
CHEMOLITHOTROPHIC BACTERIA			
T. denitrificans (350,000)	--	--	no
T. intermedius (455,000)	8L	--	yes
T. novellus (498,000)	--	--	yes
Thiobacillus A2 (512,000)	8L, 8S	--	yes
H. eutropha (516,000)	8L, 8S	yes	yes
BLUE-GREEN ALGAE			
Agm. quadruplicatum (456,000)	8L	--	--
Anab. cylindrica (452,000)	8L (8S)	--	--
Anab. variabilis (18 S)	8L, 8S	--	--
Plect. boryanum (18 S)	8L, 8S	--	--
Aphanocapsa (525,000)	8L, 8S	yes	yes
GREEN ALGAE			
E. gracilis	8L, 8S	yes	yes
Chlam. reinhardi (530,000)	8L, 8S	--	--
Chlorella fusca (530,000)	8L, 8S	yes	yes
Chlorella ellipsoidea (19 S)	8L, 8S	--	--
Halimeda cylindracea (18 S)	8L, 8S	yes	--
HIGHER PLANTS			
spinach (560,000)	8L, 8S	yes	yes
spinach beet (560,000)	8L, 8S	--	--
tobacco (525,000)	8L, 8S	yes	--
French bean (17.9 S)	8L, 8S	--	--

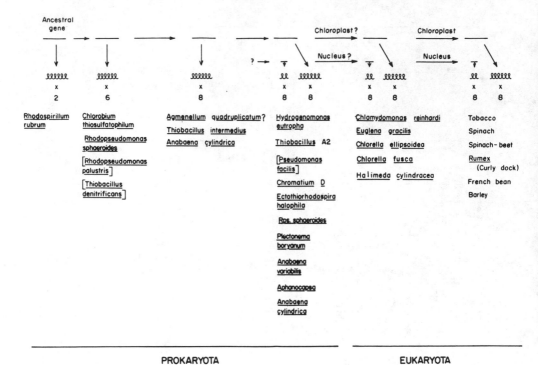

PROKARYOTA   EUKARYOTA

Figure 7. Tentative evolutionary scheme for RuBP carboxylase/
oxygenase. The quaternary structure of the enzyme from the
bracketed microorganisms is unknown, but placement is based on the
molecular weight of the native enzyme.

latter were detectable only after electrophoresis of the dissociation products which had been applied at lower concentrations to gels
(12). In contrast, examination of the enzyme from T. intermedius
revealed dissociation into a single 54,500-dalton electrophoretic
species (46). In light of this variability of large subunit structure, we urge the reexamination of the quaternary structure of other
RuBP carboxylases. This should be done after isolation of enzymes
homogeneous to gel electrophoresis, and under conditions that minimize proteolysis. It is relevant that heterogeneity of both large
and small subunits of the higher plant enzyme has been detected
by isoelectric focusing, as mentioned previously (41). If the
large subunits are indeed chloroplast DNA-encoded and heterogeneous
in this enzyme from all higher plants, then heterogeneity of analogous subunits from prokaryota would be compatible with the endo-
symbiont theory of chloroplast origin. Of interest in this regard
would be a reinvestigation of the quaternary structure of RuBP carboxylases from blue-green and unicellular green algae. The enzyme
of one of the cyanobacteria (blue-green algae), Oscillatoria
limnetica, is of particular interest. This organism can conduct

either oxygenic or anoxygenic photosynthesis, using H_2S in the
latter case, and thus meets one expectation of an organism linking
anoxygenic and oxygenic microorganisms in evolution (133).

To test the postulated evolution of RuBP carboxylase described
here (Figure 7) and in more detail elsewhere (11), immunochemical
studies, tryptic fingerprinting, and, finally, amino acid sequencing
will be required for both large and small subunits. X-ray crystallography should yield information about the arrangement of the subunits and other aspects of three-dimensional structure. In the
final analysis, it is evident that knowledge of the structure,
function, and regulation of RuBP carboxylase may yield considerable
information about the evolution of autotrophism, especially when it
is coupled with information about the primary structures of other
proteins found in autotrophic species.

Acknowledgments: We acknowledge support in part through
grants from the NIH (GM-19,972) and Frasch Foundation. We
thank Dr. R. E. Hurlbert for the generous supply of T. roseopersicina cells and V. B. Lawlis for providing data from the
butadione experiments.

REFERENCES

1. Urey, H. C., Proc. Natl. Acad. Sci. USA 38, 351-63 (1952).
2. Miller, S. L. and Orgel, L. E., The Origins of Life on the
    Earth, Prentice-Hall, Englewood Cliffs, New Jersey, 1974.
3. Oparin, A. I., The Origin of Life, Macmillan, New York, 1938.
4. Haldane, J. B. S., Rationalist Ann. 1-10 (1929).
5. McFadden, B. A., Bacteriol. Rev. 37, 289-319 (1973).
6. McFadden, B. A. and Tabita, F. R., Biosystems 6, 93-112 (1974).
7. Horowitz, N. H., Proc. Natl. Acad. Sci. USA 31, 153-7 (1945).
8. McFadden, B. A., in Proc. Symp. on Microbial Production and
    Utilization of Gases, pp. 267-80, H. G. Schlegel et al.,
    Editors, E. Golze, Göttingen, 1976.
9. McFadden, B. A., in Proc. 2nd Int. Symp. on Photosynthetic
    Procaryotes, pp. 190-2, G. A. Codd and W. D. P. Stewart,
    Editors, U. of Dundee, 1976.
10. McFadden, B. A., in Proc. 2nd Int. Symp. on Microbial Growth
    on C_1-compounds, pp. 106-7, G. K. Skryabin et al., Editors,
    Academy of Sciences, Pushchino, USSR, 1977.
11. McFadden, B. A., in The Bacteria, Vol. 6, L. N. Ornston and
    J. R. Sokatch, Editors, Academic Press, New York, in press,.
    1978.
12. Purohit, K. and McFadden, B. A., J. Bacteriol. 129, 415-21
    (1977).
13. Tabita, F. R. and McFadden, B. A., J. Bacteriol. 126, 1271-7
    (1976).
14. Akazawa, T., Kondo, H., Shimazue, T., Nishimura, M., and
    Sugiyama, T., Biochemistry 11, 1298-1303 (1972).

15. Purohit, K. and McFadden, B. A., Biochem. Biophys. Res. Commun. 71, 1220-7 (1976).
16. Bowien, B., Mayer, F., Codd, G. A., and Schlegel, H. G., Arch. Microbiol. 110, 157-66 (1976).
17. Charles, A. M. and White, B., Arch. Microbiol. 108, 195-202 (1976).
18. Charles, A. M. and White, B., Arch. Microbiol. 108, 203-9 (1976).
19. McCarthy, J. T. and Charles, A. M., Arch. Microbiol. 105, 51-9 (1975).
20. Shively, J. M., Saluja, A., and McFadden, B. A., J. Bacteriol. 134, 1123-32 (1978).
21. Takabe, T., Nishimura, M., and Akazawa, T., Biochem. Biophys. Res. Commun. 68, 537-44 (1976).
22. Codd, G. A. and Stewart, W. D. P., Arch. Microbiol. 113, 105-10 (1977).
23. Stewart, R., Auchterlonie, C. C., and Codd, G. A., Planta 136, 61-4 (1977).
24. Sugiyama, T., Ito, T., and Akazawa, T., Biochemistry 10, 3406-11 (1971).
25. Lord, J. M. and Brown, R. H., Plant Physiol. 55, 360-4 (1975).
26. Givan, A. L. and Criddle, R. S., Arch. Biochem. Biophys. 149, 153-63 (1972).
27. McFadden, B. A., Lord, J. M., Rowe, A., and Dilks, S., Eur. J. Biochem. 54, 195-206 (1975).
28. Akazawa, T. and Osmond, C. B., Aus. J. Plant Physiol. 3, 93-103 (1976).
29. Rutner, A. C. and Lane, M. D., Biochem. Biophys. Res. Commun. 28, 531-7 (1967).
30. Sugiyama, T. and Akazawa, T., J. Biochem. Tokyo 62, 474-82 (1967).
31. Moon, K. E. and Thompson, E. O. P., Aus. J. Biol. Sci. 22, 463-70 (1969).
32. Kawashima, N., Plant Cell Physiol. 10, 31-40 (1969).
33. Kawashima, N. and Wildman, S. G., Annu. Rev. Plant Physiol. 21, 325-58 (1970).
34. Kawashima, N., Kwok, S., and Wildman, S. G., Biochim. Biophys. Acta 236, 578-86 (1971).
35. Goldthwaite, J. J. and Bogorad, L., Anal. Biochem. 41, 57-66 (1971).
36. Gray, J. C. and Kekwick, R. G. O., Eur. J. Biochem. 44, 481-9 (1974).
37. Blair, G. E. and Ellis, R. J., Biochim. Biophys. Acta 319, 223-34 (1973).
38. Strøbaek, S. and Gibbons, G. C., Carlsberg Res. Commun. 41, 57-72 (1976).
39. Holder, A. A., Carlsberg Res. Commun. 41, 321-34 (1976).
40. Börner, T., Jahn, G., and Hagemann, R., Biochem. Physiol. Pflanzen 169, 179-81 (1976).

41. Chen, K., Kung, S. D., Gray, J. C., and Wildman, S. G., Plant
     Sci. Lett. 7, 429-34 (1976).
42. Patel, T. R. and McFadden, B. A., Exp. Parasitol. 44, 72-81
     (1978).
43. Gray, J. C., Kung, S. D., and Wildman, S. G., Arch. Biochem.
     Biophys. 185, 272-81 (1978).
44. Tabita, F. R. and McFadden, B. A., J. Biol. Chem. 249, 3459-
     64 (1974).
45. Tabita, F. R., McFadden, B. A., and Pfennig, N., Biochim.
     Biophys. Acta 341, 187-94 (1974).
46. Purohit, K., McFadden, B. A., and Cohen, A. L., J. Bacteriol.
     127, 505-15 (1976).
47. Tabita, F. R., Stevens, S. E. Jr., and Quijano, R., Biochem.
     Biophys. Res. Commun. 61, 45-52 (1974).
48. Tabita, F. R., Stevens, S. E. Jr., and Gibson, J. L., J.
     Bacteriol. 125, 531-9 (1976).
49. Takabe, T., Agric. Biol. Chem. 41, 2255-60 (1977).
50. Gibson, J. L. and Tabita, F. R., J. Bacteriol. 132, 818-23
     (1977).
51. Gibson, J. L. and Tabita, F. R., J. Biol. Chem. 252, 943-9
     (1977).
52. Gibson, J. L. and Tabita, F. R., J. Bacteriol. 131, 1020-2
     (1977).
53. Marchalonis, J. J. and Weltman, J. K., Comp. Biochem. Physiol.
     38B, 609-25 (1971).
54. Takabe, T. and Akazawa, T., Plant Cell Physiol. 16, 1049-60
     (1975).
55. Siegel, M. I. and Lane, M. D., Biochem. Biophys. Res. Commun.
     48, 508-16 (1972).
56. Takabe, T. and Akazawa, T., Arch. Biochem. Biophys. 157, 303-
     8 (1973).
57. Nishimura, M., Takabe, T., Sugiyama, T., and Akazawa, T.,
     J. Biochem. Tokyo 74, 945-96 (1973).
58. Nishimura, M. and Akazawa, T., Biochemistry 13, 2277-81 (1974).
59. Takabe, T. and Akazawa, T., Biochemistry 14, 46-50 (1974).
60. Lord, J. M., Codd, G. A., and Stewart, W. D. P., Plant Sci.
     Lett. 4, 377 83 (1975).
61. Gray, J. C. and Kekwick, R. G. O., Eur. J. Biochem. 44, 481-9
     (1974).
62. McFadden, B. A., Tabita, F. R., and Kuehn, G. D., Methods
     Enzymol. 42, 461-72 (1975).
63. Tabita, F. R. and McFadden, B. A., Biochem. Biophys. Res.
     Commun. 48, 1153-60 (1972).
64. Tabita, F. R. and McFadden, B. A., Arch. Microbiol. 99, 231-
     40 (1974).
65. Kawashima, N. and Wildman, S. G., Biochim. Biophys. Acta
     229, 240-9 (1971).
66. Kwok, S., Kawashima, N., and Wildman, S. G., Biochim. Biophys.
     Acta 234, 293-6 (1971).

67. Chan, P. H., Sakano, K., Singh, S., and Wildman, S. G.,
     Science 176, 1145-6 (1972).
68. Pardies, H. H., Zimmer, B., and Werz, G., Biochem. Biophys.
     Res. Commun. 74, 397-404 (1977).
69. Zimmer, B., Pardies, H. H., and Werz, G., Biochem. Biophys.
     Res. Commun. 74, 1496-500 (1977).
70. Baker, T. S., Suk, S., and Eisenberg, D., Proc. Natl. Acad.
     Sci. USA 74, 1037-41 (1977).
71. Baker, T. S., Eisenberg, D., and Eiserling, F., Science 196,
     293-5 (1977).
72. Cooper, T. G., Filmer, S., Wishnick, M., and Lane, M. D.,
     J. Biol. Chem. 244, 1081-3 (1969).
73. Siegel, M. I. and Lane, M. D., J. Biol. Chem. 248, 5486-98
     (1973).
74. Sjödin, B. and Vestermark, A., Biochim. Biophys. Acta 297,
     165-73 (1973).
75. Miziorko, H. M. and Mildvan, A. S., J. Biol. Chem. 249, 2743-
     50 (1974).
76. Lorimer, G. H., Andrews, T. J., and Tolbert, N. E., Biochemis-
     try 12, 18-23 (1973).
77. Andrews, T. J., Lorimer, G. H., and Tolbert, N. E., Biochemis-
     try 12, 11-18 (1973).
78. Ogren, W. L. and Bowes, G., Nature London 230, 159-60 (1971).
79. Zelitch, I., Science 188, 626-33 (1975).
80. Bassham, J. A. and Kirk, M., Plant Physiol. 52, 407-11 (1973).
81. Davis, B., and Merrett, M. J., Plant Physiol. 55, 30-4 (1975).
82. Nelson, P. E. and Surzycki, S. J., Eur. J. Biochem. 61,
     465-74 (1976).
83. Nelson, P. E. and Surzycki, S. J., Sur. J. Biochem. 61,
     475-80 (1976).
84. Lorimer, G. H., Krause, G. H., and Berry, J. A., FEBS Lett.
     78, 199-202 (1977).
85. Andrews, T. J., Lorimer, G. H., and Tolbert, N. E., Biochemis-
     try 10, 4777-82 (1971).
86. Dimon, B. and Gerster, R., C. R. Hebd. Seances Acad. Sci.
     Ser. D. 283, 507-10 (1976).
87. Dimon, B., Gerster, R., and Tournier, P., C. R. Hebd. Seances
     Acad. Sci. Ser. D. 284, 297-9 (1977).
88. Asami, S. and Akazawa, T., Biochemistry 16, 2202-7 (1977).
89. Krause, G. H., Thorne, S. W., and Lorimer, G. H., Arch.
     Biochem. Biophys. 183, 471-9 (1977).
90. Asami, S. and Akazawa, T., Plant Cell Physiol. 15, 571-6
     (1974).
91. Codd, G. A. and Smith, B. M., FEBS Lett. 48, 105-8 (1974).
92. McFadden, B. A., Biochem. Biophys. Res. Commun. 60, 312-17
     (1974).
93. Takabe, T. and Akazawa, T., Biochem. Biophys. Res. Commun.
     53, 1173-9 (1973).
94. Lorimer, G. H., Osmond, C. B., Akazawa, T., and Asami, S.,
     Arch. Biochem. Biophys. 185, 49-56 (1978).

95. Codd, G. A., Bowien, B., and Schlegel, H. G., Arch. Micro-
     biol. 110, 167-71 (1976).
96. Kuehn, G. D. and McFadden, B. A., Biochemistry 8, 2394-402
     (1969).
97. Tabita, F. R. and McFadden, B. A., J. Biol. Chem. 249,
     3453-8 (1974).
98. Pon, N. G., Rabin, B. R., and Calvin, M., Biochem. Z. 338,
     7-19 (1963).
99. Lorimer, G. H., Badger, M. R., and Andrews, T. J., Biochemis-
     try 15, 529-36 (1976).
100. Laing, W. A. and Christeller, J. T., Biochem. J. 159, 563-
     70 (1976).
101. Bahr, J. T. and Jensen, R. G., Plant Physiol. 53, 39-44
     (1974).
102. Bahr, J. T. and Jensen, R. G., Arch. Biochem. Biophys. 164,
     408-13 (1974).
103. Bahr, J. T. and Jensen, R. G., Biochem. Biophys. Res. Commun.
     57, 1180-5 (1974).
104. Laing, W. A., Ogren, W. L., and Hageman, R. H., Plant Physiol.
     54, 678-85 (1974).
105. Laing, W. A., Ogren, W. L., and Hageman, R. H., Biochemistry
     14, 2269-75 (1975).
106. Badger, M. R. and Andrews, J. T., Biochem. Biophys. Res.
     Commun. 60, 204-10 (1974).
107. Jackson, W. A. and Volk, R. J., Annu. Rev. Plant Physiol.
     21, 385-432 (1970).
108. Jollife, P. A. and Tregunna, E. B., Can. J. Botany 51, 841-
     53 (1973).
109. Badger, M. R. and Lorimer, G. H., Arch. Biochem. Biophys.
     145, 723-9 (1976).
110. Bassham, J. A. and Kirk, M., in Comparative Biochemistry and
     Biophysics of Photosynthesis, pp. 365-78, K. Shibata et al.,
     Editors, U. of Tokyo Press, 1968.
111. Chu, D. K. and Bassham, J. A., Plant Physiol. 50, 224-7
     (1972).
112. Singer, S. J. and Nicholson, G. L., Science 175, 720-31
     (1972).
113. Kawashima, N. and Wildman, S. G., Biochim. Biophys. Acta 262,
     42-9 (1972).
114. Chan, P. H., Sakano, K., Singh, S., and Wildman, S. G.,
     Science 176, 1145-6 (1972).
115. Criddle, R. S., Daw, B., Kleinkopf, G. E., and Huffaker, R. C.,
     Biochem. Biophys. Res. Commun. 41, 621-7 (1970).
116. Alscher, R., Smith, M. A., Petersen, L. W., Huffaker, R. C.,
     and Criddle, R. S., Arch. Biochem. Biophys. 174, 216-25 (1976).
117. Roy, H., Patterson, R., and Jagendorf, A. T., Arch. Biochem.
     Biophys. 172, 64-73 (1976).
118. Hartley, M. R., Wheeler, A., and Ellis, R. J., J. Mol. Biol.
     91, 67-77 (1975).

119. Sagher, D., Grosfeld, H., and Edelman, M., Proc. Natl. Acad.
      Sci. USA 73, 722-6 (1976).
120. Dobberstein, B., Blobel, G., and Chua, N., Proc. Natl. Acad.
      Sci. USA 74, 1082-5 (1976).
121. Shiveley, J. M., Ball, F. L., Brown, D. H., and Saunders,
      R. E., Science 182, 584-6 (1973).
122. Shiveley, J. M., Bock, E., Westphal, K., and Cannon, G. C.,
      J. Bacteriol. 132, 673-5 (1977).
123. Shiveley, J. M., Ball, F. L., and Kline, B. W., J. Bacteriol.
      116, 1405-11 (1973).
124. Purohit, K., McFadden, B. A., and Shaykh, M. M., J. Bacteriol.
      127, 516-22 (1976).
125. Codd, G. A. and Stewart, W. D. P., Planta 130, 323-6 (1976).
126. Norton, I. L., Welch, M. H., and Hartman, F. C., J. Biol.
      Chem. 250, 8062-8 (1975).
127. Schloss, J. V. and Hartman, F. C., Biochem. Biophys. Res.
      Commun. 75, 320-8 (1977).
128. Whitman, W. and Tabita, F. R., Biochem. Biophys. Res. Commun.
      71, 1034-9 (1976).
129. Lawlis, V. B. and McFadden, B. A., Biochem. Biophys. Res.
      Commun. 80, 580-5 (1978).
130. Baker, T. S., Eisenberg, D., Eiserling, F. A., and Weissman,
      L., J. Mol. Biol. 91, 391-9 (1975).
131. Orr, M. W., Blakeley, R. L., and Panagou, D., Anal. Biochem.
      45, 68-85 (1972).
132. Jensen, R. G. and Bahr, T. J., Annu. Rev. Plant Physiol.
      28, 379-400 (1977).
133. Cohen, Y., Padan, E., and Shilo, M., J. Bacteriol. 123, 855-
      61 (1975).

DISCUSSION

HUNER: In my work with RuBP carboxylase from rye plants, I
have also observed heterogeneity in the region of the large subunit.
I observed polypeptides of MW 54,000 and 47,000 in the ratio of
5:3 in the NaCl-inactivated enzyme and in the active enzyme prepared for electrophoresis at slightly acidic pH but not at alkaline
pH. Have you considered the possibility that the heterogeneity you
observed may be an artifact due to your method of preparing samples
for electrophoresis?

Mc FADDEN: The samples for SDS-PAGE were prepared by the method
of Lemmli (Nature, 1970), and the SDS dissociation was done at
pH 6.8. We varied the pH of electrophoresis in Tris/glycine from
8.3 to 9.3 and observed the same heterogeneity of large subunits
derived from RuBP carboxylase from H. eutropha. The Lemmli procedure has been extensively used during the last 8 years and to our
knowledge gives good resolution. We always used active enzyme and
did not dissociate at an alkaline pH. Variation of the heat treatment (with and without β-mercaptoethanol) had no effect on the SDS

dissociation products we described in Biochem. Biophys. Res. Commun.
(1976) and in J. Bacteriol. (1977). Isolated mixed large subunits
gave the same banding pattern upon gel electrophoresis as described
above. I stress, once again, that we observed heterogeneity only
when very small amounts of protein (2 to 5 µg) were loaded onto the
gels (after dissociation). We do not see how these procedures could
have resulted in artifacts.

    LORIMER: What is the ratio of carboxylase to oxygenase activity in the procaryotic organism?

    McFADDEN: About 6 to 10.

    LORIMER: Is that a ratio of V_{max} activity?

    McFADDEN: Yes.

    HOWELL: Is there any evidence suggesting that either one of
these subunits might be borne on a nonchromosomal piece of genetic
information?

    McFADDEN: Schlegel reported at a conference in Pushchino,
Russia (sponsored last fall by the Soviet Academy of Science), that
among the hydrogen bacteria the genes for autotrophic CO_2 fixation,
in particular hydrogenase, RuBP carboxylase, and phosphoribulo-
kinase, are apparently located on plasmids. This has not been published yet, so I have not been able to evaluate it.

    HARTMAN: On the gels you showed to present evidence of an
enzymically active monomeric subunit, I was surprised to see the
activity contained in a single slice.

    McFADDEN: The gel slices are 1.46 mm thick. We have done four
other experiments, and we found activity over two slices in that
region of the gel, so I am quite satisfied that this is reproducible.

    CHOLLET: Regarding the differential regulation, have you made
simultaneous runs to make sure you can directly confirm it?

    McFADDEN: No. We did the experiments one day apart, always
on a single enzyme preparation, and we took the precautions I indicated. We matched temperature, pH, and all buffer components. The
only difference was the carry-over of bicarbonate into the oxygenase assay.

    YADOV: Can you dissociate the protein outside the gel with
glycerol and then see whether there is any activity associated with
it?

    McFADDEN: We have not taken a careful look at the dependence
of dissociation on glycerol concentration. In sedimentation velocity studies there are two limiting structures of 16 S and 19.3 S
at 10% and 4% (v/v) glycerol, respectively. We see a single symmetrical peak in each case, but we have not studied the intermediate glycerol concentrations, 2, 4, or 10% by volume.

RIBULOSE BISPHOSPHATE CARBOXYLASES FROM Chromatium vinosum
AND Rhodospirillum rubrum AND THEIR ROLE IN PHOTOSYNTHETIC
CARBON ASSIMILATION*

Takashi Akazawa, Tetsuko Takabe, Sumio Asami,

and Hirokazu Kobayashi

Research Institute for Biochemical Regulation
School of Agriculture, Nagoya University
Chikusa, Nagoya 464, Japan

Ribulose bisphosphate (RuBP) carboxylase is widely distributed
in both prokaryotic (Bacteriophyta and Cyanophyta) and eukaryotic
photoautotrophs (Euglenophyta, and Chlorophyta such as Chlorophytina
and Tracheophytina). It is also present in some chemosynthetic
bacteria. RuBP carboxylase from higher plants and green algae has
a molecular weight of (5 to 5.5) x 10^5 and contains two types of
subunits, A (MW 5.5 x 10^4) and B (MW 1.2 x 10^4). On the basis of
biochemical and immunochemical experiments, it is now well established that the large subunit moiety (A) of the enzyme molecule
carries the catalytic site, whereas the small subunit (B) is probably involved in the regulatory function, discussed below. Although
the spinach enzyme has been used in the most thorough investigations,
in various laboratories, of the structure and function of the carboxylase molecule, other biochemical and biophysical studies have
supported the idea that the plant-type enzyme molecule in general
has a symmetrical structure consisting of an octamer of the two
subunits, A_8B_8 (1).

---

*Research supported by grants from the Ministry of Education
of Japan (211113), the Toray Science Foundation (Tokyo), and the
Nissan Science Foundation (Tokyo).
    Abbreviations: PEP, phosphoenolpyruvate; 3-PGA, 3-phosphoglyc-
erate; RuBP, ribulose 1,5-bisphosphate; A and B, large and small
subunits of RuBP carboxylase, respectively.

A.   INTERSPECIFIC HOMOLOGIES OF SUBUNIT A AND MOLECULAR
EVOLUTION OF RuBP CARBOXYLASE

It is intriguing to try to elucidate the molecular mechanism
of the evolution of complex biopolymers having different functions,
and RuBP carboxylase is of particular interest from the standpoint
of the origin of autotrophy (2-5).   One can speculate that during
the developmental stage of RuBP carboxylase the catalytic subunit
A evolved first, followed by subunit B.   However, the fact that
the carboxylase from the prokaryotic bacterium Chromatium vinosum
resembles the plant-type enzyme and contains subunit B (6, 7) makes
it evident that the small subunit molecule evolved before the appearance of eukaryotic organisms.   Establishment of the primary
structures of RuBP carboxylases from divergent origins for comparison is crucial to any discussion of the phylogenetic development of autotrophic organisms.   Structural investigations have been
initiated for determining the primary sequences of subunit B from
several higher plant carboxylases (8), but knowledge is still frag-
mentary.   Although determination of the total amino acid sequences
of subunit A will be difficult because of its large molecular size,
it is noteworthy that Hartman and associates (9, 10) have found
lysine in the primary sequence at the RuBP-binding site of the
enzyme from both spinach and Rhodospirillum rubrum.
        Sizable amounts of data have accumulated regarding the qua-
ternary structures, amino acid compositions, and immunochemical
resemblances among RuBP carboxylases from various sources, which
may provide some useful information about the molecular evolution
of this enzyme.   Experimental results from different laboratories
have indicated that the amino acid composition of subunit A is
similar for enzymes of widely different origins, but that of subunit B is more variable.   To assess the interspecific structural
homologies for subunit A and dissimilarities for subunit B, statis-
tical SΔQ values were calculated from published analytical data by
the method of Marchalonis and Weltman (11):

$$S\Delta Q \equiv \sum_j (X_{i,j} - X_{k,j})^2 \tag{1}$$

where i and k denote the proteins being compared, and $X_{i,j}$ is the
mole % of amino acid j in protein i.   A close relationship between
SΔQ value and difference in amino acid sequences was reported (11)
for hemoglobin (141 to 146 residues), the light chain of immuno-
globin (213 to 221 residues), and cytochrome c (103 to 104 residues)
of vertebrate origin.   However, our calculation (4) from Dayhoff's
amino acid sequence data (12) indicated that SΔQ values are not
directly proportional to % differences in amino acid sequence.
From available data, the ratio f,

$$f \equiv D/S\Delta Q \tag{2}$$

where D = % sequence differences, was plotted as a function of the
number of amino acid residues in protein molecules, and a hyperbolic
curve was obtained. By using f values from this calibration curve,
one can calculate $D_{S\Delta Q}$ (% probable sequence differences) for protein
molecules of unknown sequence:

$$D_{S\Delta Q} = f \cdot S\Delta Q \quad . \tag{3}$$

Although the number of amino acid residues in subunit A or B varies
slightly for RuBP carboxylase molecules from various sources, it
does not vary drastically because the molecular weights are nearly
the same. We obtained f values of 2.1 for subunit A and 0.66 for
subunit B, respectively. In Table 1, $S\Delta Q$ and $D_{S\Delta Q}$ values calculated
by this technique are presented. Average $D_{S\Delta Q}$ values are quite
small (16%) for subunit A among six eukaryotic origins, whereas
values between prokaryotes and eukaryotes are much larger. RuBP
carboxylase from Rhodospirillum rubrum appears to be unique; $D_{S\Delta Q}$
values are large between R. rubrum and other organisms, reflecting
the characteristic structure of the R. rubrum enzyme molecule.
Tabita and McFadden (13) reported this structure as being dimeric
in subunit A (A_2) (MW of monomer 5.2×10^4) and lacking the small
subunit (B); unpublished results of our own analytical study are
in agreement.

Table 1

Comparison of $S\Delta Q$ and $D_{S\Delta Q}$ Values of Subunit A of RuBP Carboxylase
From Different Sources

Data from Takabe (5). Values of $S\Delta Q$ and $D_{S\Delta Q}$ between Rhodospirillum rubrum and other organisms are underlined.

	SΔQ	$D_{S\Delta Q}$								
		I	II	III	IV	V	VI	VII	VIII	IX
	PROKARYOTES									
I	Rhodospirillum rubrum		59	76	42	38	61	53	57	82
II	Chromatium vinosum	28		53	44	19	36	34	32	32
III	Anabaena cylindrica	36	25		36	42	25	23	25	34
	EUKARYOTES									
IV	Chlorella ellipsoidea	20	21	17		23	13	34	25	27
V	Chlamydonomas reinhardi	18	9	20	11		17	19	13	19
VI	Nicotiana tabacum	29	17	12	6	8		17	8	11
VII	Hordeum vulgare	25	16	11	16	9	8		4	21
VIII	Beta vulgaris	27	15	12	12	6	4	2		11
IX	Spinacia oleracea	39	15	16	13	9	5	10	5	

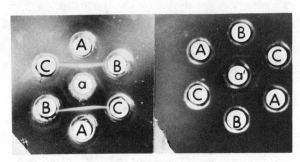 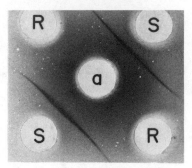

a: anti-[A](spinach) a′: anti- [A](spinach)   a: anti- [N] (spinach)

A: spinach (N)   A: <u>Chromatium</u>(N)   S: spinach (N)

B: <u>Chromatium</u>(A$_8$)   B: <u>Chromatium</u>(A$_8$)   R: <u>R.rubrum</u>(N)

C: <u>Chromatium</u>(B)   C: <u>Chromatium</u>(B)

Figure 1. Ouchterlony double immunodiffusion experiments. Left: Cross-reaction between anti-[A] (spinach) serum and spinach RuBP carboxylase and <u>Chromatium</u> RuBP carboxylase and its subunits (A$_8$ or B). After Takabe and Akazawa (6). Right: Cross-reaction between anti-[N] (spinach) serum and spinach and <u>R. rubrum</u> RuBP carboxylases. From Kobayashi and Akazawa (unpublished).

RuBP carboxylase from <u>Chromatium</u> is analogous to that from spinach in many respects, and the RuBP carboxylase/oxygenase activities displayed by purified preparations of it are strongly inhibited by rabbit antiserum against the spinach native enzyme, anti-[N], or its catalytic oligomer (A$_8$), anti-[A] (6). Results of the immunological double diffusion test (Figure 1) indicate close structural homologies between the catalytic subunit A segments of the spinach and <u>Chromatium</u> enzymes (6). In contrast, the activity of RuBP carboxylase from <u>R. rubrum</u> was totally indifferent to both anti-[N] and anti-[A], and showed no immunological cross-reactivity with the spinach enzyme (14) (Figure 1).

We have discussed previously that the catalytically important subunit A core is genetically conserved during the evolutionary history of the enzyme molecules, whereas conceivably the rate of mutational amino acid substitution in the noncatalytic subunit B moiety is large (5). On the same grounds it can be theorized that the <u>R. rubrum</u> carboxylase molecule, having less structural constraint because it lacks intersubunit interaction, should be subject to greater mutational change. In addition, the functionally less important role of RuBP carboxylase in carbon assimilation of bacteria growing principally under heterotrophic conditions (see below) should increase the tendency toward rapid molecular mutation. Conversely, as to whether molecular structure A$_2$ of <u>R. rubrum</u> carboxylase is an artifact of the enzyme isolation procedure, one can

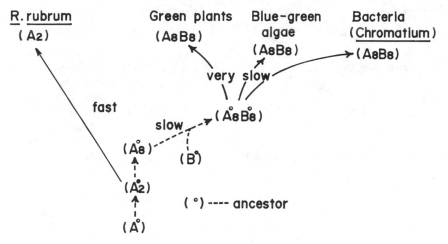

Figure 2.   Molecular evolution of RuBP carboxylase.

consider that this enzyme exists and functions in vivo in such simple macromolecular form because of its rapid mutation rate. Figure
2 shows a hypothetical scheme for the molecular evolution of RuBP
carboxylases from the ancestor polypeptide molecule of the large
subunit (A^0), in which the R. rubrum carboxylase molecule has a
unique situation. Various research approaches, such as comparative
and systematic cataloging of the ribosomal RNA species of bacteria,
as done by Fox et al. (15), may provide more definite information.

B.   FUNCTIONAL ROLE OF RuBP CARBOXYLASE/OXYGENASE
IN CARBON ASSIMILATION OF C. vinosum AND R. rubrum

Experimental evidence from various lines of research indicates
operation of the Calvin-Benson cycle in the carbon assimilation of
Chromatium (16-18). It has been shown, however, that the primary
products of the CO_2 fixation reaction are 3-PGA and aspartic acid;
this, in combination with experimental results of the $\delta^{13}C$ determination, led to proposal of the double carboxylation pathway (19):

$$RuBP \xrightarrow{CO_2} 3\text{-PGA} \rightarrow PEP \xrightarrow{CO_2} oxalacetate \rightarrow aspartate \ . \qquad (4)$$

The results of our own analytical studies on the time-dependent
incorporation of $^{14}CO_2$ into various compounds are consistent with
such a mechanism (18). The inhibitory effect of KCN on photosynthetic CO_2 fixation by intact bacterial cells also supports an important role of RuBP carboxylase in carbon assimilation.
The mechanism of the photosynthetic carbon pathway in R. rubrum,
on the other hand, has been a matter of debate for many years (20-
23). R. rubrum can grow both autotrophically and heterotrophically,

Table 2

$K_m(CO_2)$, $K_m(O_2)$, and V_{max} Values of RuBP Carboxylase/
Oxygenase Reactions by Purified Enzyme Preparations
From R. rubrum and Spinach

Unpublished data of Kobayashi and Akazawa. Enzyme
assay based on method of Lorimer et al. (60).

	Carboxylase		Oxygenase	
	$K_m(CO_2)$ (μM)	V_{max} (μmoles CO_2 per mg protein per min)	$K_m(O_2)$ (μM)	V_{max} (μmole O_2 per mg protein per min)
R. rubrum	1000	2.2	350	0.8
Spinach	17	1.1	200	0.4

and many studies have been done with cells cultured on malate as
the carbon source. Photosynthesis by malate-grown R. rubrum cells
is totally indifferent to KCN, whereas that by butyrate-grown cells
is appreciably inhibited (47%) by 1 mM KCN (unpublished results)
although less so than that by Chromatium cells (cf. ref. 18).
These results can be interpreted on the basis of the Calvin-Benson
cycle having a relatively minor role in carbon assimilation. In
addition, purified RuBP carboxylase from butyrate-grown R. rubrum
cells has been found to have a very large $K_m(CO_2)$ value (1 mM) in
spite of its high specific carboxylase activity; this kinetic property is in sharp contrast to that of the enzyme from spinach or
Chromatium (Table 2).
     Since the discovery of the RuBP oxygenase activity of the car-
boxylase molecule isolated from various sources (cf. ref. 1), the
role of this activity in photorespiration has been increasingly
substantiated. It is now generally agreed that the competitive
nature of CO_2 and O_2 in the RuBP carboxylase/oxygenase reactions
indicates that these two enzyme reactions take place at the same
catalytic site of the enzyme molecule and control photosynthesis
and photorespiration (24-28), and that the basic mechanism underlying the Warburg effect is probably governed by this bifunctional
enzyme reaction. Lorimer et al. (29) have proposed that the photosynthetic carbon oxidation cycle is the key mechanism controlling
the carbon assimilation reaction of photosynthetic organisms in
photorespiratory environments.
     The O_2 sensitivity of photosynthesis by Chromatium, particu-
larly the competitive inhibitory effect of O_2 with respect to CO_2,
is basically analogous to that in green plants (18). In contrast,
the O_2 response of the photosynthetic reaction by R. rubrum is
quite different. The O_2 effect was determined with R. rubrum cells

grown on either malate or butyrate as the carbon source. The CO_2
fixation reaction of both types of cells was drastically suppressed
in a 100% O_2 atmosphere, and in the case of malate-grown cells the
inhibition was irreversible, although in the case of butyrate-grown
cells partial restoration occurred upon substitution of N_2 for O_2
(unpublished results). These findings, together with the pattern
of $^{14}CO_2$ incorporation, make it appear likely that the TCA cycle
and the reductive carboxylic acid cycle, which presumably plays a
role in the carbon metabolism of Chlorobium thiosulfatophilum and
is sensitive to O_2 (18), operate in the photosynthetic carbon assimilation of R. rubrum (Figure 3). It is noteworthy that analyses of
isotope discrimination, by Quandt et al. (30), showed the $\delta^{13}C$ value
for R. rubrum to be much smaller than that for Chromatium.

Another important facet of the photorespiratory process is
the formation of glycolate, which is drained off from the inter-
mediate(s) of the Calvin-Benson cycle, and its subsequent transformation (the glycolate pathway) (cf. ref. 27). The nature of the
enzymic mechanism responsible for glycolate biosynthesis has been

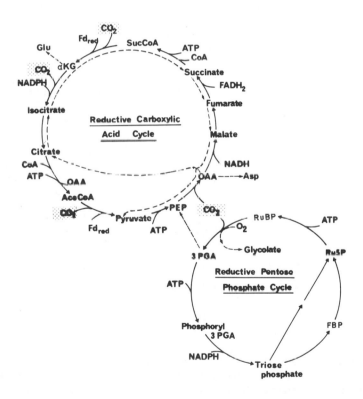

Figure 3. Illustrative representation of reductive pentose phosphate (Calvin-Benson) cycle linked to reductive carboxylic acid
cycle. After Takabe and Akazawa (18).

long argued, but recent experiments support the idea that the RuBP
oxygenase reaction is the principal one (28). The enrichment of
^{18}O in the carboxyl carbon of glycolate produced by <u>Chromatium</u>
and <u>Chlorella</u> cells (31) as well as in isolated spinach chloroplasts (32) exposed to $^{18}O_2$ can be satisfactorily explained by this
enzymic mechanism. It is therefore likely that in a photorespiratory environment phosphoglycolate derived from RuBP by the RuBP
oxygenase reaction will be enzymically hydrolyzed to glycolate and
further metabolized by the glycolate pathway. RuBP carboxylase
from <u>R. rubrum</u> was shown to have oxygenase activity (Table 2). We
also found that detectable amounts of $^{14}CO_2$ are incorporated into
glycolate produced by the cells under air (21% O_2) or 100% O_2, but,
in contrast to the original report by Anderson and Fuller (33), we
found no glycolate production under anaerobic conditions (unpublished results). It remains to be established whether or not RuBP
oxygenase alone is responsible for the formation of glycolate in
<u>R. rubrum</u>, and we hope to do this by ^{18}O enrichment experiments.

C.   REGULATORY ROLE OF SUBUNIT B IN RuBP CARBOXYLASE

Recent investigations by Lorimer et al. (34) have disclosed
the mechanism of activation of the RuBP carboxylase molecule:

$$\text{Enzyme} + CO_2 \rightarrow \text{enzyme-}CO_2 + Mg^{2+} \rightarrow \text{enzyme-}CO_2\text{-}Mg^{2+} \quad . \quad (5)$$
$$\text{(inactive)} \qquad\qquad\qquad\qquad\qquad\qquad \text{(active)}$$

It is now evident that there are two separate steps in the RuBP
carboxylase/oxygenase reactions, i.e., activation and catalysis,
and the kinetically active enzyme molecule, having low $K_m(CO_2)$,
exists in the form of enzyme-CO_2-Mg^{2+}. In another reaction model,
Laing and Christeller (35), using the soybean enzyme, suggested
that Mg^{2+} is not directly involved in the enzyme catalysis:

$$\text{Enzyme} \xrightarrow{CO_2,\ Mg^{2+}} \text{enzyme-}CO_2\text{-}Mg^{2+} \xrightarrow{RuBP,\ CO_2} \qquad (6)$$
$$\text{(inactive)} \qquad\qquad \text{(active)}$$

$$(\text{enzyme-}CO_2\text{-}Mg^{2+})\text{-}RuBP\text{-}CO_2 \rightarrow \text{enzyme} + \text{product} \quad .$$

Experimental results obtained by Bahr and Jensen (36) with intact
chloroplasts suggest that activation of the RuBP carboxylase molecule, which depends on Mg^{2+}, CO_2 partial pressure, and pH, constitutes an important part in the regulation of photosynthesis
<u>in vivo</u>.
        The requirement for Mg^{2+} in the RuBP carboxylase reaction
observed in the initial research by Weissbach et al. (37) has
been constantly confirmed by later workers, and the mode of action
of Mg^{2+} appears to be partly ascribable to the enzyme activation
process described above. However, the exact mechanism of Mg^{2+}
interaction with the enzyme molecule, particularly at the level of

the quaternary enzyme structure, remains unknown. Experimental
data have accumulated showing that subunit B is probably involved
in the Mg^{2+} effect on the enzyme reaction. Increasing Mg^{2+} concentrations in the assay mixture from 0 to 10 mM not only enhanced
the carboxylase activities of enzymes from different sources but,
more notably, shifted the optimal pH of the reaction from alkaline
(pH 9.0) to near neutral (about pH 7.5) (38). This can be observed by using different types of buffers and even by assaying
enzyme activity after activation.

The pH shift is not seen, however, in the enzyme reaction of
the catalytic oligomer, A_8, isolated from either spinach (39) or
Chromatium enzyme (40), depleted of subunit B. Furthermore, it
cannot be detected in the R. rubrum enzyme (A_2) (14). The addition
of anti-[B] to the native spinach enzyme prevents the pH shift,
optimal pH remaining on the alkaline side (pH 8.5) regardless of
the presence or absence of Mg^{2+} (41). The reconstituted enzyme
molecule (with subunit B added to the catalytic oligomer, A_8) again
shows the Mg^{2+}-induced shift in optimal pH (42). The possible
significance of this phenomenon, observed in vitro, in the control
mechanism of photosynthesis in vivo has been discussed in relation
to the light-induced uptake of Mg^{2+} accompanied by an H^+ influx in
the thylakoid membranes of chloroplasts (cf. ref. 1). Although
the exact nature of the Mg^{2+}-dependent shift in optimal pH at the
molecular level is not known, a recent finding by Schloss and
Hartman (43) on affinity labeling of the active site is of interest.
They reported that in the presence of Mg^{2+}, the binding site of N-
bromoacetylethanolamine phosphate in spinach RuBP carboxylase appears to change from cysteinyl to lysyl residues.

Some implication regarding the possible role of subunit B in
the enzyme activation process can be drawn from the following experimental findings. As shown in Table 3, the $K_m(CO_2)$ of RuBP carboxylase from R. rubrum, which lacks subunit B, is large (1 mM).

Table 3

$K_m(CO_2)$ Values (μM) of RuBP Carboxylase in Various Forms
Determined With or Without Activation

Unpublished data of Kobayashi and Akazawa. Enzyme assay
based on method of Lorimer et al. (60).

| | Spinach | | | | R. rubrum |
	Native (A_8B_8)	A_8	Native + anti-[B]	A_8 + [B]	Native (A_2)
Activation	17	140	125	70	1000
No activation	100	-	160	-	3900

Treatment of spinach RuBP carboxylase with anti-[B] caused the $K_m(CO_2)$ to increase appreciably. The $K_m(CO_2)$ of partially recon-stituted enzyme was significantly lower than that of A_8. However, since the catalytic oligomer (A_8) of spinach RuBP carboxylase or R. rubrum carboxylase (A_2) becomes activated upon incubation with Mg^{2+} and CO_2, apparently both Mg^{2+} and CO_2 bind to subunit A. Therefore, one plausible role for subunit B could be enhancement of the activation process after the reaction shown in equation 5 or 6 (cf. ref. 44). The finding that enzymically inactive subunit B isolated from spinach enzyme emits measurable fluorescence upon incubation with Mg^{2+}, independent of CO_2, suggests a possible mechanism, i.e., that the conformational change in the small sub-unit induced by Mg^{2+} binding is related to this enhancement (unpublished results).

D.   GLYCOLATE METABOLISM IN CHROMATIUM VINOSUM:\nPROTOTYPE OF PHOTORESPIRATION

1.   Warburg Effect

Understanding the mechanism of the formation and extracellular excretion of glycolate by the obligate photoanaerobe Chromatium in O_2-containing atmospheres would help clarify the origin of photo-respiration. Hurlbert (45) first reported that Chromatium can survive in an O_2-containing atmosphere for an extended period. He also noted that, although a truly aerobic atmospheric environment is foreign to this bacterium, molecular oxygen is frequently en-countered in its ecosystem, e.g., in air-agitated sewage or oceanic mud. It is thus conceivable that the bacterium has acquired a potential adaptability towards O_2 in its natural environment, in which the RuBP carboxylase molecule with its oxygenase activity has played a crucial role. Upon exposure to an O_2 atmosphere, photosynthesizing cells of Chromatium produce large amounts of glycolate and glycine and excrete them extracellularly (46). As described above, Chromatium RuBP carboxylase exhibits oxygenase activity (6, 7, 47), and experiments on ^{18}O-labeling of the glyco-late carboxyl group give results in accord with RuBP oxygenase having the principal role in glycolate production (31).

Lorimer and Andrews (48) proposed that oxygenolytic cleavage of the enzyme-bound enolate anion, liberating phosphoglycolate in the presence of O_2, is inherent to the RuBP carboxylase molecule. It can be postulated that this bifunctional enzyme reaction may serve as a safety mechanism protecting the bacterial cells against the O_2-containing environment. During short-term incubation, O_2 was found to exhibit a competitive type of inhibitory effect on photosynthetic CO_2 fixation by autotrophically grown bacterial cells, accompanied by the formation of glycolate (18). However, in bacteria grown on acetate and malate as carbon source, O_2 has a type of in-hibitory effect that is noncompetitive, and the true photosynthesis

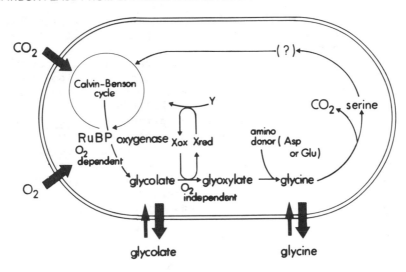

Figure 4. Glycolate pathway in <u>Chromatium</u>, after Asami et al. (50), modified on the basis of recent experiments (Asami and Akazawa, unpublished) dealing with the glycine-serine conversion reaction with concomitant evolution of CO_2. Cyclic return of metabolite(s) to the Calvin-Benson cycle is hypothetical.

rate is much lower (49). For autotrophically grown cells in an O_2-containing atmosphere, the larger the O_2/CO_2 ratio, the more marked is glycolate production, the overall situation being analogous to that in green plants and in green algae. These results also support the concept that the Warburg effect is a ubiquitous phenomenon, occurring in photosynthetic organisms, including prokaryotes, in which the Calvin-Benson cycle has a predominant function.

The general features of glycolate metabolism in <u>Chromatium</u> are also similar to those in green plants: glycolate is transformed to glycine via glyoxylate, and the subsequent metabolism of glycine is accompanied by the production of CO_2 (Figure 4) (49, 50). Although the overall mechanism in bacteria is similar to the glycolate pathway, it has some unique aspects: (i) Glycolate oxidation is known to be independent of O_2, although the molecular identity of the electron acceptor is unknown. (ii) Most of the glycine molecules produced are excreted from the bacterial cells. Our recent studies (unpublished) on the formation of glycolate and glycine under a diminished O_2 atmosphere (0.05 mM) showed no measurable excretion of glycolate and glycine, even though $^{14}CO_2$ is incorporated into glycolate and glycine if they are added to the incubation medium. Our results indicate that the capacity for glycolate metabolism in <u>Chromatium</u> cells is much smaller than in green plants, and that excess glycolate and glycine molecules produced in a high O_2 atmosphere (beyond the intracellular pool size) are inevitably excreted. Note that in green plants the whole process of glycolate metabolism proceeds by multi-organellar channeling (chloroplasts, peroxisomes, and mitochondria), and partial recovery of a carbon compound (3-PGA) by chloroplasts after the turn in the glycolate pathway constitutes a mechanism for conserving the carbon skeleton (29). Thus, one can imagine that in the prokaryotic photoanaerobe, Chromatium, only a part of the glycolate pathway inside the cell is a prototype of the photorespiratory system, and that the mechanism developed evolutionarily in aerobic green plant tissues (C_3 plants) effectively adapting to the O_2-containing environment (cf. Figure 4). Naturally the mechanism has been further advanced and elaborated in C_4 plants, both anatomically (Kranz structure) and enzymically.

2.   Irreversible Inhibitory Effect of O_2

Apart from the reversible O_2 inhibition of photosynthesis (Warburg effect), there is an irreversible O_2 inhibition, sometimes called photodynamic action (51, 52), which is thought to be closely related to deterioration of membrane structure in the photosynthetic apparatus. Recent biochemical experiments have provided evidence that the active O_2 species produced intracellularly (O_2^-, $^1O_2$, ·OH, and H_2O_2) have a toxic effect on several biological systems (52, 53). One can theorize that production of these molecular species, coupled with photosystem I, in a light/O_2 environment, may eventually result in photooxidative damage to photosynthesis systems:

$$O_2 + e^- \rightarrow O_2^-, \tag{7}$$

$$2O_2^- + 2H^+ \rightarrow {}^1O_2 + H_2O_2 . \tag{8}$$

Generally, however, chlorophyllous aerobic photoautotrophs (green plants) are provided with protective devices against harmful effects of active O_2 molecules, and conspicuous photoinhibitory effects are not observed. Most of the O_2^- and H_2O_2 molecules are removed by superoxide dismutase and catalase, respectively, and carotenoids are known to scavenge $^1O_2$ (54). It has been established that the herbicidal action of methylviologen on foliage is due to excessive production of O_2^- in chloroplasts under light/O_2 conditions (55). Thus, detailed analysis of O_2 toxicity in the obligate anaerobic bacterium, Chromatium, should provide an ideal model for elucidating the mechanism underlying photodynamic action.

Exposure of Chromatium cells to strong light and prolonged bubbling with O_2 in the absence of $NaHCO_3$ results in gradual loss of photosynthetic CO_2 fixation (Figure 5), but the presence of $NaHCO_3$ markedly delays the decline. Under dark/O_2 or light/N_2 conditions, no such effect was observed. Before the drop in CO_2

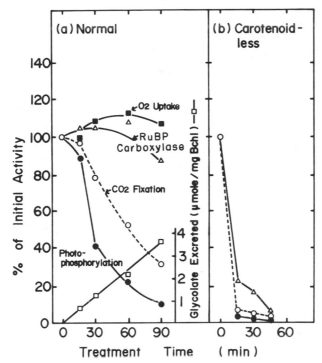

Figure 5.   Photosynthetic CO_2 fixation, RuBP carboxylase activity,
O_2 uptake, and photophosphorylation of normal and carotenoidless
Chromatium cells subjected to light/O_2 treatment. Portions of the
treated cells were withdrawn for use in measurements. For normal
cells, glycolate formation in the extracellular fraction was also
determined. Data from Asami and Akazawa (unpublished).

fixation, light/O_2 caused inactivation of photophosphorylation according to an antisigmoidal curve. RuBP carboxylase activity was
not much affected. Note that in carotenoidless Chromatium cells
prepared by the method of Fuller and Anderson (56), the photooxidative damage is almost instantaneous.

Photophosphorylation by bacterial chromatophores was similarly
impaired by light/air treatment, but the inclusion of either Tiron
or α-tocopherol had a marked preventive effect (Table 4). The
overall results suggest that O_2^- or $^1O_2$ is the actual agent involved in the photooxidative damage both in vitro and in vivo.
The inhibitory effect of light/O_2 was prominently stimulated by
simultaneous addition of either KCN (which inhibits RuBP carboxylase)
or methylviologen (which produces O_2^-) to bacterial suspension medium
without $NaHCO_3$, but addition of exogenous $NaHCO_3$ to the system effectively counteracted the inhibition. In green plants one role of
the glycolate pathway is thought to be maintenance of intracellular
CO_2 tension, which serves as a supplementary protective device in

Table 4

Photosensitivity of Photophosphorylation by
Chromatium Chromatophores

Unpublished data of Asami and Akazawa. Chromatophores
prepared from normal Chromatium cells were treated with
light (60 klux) for 13 min. Activity of air/dark treated chromatophores was 6.9 μmoles ATP formed per mg Bchl
per hr.

Treatment conditions of chromatophores	Photophosphorylation (%)
Air/dark	100
N₂/light	48.8
Air/light	31.1
Air/light + 10 mM Tiron	68.8
Air/light + 0.2 mM α-tocopherol	54.2

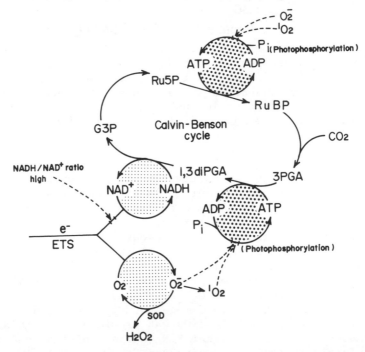

Figure 6. Hypothetical mechanism of "cyclic and amplified" type of
photooxidative damage to photosynthetic activities of Chromatium
cells under light/O_2 conditions.

photorespiratory environments. The CO_2 evolved in <u>Chromatium</u> cells
during the glycine-serine transformation step may also participate
in a similar function. It should be recognized, however, that,
even with a high level of exogenously supplied $NaHCO_3$, photoanaerobic bacteria cannot withstand photooxidative damage because they
lack the basic machinery for effective protection of the photophosphorylation site from photodynamic action. Björkman (57, 58)
proposed that photorespiration represents a mechanism for dissi-
pating excess light energy under conditions of CO_2 deficiency in
green plants. The experimental results described above lead us
to agree. This basic mechanism is most clearly shown in cells of
the photoanaerobic bacterium <u>Chromatium</u> when placed in a light/O_2
environment. The primary target of the active O_2 molecules (O_2^-
and $^1O_2$) produced in competition with the reduction of NAD is the
photophosphorylation site, and the lowered rate of ATP formation
may slow down the turnover of the Calvin-Benson cycle. The ensuing
higher NADH/NAD ratio will accelerate the formation of O_2^- and $^1O_2$
in turn. Such a "cyclic and amplified" type of mechanism for
photooxidative damage is sketched in Figure 6.

    Finally it must be emphasized that, in spite of the dominant
role of the RuBP oxygenase reaction in the production of glycolate
in <u>Chromatium</u> cells, a possible role of a transketolase mechanism
(with O_2^- detoxifying O_2 in photorespiratory environments) should
not be overlooked (59). The results in Figure 5 show that, under
conditions causing marked photooxidative damage, glycolate formation
continues to rise linearly. It is likely that the glycolate oxidation pathway is also damaged by O_2^- or $^1O_2$ molecules.

REFERENCES

1. Akazawa, T., in Photosynthesis II, Encyclopedia of Plant
   Physiology, M. Gibbs and E. Latzko, Editors, Springer,
   Berlin, in press, 1978.
2. McFadden, B. A., Bacteriol. Rev. 37, 289-319 (1973).
3. McFadden, B. A. and Tabita, F. R., BioSystems 6, 93-112 (1974).
4. Takabe, T. and Akazawa, T., Plant Cell Physiol. 16, 1049-60
   (1975).
5. Takabe, T., in Molecular Evolution and Protein Morphism,
   pp. 296-312, M. Kimura, Editor, Natl. Inst. Genetics, Mishima,
   1978.
6. Takabe, T. and Akazawa, T., Biochemistry 14, 46-50 (1975).
7. Takabe, T. and Akazawa, T., Arch. Biochem. Biophys. 169, 686-
   94 (1975).
8. Poulsen, C., Ströbaek, S., Haslett, B. G., in Genetics and
   Biogenesis of Chloroplasts and Mitochondria, pp. 17-24,
   Th. Bücher, Editor, Elsevier/North Holland Biomedical Press,
   Amsterdam, 1976.

9. Norton, I. L., Welch, M. H., and Hartman, F. C., J. Biol. Chem. 250, 8062-8 (1975).
10. Schloss, J. V. and Hartman, F. C., Biochem. Biophys. Res. Commun. 75, 320-8 (1977).
11. Marchalonis, J. J. and Weltman, J. K., Comp. Biochem. Biophys. 38B, 609-25 (1971).
12. Dayhoff, M. O., in Atlas of Protein Sequence and Structure, Vol. 5, National Biomedical Research Foundation, Washington, DC, 1972.
13. Tabita, F. R. and McFadden, B. A., J. Biol. Chem. 249, 3459-64 (1974).
14. Akazawa, T., Sato, K., and Sugiyama, T., Arch. Biochem. Biophys. 132, 255-61 (1969).
15. Fox, G. E., Pechman, K. R., and Woese, C. R., Int. J. Syst. Bacteriol. 27, 44-57 (1977).
16. Fuller, R. C., Smillie, R. M., Sisler, E. C., and Kornberg, H. L., J. Biol. Chem. 236, 2140-9 (1961).
17. Gibson, J. and Hart, B. A., Biochemistry 8, 2737-41 (1969).
18. Takabe, T. and Akazawa, T., Plant Cell Physiol. 18, 753-65 (1977).
19. Wong, W., Sackett, W. M., and Benedict, C. R., Plant Physiol. 55, 475-9 (1975).
20. Gest, H., Ormerod, J. G., and Ormerod, K. S., Arch. Biochem. Biophys. 97, 21-3 (1962).
21. Anderson, L. and Fuller, R. C., Plant Physiol. 42, 487-90 (1967).
22. Anderson, L. and Fuller, R. C., Plant Physiol. 42, 497-502 (1967).
23. Slater, J. H. and Morris, I., Arch. Microbiol. 88, 213-23 (1973).
24. Bahr, J. T. and Jensen, R. G., Arch. Biochem. Biophys. 164, 408-13 (1974).
25. Andrews, T. J., Badger, M. R., and Lorimer, G. H., Arch. Biochem. Biophys. 171, 93-103 (1975).
26. Laing, W. A., Ogren, W. L., and Hageman, R. H., Biochemistry 14, 2269-75 (1975).
27. Chollet, R. and Ogren, W. L., Bot. Rev. 41, 137-79 (1975).
28. Chollet, R., Trends Biochem. Sci. 2, 155-9 (1977).
29. Lorimer, G. H., Woo, K. C., Berry, J. A., and Osmond, C. B., in Proc. 4th Int. Congr. Photosynthesis, pp. 311-22, D. O. Hall et al., Editors, Biochem. Soc., London, 1977.
30. Quandt, L., Gottschalk, G., Ziegler, H., and Stichler, W., FEBS Lett. 1, 125-8 (1977).
31. Lorimer, G. H., Osmond, C. B., Akazawa, T., and Asami, S., Arch. Biochem. Biophys. 185, 49-56 (1978).
32. Lorimer, G. H., Krause, G. H., and Berry, J. A., FEBS Lett. 78, 199-202 (1977).
33. Anderson, L. and Fuller, R. C., Biochim. Biophys. Acta 131, 198-201 (1967).

34. Lorimer, G. H., Badger, M. R., and Andrews, T. J., Biochemistry 15, 529-36 (1976).
35. Laing, W. A. and Christeller, J. T., Biochem. J. 159, 563-70 (1976).
36. Bahr, J. T. and Jensen, R. G., Arch. Biochem. Biophys. 185, 39-48 (1978).
37. Weissbach, A., Horecker, B. L., and Hurwitz, J., J. Biol. Chem. 218, 795-810 (1956).
38. Sugiyama, T., Nakayama, N., and Akazawa, T., Arch. Biochem. Biophys. 126, 737-45 (1968).
39. Nishimura, M. and Akazawa, T., Biochem. Biophys. Res. Commun. 54, 842-8 (1973).
40. Takabe, T. and Akazawa, T., Arch. Biochem. Biophys. 157, 303-8 (1973).
41. Nishimura, M. and Akazawa, T., Biochemistry 13, 2277-81 (1974).
42. Nishimura, M. and Akazawa, T., Biochem. Biophys. Res. Commun. 59, 584-90 (1974).
43. Schloss, J. V. and Hartman, F. C., Biochem. Biophys. Res. Commun. 77, 230-6 (1977).
44. Akazawa, T., pp. 447-56 in ref. 29.
45. Hurlbert, R. E., J. Bacteriol. 93, 1346-52 (1967).
46. Asami, S. and Akazawa, T., Plant Cell Physiol. 15, 571-6 (1974).
47. Takabe, T. and Akazawa, T., Biochem. Biophys. Res. Commun. 53, 1173-9 (1973).
48. Lorimer, G. H. and Andrews, T. J., Nature London 243, 359-60 (1973).
49. Asami, S. and Akazawa, T., Plant Cell Physiol. 16, 631-42 (1975).
50. Asami, S., Takabe, T., and Akazawa, T., Plant Cell Physiol. 18, 149-59 (1977).
51. Spikes, J. D. and Straight, R., Annu. Rev. Phys. Chem. 18, 409 (1967).
52. Asada, K., Takahashi, M., Tanaka, K., and Nakano, Y., in Biochemical and Medical Aspects of Active Oxygen, pp. 45-63, O. Hayaishi and K. Asada, Editors, U. of Tokyo Press, 1977.
53. Morris, J. G., Trends Biochem. Sci. 2, 81-4 (1977).
54. Mathews-Roth, M. M., Wilson, T., Fujimori, E., and Krinsky, N. I., Photochem. Photobiol. 19, 217 (1974).
55. Asada, K., Kiso, K., and Yoshikawa, K., J. Biol. Chem. 249, 2175-81 (1974).
56. Fuller, R. C. and Anderson, I. C., Nature London 181, 252-4 (1958).
57. Osmond, C. B. and Björkman, O., Carnegie Inst. Washington Yearb. 71, 141-8 (1972).
58. Björkman, O., in Photobiology, pp. 1-63, A. Geise, Editor, Academic Press, New York, 1973.
59. Asami, S. and Akazawa, T., Biochemistry 16, 2202-7 (1977).
60. Lorimer, G. H., Badger, M. R., and Andrews, T. J., Anal. Biochem. 78, 66-75 (1977).

DISCUSSION

JENSEN: What is the role of the small subunit? Do you assume that the small subunit regulates the shift in pH optimum at
various magnesium concentrations and also the apparent change in
$K_m(CO_2)$?

AKAZAWA: Yes.

JENSEN: Do you mean that magnesium is still bound to the large
subunit but could also be bound to the small subunit?

AKAZAWA: We have no experimental evidence, but our kinetic
data suggest that the cycle of binding for both CO_2 and magnesium
is to the large subunit. We think the small subunit is involved
with the activation process but we have no clear idea of its function.

JENSEN: Do organisms having only a large subunit also have a
high $K_m(CO_2)$?

AKAZAWA: Yes, as far as our experiments show. Only enzyme
from R. rubrum was extensively studied. The large subunit of
Chromatium is not as active as the native enzyme, only partially
retaining the original activity.

ACTIVE SITE OF RIBULOSE 1,5-BISPHOSPHATE CARBOXYLASE/OXYGENASE[*]

Christian Paech, Stephen D. McCurry, John Pierce, and N. E. Tolbert

Department of Biochemistry, Michigan State University
East Lansing, Michigan 48824

The properties, distribution, biogenesis, function, and regulation of ribulose bisphosphate carboxylase/oxygenase, as described elsewhere in this Symposium, are mainly integrated around the mechanism of action of the carboxylase and oxygenase reactions. Recently, interest has been stimulated by the discovery of the oxygenase reaction (1, 2) and its role in the glycolate pathway of photorespiration (3), and by the desirability of differential regulation of the carboxylase and oxygenase activities, if possible, in favor of the former in order to increase photosynthetic productivity. A prerequisite for such regulation is a thorough knowledge of the relationships between structure and function in this protein.

RuBP carboxylase, in various molecular forms (4), is present in all photosynthetic organisms. To date, most of the work has been done either with the crystalline enzyme from tobacco leaves or with an enzyme from spinach whose ease of purification to homogeneity in large amounts has often been reported. Purified preparations can be obtained in one day by precipitation from the homogenate with polyethylene glycol followed by DEAE cellulose chromatography (N.P. Hall, unpublished). However, the complexity of this large protein and the inability to dissociate it into active subunits have been intimidating, and unfortunately little is known about RuBP carboxylase/oxygenase relative to other key enzymes.

---

[*]Supported in part by NSF grant PCM 7815891.
Abbreviations: RuBP, ribulose 1,5-bisphosphate; XuBP, xylulose 1,5-bisphosphate; xylitol BP, xylitol 1,5-bisphosphate; FBP, fructose 1,6-bisphosphate; CRBP, 2-carboxy-D-ribitol 1,5-bisphosphate; bromodihydroxybutanone BP, 3-bromo-1,4-dihydroxy-2-butanone 1,4-bisphosphate.

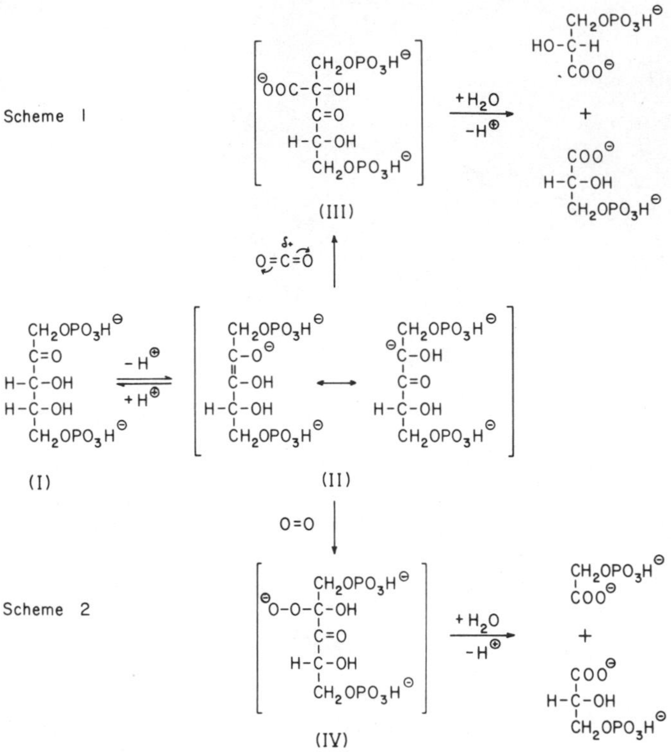

Scheme 1

$$
\left[\begin{array}{l}
CH_2OPO_3H^{\ominus} \\
{}^{\ominus}OOC-C-OH \\
C=O \\
H-C-OH \\
CH_2OPO_3H^{\ominus}
\end{array}\right]
\xrightarrow[-H^{\oplus}]{+H_2O}
$$

(III)

$$
\begin{array}{l}
CH_2OPO_3H^{\ominus} \\
HO-C-H \\
.COO^{\ominus} \\
\\
+ \\
\\
COO^{\ominus} \\
H-C-OH \\
CH_2OPO_3H^{\ominus}
\end{array}
$$

$$O \overset{\delta+}{=} C \overset{\longrightarrow}{=} O \quad \uparrow$$

Scheme 2

$$
\begin{array}{l}
CH_2OPO_3H^{\ominus} \\
C=O \\
H-C-OH \\
H-C-OH \\
CH_2OPO_3H^{\ominus}
\end{array}
\quad
\overset{-H^{\oplus}}{\underset{+H^{\oplus}}{\rightleftharpoons}}
\quad
\left[\begin{array}{l}
CH_2OPO_3H^{\ominus} \\
C-O^{\ominus} \\
\parallel \\
C-OH \\
H-C-OH \\
CH_2OPO_3H^{\ominus}
\end{array}
\longleftrightarrow
\begin{array}{l}
CH_2OPO_3H^{\ominus} \\
{}^{\ominus}C-OH \\
C=O \\
H-C-OH \\
CH_2OPO_3H^{\ominus}
\end{array}\right]
$$

(I)   (II)

$$O=O \quad \downarrow$$

$$
\left[\begin{array}{l}
CH_2OPO_3H^{\ominus} \\
{}^{\ominus}O-O-C-OH \\
C=O \\
H-C-OH \\
CH_2OPO_3H^{\ominus}
\end{array}\right]
\xrightarrow[-H^{\oplus}]{+H_2O}
$$

(IV)

$$
\begin{array}{l}
CH_2OPO_3H^{\ominus} \\
COO^{\ominus} \\
\\
+ \\
\\
COO^{\ominus} \\
H-C-OH \\
CH_2OPO_3H^{\ominus}
\end{array}
$$

The chemical mechanism of the carboxylation reaction predicted by Calvin (5) (Scheme 1) even before discovery of the enzyme is soundly based on organic chemistry, but progress has been slow in showing how the functional groups at the active site enable the enzyme to catalyze this reaction. The initial step is a base-catalyzed enolization (II) at C3 of RuBP (I), which could be the rate-limiting step of the overall reaction (6). After addition of CO_2, which is the actual substrate for the carboxylase (7), to the intermediary carbanion, at C2 of RuBP (II), C-C bond cleavage of the intermediate β-keto acid (III) occurs between C2 and C3 (8). On the basis of the enzyme's sensitivity to sulfhydryl-modifying reagents, and in order to explain unsuccessful attempts to isolate the hypothetical β-keto acid (III), a catalytic mechanism involving hemimercaptal formation of a sulfhydryl group with RuBP at carbon 2 was proposed (9). Later, support for the existence of the β-keto intermediate (III) was sought by attempts at its synthesis (10, 11). The finding that the stable transition-state analog of (III), 2-carboxy-D-ribitol 1,5-bisphosphate (CRBP) inhibits by competition with RuBP (12, 13) has been considered proof for intermediate (III) in the reaction.

Calvin's original hypothesis for the mechanism of the carboxylase reaction was also used to explain the ^{18}O-labeling pattern in phosphoglycolate produced by the oxygenase reaction (2) (Scheme 2). Although this concept is appealing, the enzyme does not contain any of the cofactors normally associated with oxygenase essential for activating the O_2 (2, 13), such as metal ions or flavins. O_2 is normally in the triplet state, which is 22 kcal/mol lower in energy than singlet O_2. A direct reaction of triplet O_2 with a singlet molecule (II) is a spin-forbidden process. Therefore, the enzymic mechanism for the oxygenation of RuBP must circumvent the mechanism described by Scheme 2, or use some other one, and it remains unknown.

Pon et al. (14) showed in 1963 that preincubation with CO_2 and Mg^{2+} was essential to the enzyme assay, and this requirement has been further ascribed to a homotropic effect of CO_2 (15, 16). Pon et al. also noted that the reaction had to be initiated with RuBP in order to get a maximal rate. But it was not clearly established until 1976 that CO_2 and Mg^{2+} activation is a necessity for obtaining a realistic $K_m(CO_2)$ of 10 to 20 μM and a $K_m(O_2)$ of about 200 μM, and that without CO_2 and Mg^{2+} pretreatment the enzyme is active (17, 18).

The activity of RuBP carboxylase in vivo is thought to be controlled also by several other mechanisms such as shifts in stromal pH and Mg^{2+} concentration due to the H^+ gradient created by electron transport in the light. In addition, some of the intermediates of the reductive and oxidative pentose phosphate pathways are effectors for isolated RuBP carboxylase/oxygenase. Among these are 6-phosphogluconate (see Bassham et al., this volume), fructose 1,6-bisphosphate (FBP), ribulose 5-phosphate, ribose 5-phosphate, and NADPH. The physiological function of these effectors is uncertain, and their mechanism and site of action on the carboxylase are unknown.

A working model for the active site (Figure 1) of RuBP carboxylase/oxygenase must take into account a binding site for RuBP and the mechanism of the base-catalyzed enolization, a CO_2 substrate site near carbon 2 of RuBP, and the possiblity of an O_2 site, as well as amino acid residues essential for catalysis. The CO_2 activator site must bind both CO_2 and Mg^{2+} in a manner to activate or affect the catalytic site. In addition, there is a need for effector sites.

ESSENTIAL LYSYL RESIDUES AND SULFHYDRYL GROUPS

Pyridoxal 5'-phosphate has been used as a probe to demonstrate that ε-amino groups of lysyl residues are important for the mechanism of RuBP carboxylase/oxygenase from spinach (19-21) and Rhodospirillum rubrum (22-24). Pyridoxal phosphate has some advantage over other affinity labels employed so far (25-29) (see Hartman et al., this volume) in that it not only reacts reversibly with RuBP carboxylase forming a Schiff base, but also can be used as an irreversible inhibitor through subsequent $NaBH_4$ reduction. In addition to the characteristic spectral properties of both the Schiff base and its reduced form, a radioactive label can be easily introduced

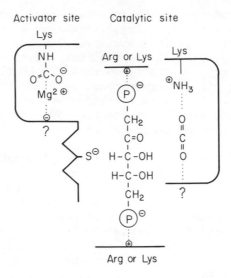

Figure 1.   Working model for active site of RuBP carboxylase/oxygenase.

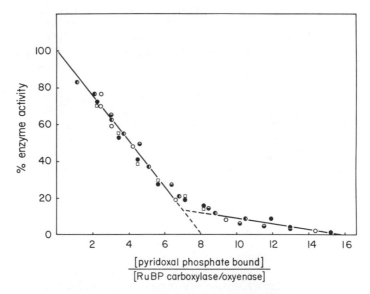

Figure 2.   Activity of RuBP carboxylase/oxygenase with increasing
amounts of bound pyridoxal phosphate after $NaBH_4$ reduction.   Open
circles represent carboxylase activity and squares, oxygenase
activity.   The enzyme was incubated with increasing concentrations of
pyridoxal phosphate and reduced with $NaBH_4$, and the amount of bound
pyridoxal phosphate was determined spectrophotometrically (21).

by NaB^3H_4 reduction, but the stoichiometry is su[l]
effect (C. Paech, unpublished). Pyridoxal phosp[h]
have an exclusive specificity for primary amino [g]
the affinity labels 3-bromo-1,4-dihydroxy-2-buta[n]
phate and N-bromoacetylethanolamine phosphate ex[h]
activity with sulfhydryl groups (26, 28).

RuBP carboxylase/oxygenase from spinach, wh[i]
structure, was completely inactivated when 16 ly̲s̲y̲l̲ ̲r̲e̲s̲i̲d̲u̲e̲s̲ ̲w̲e̲r̲e̲
blocked by pyridoxal phosphate (21). The stoichiometry of pyridoxal
phosphate incorporation was determined spectrophotometrically with
an extinction coefficient of 4800 M^{-1} cm^{-1} for the reduced Schiff
base, as derived for this system (21). This is different from the
coefficient used for this purpose as found in the literature, but
it is in the same range as that derived for aspartate transcarbamyl-
ase (30). This value for the reduced Schiff base should not be con-
fused with the value 5800 M^{-1} cm^{-1} derived for the Schiff base between RuBP carboxylase and pyridoxal phosphate. Loss of enzyme
activity with increasing amounts of irreversibly bound pyridoxal
phosphate was linear but biphasic (Figure 2). The two linear portions extrapolated to 8 and 16 mols of bound pyridoxal phosphate
per mol of enzyme. Complete inactivation was not reached until all
16 mols of pyridoxal phosphate were bound. The carboxylase and the
oxygenase activities declined in parallel.

SDS gel electrophoresis of enzyme-pyridoxal phosphate complexes
reduced with NaB^3H_4 revealed no tritium incorporation into the small
subunit under conditions producing complete inactivation. Introduction of covalently bound pyridoxal phosphate resulted in a net charge
change on the subunits, so that specificity of pyridoxal phosphate incorporation could also be monitored directly by gel electrophoresis
in 8 M urea. Even under conditions such that 16 pyridoxal phosphate
molecules were bound, only the mobility of the large subunit was
changed. For both activities, modification of the enzyme decreased
V_{max} values but did not change K_m values, which indicates only a
decrease in the amount of catalytically active enzyme and not a
change in the catalytic properties of the enzyme.

Bicarbonate and Mg^{2+} competitively reduced the pyridoxal phosphate inhibition of the spinach carboxylase (Figure 2 of ref. 20).
This is in contrast to the enhancement of inhibition reported for
the enzyme from R. rubrum (23). In the absence of Mg^{2+}, RuBP, alone
or with bicarbonate, provided full protection against pyridoxal
phosphate inactivation. Although the number of pyridoxal phosphate
molecules incorporated was reduced by 16, a small amount of nonspecific binding of pyridoxal phosphate still occurred (21). RuBP
was at first reported to protect about half the reactive lysyl
residues against pyridoxal phosphate (20). This was an error
caused by the experiments being done in the presence of Mg^{2+} and
incubated for 20 min to establish the Schiff base equilibrium.
Under these conditions the RuBP was slowly lost through catalytic
reaction with O_2 and CO_2, which had not been excluded from the incubation mixture, and therefore the protective effect was lost.

On the basis of these data, the simplest explanation is that
e 16 essential amino groups are located in the active-site region. A more complicated explanation would be that RuBP exerts
its protective effect both at the catalytic site directly and at
an allosteric site through a proposed RuBP-induced conformational
change of the enzyme.

To gain further information about the function of the two lysyl
groups, the reactivation pattern of the pyridoxal phosphate-inhibited
enzyme during dissociation of the Schiff base complex was studied
(21). From low initial inhibition, reactivation proceeded in a first-
order reaction, independent of whether enzyme activity was measured
directly during reactivation or after samples taken at different
times had been fixed by $NaBH_4$ reduction. From high levels of pyridoxal phosphate inhibition, the reactivation followed first-order
kinetics with the same rate constant only in the experiment that
included $NaBH_4$ reduction. Direct assay showed a clear delay in the
reactivation process, as if a rate-limiting step were imposed on
the overall process. This suggested that the lysyl groups reacting
only at high pyridoxal phosphate concentrations are at the activator
sites, and that the delay in regaining activity is due to slow activation of the enzyme with CO_2 and Mg^{2+} after liberation of the pyridoxal phosphate. This has to take place before any particular subunit is catalytically active. The 8 amino groups reacting at low
pyridoxal phosphate concentration appear to be at the catalytic
sites. This is consistent with the reactivation kinetics and with
the unaltered K_m values of the pyridoxal phosphate-modified enzyme.
The structural similarity between RuBP and pyridoxal phosphate,
i.e., that the carbonyl group and one phosphate are 3 carbons apart,
may account for a higher affinity for pyridoxal phosphate at the
substrate site.

Differential labeling of the enzyme first with p-chloromercuribenzoic acid and then with pyridoxal phosphate indicated that one
group of 8 of the lysyl residues was protected from the pyridoxal
phosphate. Removal of the mercuribenzoate group by dithiothreitol
left the enzyme inactive with only 8 pyridoxal phosphate molecules
bound to it. Although both p-chloromercuribenzoic acid and iodoacetamide inactivate the enzyme, and the binding of each can be prevented by RuBP, only the large p-chloromercuribenzoate group blocks
pyridoxal phosphate binding. Since modification of the sulfhydryl
group(s) prevents binding of RuBP, a sulfhydryl group(s) must be very
close to the binding site for RuBP. Therefore, the 8 lysyl groups
excluded from reaction with pyridoxal phosphate in the presence of
p-chloromercuribenzoic acid are thought to be located at the catalytic site, and the other group of 8 that are still available for
Schiff base formation are possibly located in the vicinity of the
activator site.

The function of the ε-amino group of lysine at the catalytic
site is most likely noncovalent binding of CO_2, since pyridoxal
phosphate appears as a competitive inhibitor for CO_2 (20). However,
lysyl groups, there, may also be involved in proton transfer and

binding of phosphate groups. An ϵ-amino group of lysine at the
activator sites could provide the requirements for allosteric
binding of CO_2 and Mg^{2+} that would be exothermic, reversible,
and pH dependent according to the model in Scheme 3, which is
based on the model for CO_2 activation (17):

$$Enz\text{-}NH_2 + CO_2 \rightleftharpoons Enz\text{-}NH\text{-}COO^- \underset{-\,Mg^{2+}}{\overset{+\,Mg^{2+}}{\rightleftharpoons}} Enz\overset{NH}{\underset{Mg^{2+}}{\diagdown\diagup}}COO^-$$

Supporting the model in Figure 1 for two types of primary
amino groups is the establishment of two different lysyl residues,
also, with the use of bromodihydroxybutanone bisphosphate (29).
Since this affinity label is one carbon atom shorter than RuBP,
it may have greater mobility at the active site and therefore
react with either lysyl group; but, once bound to one of the two,
the first molecule may preclude binding of another to the second
amino group because of a high charge density through the phosphate
groups. This does not explain the low stoichiometry of 4 to 5
moles of reagent per mole of enzyme for inhibition, when the
stoichiometry of inhibition with CRBP clearly indicates the requirement of 8 equivalents per mole of enzyme (11). The determination of bound bromodihydroxybutanone BP is based either on ^{32}P
measurement of the incorporated affinity label or on ^3H measurements after NaB^3H_4 reduction of the enzyme-inhibitor complex.
Both methods are subject to errors due to the known lability of
the affinity label (loss of one or two phosphate groups) (28) and
to possible isotopic effects during reduction with NaB^3H_4.

The A_2 enzyme from R. rubrum binds two pyridoxal phosphate
molecules per molecule or one per subunit, but is inactivated after
blocking of the first catalytic site (24). This agrees with data
obtained with bromodihydroxybutanone BP for the same enzyme (25),
but is in contrast to the finding that two pyridoxal phosphate
molecules must be bound to each subunit of the spinach enzyme.
One suggested reason for this difference is that RuBP carboxylase
from R. rubrum may behave like a half-site enzyme. Although the
amino acid compositions of the large subunits of the carboxylase
from all its various sources are apparently similar, there should
be some differences between the A_2 and the A_8B_8 enzyme, expressed
in catalytic and regulatory properties.

ARGININE AT BINDING SITE FOR RuBP

Arginine residues in enzymes serve a general function as binding sites for negatively charged groups such as phosphate or carboxylate ions (31). Similarly, RuBP carboxylase/oxygenase is

inactivated by reagents specific for arginyl residues, i.e., 2,3-butanedione and phenylglyoxal. McFadden's group (32), using 2,3-butanedione with borate to bind a third of the total arginines, noted enzyme inhibition that was partially protected against by the product 3-phosphoglycerate, as if the modified arginines were at the active site. Phenylglyoxal, which does not require borate, appears to be a better reagent for arginyl residues in this enzyme. Inactivation follows saturation kinetics, and ~ 25 to 30 arginyl residues are blocked, as judged by amino acid analysis. RuBP has given about 70% protection against phenylglyoxal inhibition (C. Paech and M. Spellman, unpublished), although Hartman's group (33) reports finding no such protection.

These preliminary results indicate that one or more arginyl residues at the active site probably function in binding the phosphate of RuBP. Further work is necessary to decide whether both phosphate groups of RuBP are bound by arginine or whether some other cationic group such as a lysyl ε-amino group may also be involved.

STUDIES OF THE RuBP BINDING SITE WITH SUBSTRATE ANALOGS

RuBP carboxylase/oxygenase is specific for RuBP as a substrate for carboxylation and oxygenation. A very low dissociation constant for the enzyme-RuBP complex has been estimated: 0.5 μM (34) or ≤1 μM (12). In fact, the RuBP-enzyme complex can be separated from excess RuBP by gel filtration, and the bound RuBP subsequently detected by the carboxylase reaction. Whereas most sugar mono- and bisphosphates are poor inhibitors, xylitol BP (S. D. McCurry, unpublished, but see ref. 35) and xylulose bisphosphate (XuBP) (36) are powerful, competitive inhibitors with respect to RuBP. The maximum inhibitory effect of either one is manifested after ~20 min, and the binding constant for xylitol BP is <1 μM (McCurry, unpublished). Like CRBP, ^{3}H-xylitol BP could not be dialyzed away from the enzyme. XuBP and xylitol BP differ from RuBP stereochemically at C3, and their severe inhibitory effect may reside in hydrogen-bonding between that hydroxyl group and the base at the catalytic site that participates in enediol formation for RuBP. The energy required to break this hydrogen bond could contribute substantially to stability of the enzyme-inhibitor complex, as evidenced by ribitol BP being a poor inhibitor (F. J. Ryan, unpublished).

XuBP has not been reported to occur in nature, but it could be readily formed in situ, and it is prepared by aldolase in vitro from dihydroxyacetone phosphate and glycolaldehyde phosphate. Apparently the formation of free glycolaldehyde phosphate would be lethal, and this could explain why all C_2 transformations in carbohydrate metabolism involve a glycolaldehyde-thiamin pyrophosphate complex with transketolase rather than free glycolaldehyde phosphate. Conceivably XuBP could be formed by epimerization from RuBP in the active site of the carboxylase from the enediol intermediate if it were allowed to revert, but the absence of such reversion is evidenced

by the lack of ^3H exchange in the absence of CO_2 with RuBP labeled
with ^3H at C3 (6). The question of whether or not such an epimerization would be permitted by the enzyme remains moot.

The most effective inhibitor known for carboxylase/oxygenase
is the transition-state analog CRBP (11, 12), a competitive, time-
dependent, irreversible inhibitor. Each enzyme molecule has 8 binding sites for this compound, and a divalent cation is required for
binding. Synthesis of CRBP from cyanide and RuBP produces two
epimers, CRBP and 2-carboxy-D-arabinitol 1,5-bisphosphate, and the
proportion of inhibition due to each is unknown. These two compounds have been synthesized and purified in our laboratory
(J. Pierce, unpublished), and their effects on RuBP carboxylase/
oxygenase are under investigation. In addition, the compounds
2-carboxy-D-xylitol 1,5-bisphosphate and 2-carboxy-D-lyxitol 1,5-
bisphosphate have been synthesized and purified. Determination
of the relative effects of these compounds on the enzyme, coupled
with knowledge of the stereochemistry of known inhibitors of the
enzyme, may allow estimation of the geometry of the active site.

It has been suggested that carboxylase inhibition by RuBP is
an allosteric effect (37) or is due to an interaction with lysine
at the catalytic site (19). Attempts to detect a Schiff base between either RuBP or XuBP at a lysine by reduction with NaB^3H_4
have not been successful (Paech and McCurry, unpublished). Summarized below is new evidence that the RuBP inhibition is caused by
other inhibitory compounds, leaving unanswered the question whether
RuBP itself has any regulatory control of enzyme activity.

RuBP DEGRADATION PRODUCTS AS INHIBITORS

Inhibition of the initial rate of carboxylase activity by
high concentrations of RuBP has been known for many years (14).
Pre-incubation of the enzyme with RuBP in the absence of CO_2 results in inhibition. Therefore the enzyme assay is initiated with
RuBP after first activating with CO_2 and Mg^{2+}. Although RuBP inhibition has been ascribed to an allosteric effect (37) or to an
interaction at the active site (19), it does not occur in assays
with whole chloroplasts (38). We have obtained data indicating
that the "substrate inhibition" phenomenon produced with solutions
of RuBP is due to the presence of inhibitors formed from the RuBP
by epimerization at C3 or by β-elimination of the C1 phosphate (39).

The idea that an inhibitory compound was present in RuBP arose
from the observation that different batches of RuBP had different
initial reaction rates although all contained nearly the same amount
of RuBP as judged by enzyme, phosphate, and carbohydrate analyses.
The initial catalysis rate, when a reaction is started with RuBP,
is almost unaffected by the suspected impurities; but, with time or
in the absence of RuBP after the substrate is used up (Figure 3),
time-dependent inhibition of these impurities becomes apparent.
Addition of a second aliquot of substrate to the same enzyme results

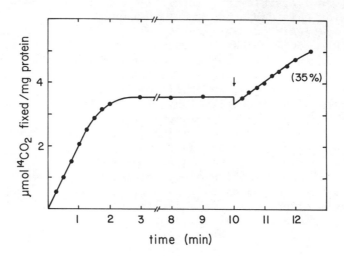

Figure 3. Time course of RuBP carboxylase reaction with a limiting amount of RuBP. After the reaction had gone to completion, a
second aliquot of substrate was added, and the change in the initial rate was recorded as percent inhibition (39).

in a much slower, or inhibited, rate of carboxylation. Product
inhibition was ruled out (39). In RuBP solutions, the inhibitors
build up with increasing storage time, particularly at pH above
8 or elevated temperatures.

Incubation of RuBP solutions at pH 11 and 30° results in total
loss of phosphate at carbon 1. Under the condition of the carboxylase assay, RuBP decomposes at the rate of 1.25% per hr. Mild
base treatment causes substantial inhibition without substantial
loss of substrate. Fresh solutions of enzymatically synthesized
RuBP have essentially no inhibitory components, but RuBP from
commercial sources is inhibitory because of degradation products
formed during the isolation steps.

Two inhibitory products have been found in RuBP solutions.
One is XuBP, which arises from nonenzymic epimerization, as shown
in reaction Scheme 4:

$$
\begin{array}{ccccc}
\begin{array}{l}
CH_2OPO_3^{2-} \\
\mid \\
C=O \\
\mid \\
H-C-OH \\
\mid \\
H-C-OH \\
\mid \\
CH_2OPO_3^{2-}
\end{array}
& \xrightleftharpoons{\;OH^-\;} &
\left[\begin{array}{l}
CH_2OPO_3^{2-} \\
\parallel \\
C-O \cdots H \\
\parallel \\
C-O \\
\mid \\
H-C-OH \\
\mid \\
CH_2OPO_3^{2-}
\end{array}\right]
& \xrightleftharpoons{\;OH^-\;} &
\begin{array}{l}
CH_2OPO_3^{2-} \\
\mid \\
C=O \\
\mid \\
HO-C-H \\
\mid \\
H-C-OH \\
\mid \\
CH_2OPO_3^{2-}
\end{array} \\[4ex]
\text{RuBP-} & & & & \text{XuBP-}
\end{array}
$$

$\searrow \quad PO_4^{3-}$

$$
\left[\begin{array}{ll}
CH_2 & CH_3 \\
\parallel & \mid \\
C-OH & C=O \\
\mid & \mid \\
C=O \;\rightleftharpoons\; & C=O \\
\mid & \mid \\
H-C-OH & H-C-OH \\
\mid & \mid \\
CH_2OPO_3^{2-} & CH_2OPO_3^{2-}
\end{array}\right]
\xrightarrow{\;+OH^-\;}
\begin{array}{l}
COO^- \\
\mid \\
H_3C-C-OH \\
\mid \\
H-C-OH \\
\mid \\
CH_2OPO_3^{2-}
\end{array}
\;+\;
\begin{array}{l}
COO^- \\
\mid \\
HO-C-CH_3 \\
\mid \\
H-C-OH \\
\mid \\
CH_2OPO_3^{2-}
\end{array}
$$

(V)	(VI)	(VII)
1-deoxy-D-glycero- 2,3-pentodiulose- 5-phosphate	2-C-methyl-D- erythrotetronic acid 4-phosphate	2-C-methyl-D- threo tetronic acid 4-phosphate

Proof of its presence and structure has been obtained by thin-layer
and gas-liquid chromatography after reduction and phosphatase treatments. XuBP preparations used as substrate with isolated enzyme
cause some initial CO_2 incorporation into 3-phosphoglycerate (39),
but this catalysis is due to the presence of a little RuBP arising
from epimerization rather than to XuBP acting as a substrate.

The other inhibitory compound arising from RuBP appears to be
a diketo degradation product formed from one or the other or both
of the bisphosphates by β-elimination. Its structure has not been
unequivocally established, but the UV absorption spectra, NMR spectra, and phosphate elimination support the idea that it is 1-deoxy-
D-glycero-2,3-pentodiulose 5-phosphate. Additional support comes
from treatment of RuBP solutions with o-phenylenediamine resulting
in partial removal of the inhibitory effect: compound V is relatively unstable and probably undergoes rearrangement to a stable
end product having properties consistent with those of a branched-
chain compound (Scheme 4), clearly recognizable by NMR spectroscopy
after extensive base treatment of RuBP. Since the diketo compound,
which appears to be the inhibitory component, is rather unstable,
it accumulates to only a small extent and does not persist in the
reaction mixture.

The physiological implication of these data is that RuBP is
an unstable substrate with toxic degradation products, and therefore

free RuBP must not occur in chloroplasts. Under normal conditions in chloroplasts the estimated concentration of RuBP is 0.2 to 0.4 mM but that of RuBP binding sites in the carboxylase is 8 times as high: 3 mM (38, 40). Therefore, because of the very low dissociation constant (0.5 μM) for the enzyme-RuBP complex (11), it is unlikely that this labile substrate ever exists free in solution in the chloroplast. In chloroplasts, only 40 to 60% of the carboxylase is estimated to be active at any one time (41). Our hypothesis is that one function for the large amounts of the carboxylase protein in chloroplasts is storage of RuBP to prevent its breakdown to inhibitory products.

CARBOXYLASE/OXYGENASE RATIO AND O_2 SITE

Since RuBP carboxylase activity initiates the photosynthetic carbon cycle and RuBP oxygenase activity diverts carbon flow into phosphoglycolate formation and photorespiration, the regulation and control of these two overlapping and competing carbon pathways appears to be a major consideration for net photosynthesis. Comparison with other systems suggests that regulation may depend on a combination of substrate availability and alteration in enzyme activity for reactions that initiate metabolic pathways or are at branch points. Indeed, compelling evidence indicates that RuBP carboxylase and oxygenase activities are competitively dependent on the CO_2 and O_2 substrate concentrations (42), but some evidence suggests that the amount of carbon flow through the glycolate pathway in the leaf is in part independent of the CO_2/O_2 ratio.

A change in enzyme activity at branch points is the general way of dealing with alterations in limiting amounts of protein, through feedback regulants or effectors. The carboxylase/oxygenase, however, is present in large excess. Effectors for this enzyme have been reported, but they alter the carboxylase and oxygenase activities similarly (43). All the inhibitors or substrate analogs so far examined also affect the carboxylase and oxygenase activities similarly. So far there is no confirmed evidence for any alteration in the carboxylase/oxygenase ratio by any mechanism other than CO_2 and O_2 competition. We cannot confirm the claim that RuBP carboxylase and RuBP oxygenase are two different proteins (44) (manuscript in preparation). There have been reports of variations in the activity ratio of preparations from different stages of leaf development, from leaves versus fruit, from different plants or mutants, after treatment with different inhibitors or effectors, and at different degrees of purification (literature not reviewed). However, many of these reports cannot be confirmed; until such reports are confirmed by several groups, a change in the carboxylase/oxygenase ratio must be considered unproven. One of the difficulties is that the oxygenase assay is relatively insensitive compared with the carboxylase assay. For instance, the enzyme must be CO_2 activated, but CO_2 is an inhibitor of the oxygenase. The CO_2 and

Mg^{2+}-activated enzyme, upon dilution into a CO_2-free oxygenase
assay medium, loses activity with a half-life of 1 min (N. P. Hall,
unpublished).

The nature of the RuBP oxygenase reaction, the possible existence of an O_2 site or an undiscovered cofactor for the oxygenase,
and whether the reaction can be regulated are critical unanswered
questions. It is very unlikely that the lysyl CO_2-binding sites
could bind or activate O_2. Despite intensive search, no cofactor
has been found that is required for the oxygenase reaction. An
early report that Cu^{2+} is present in the enzyme but has no effect
on the carboxylase activity (45) has twice been reexamined without
finding Cu^{2+} or any effect of Cu^{2+} on oxygenase activity (2, 13).
The speculation that inherent amino acid functions of the enzyme,
such as sulfhydryl groups, could serve as activator and/or binding
site for the oxygen (46) was based on a claim that differential
regulation of carboxylase and oxygenase could be achieved by modifying sulfhydryl groups with glycidate or iodoacetamide; but this
could not be confirmed by several laboratories including ours.
The basis for CO_2 and O_2 being competitive inhibitors of each other
may be competition for a reaction with the other substrate, RuBP,
rather than for a reactive-site component. In fact the most important property of this enzyme may be that it catalyzes the carboxylation reaction in the face of an unavoidable oxygenase activity.

The unknown nature of the oxygenase is of great interest mechan-
istically as well as physiologically. RuBP oxidation is an enzymic
property of the RuBP carboxylase/oxygenase protein. Without the
activated enzyme RuBP oxidation is relatively slow. Not only does
RuBP rearrange and decompose to inhibitory products as described
above, but it may also be attacked by strong oxidants equivalent
to those found in the chloroplasts (F. J. Ryan, unpublished; J.
Pierce, unpublished; N. P. Hall, unpublished). Just as we are pro-
posing that RuBP is stabilized against rearrangement in the chloroplast by being stored at the RuBP binding site on the enzyme, it
may also need protection against oxidation in the chloroplasts.
The enzyme-RuBP storage complex is inactive as either a carboxylase
or an oxygenase, but CO_2 and Mg^{2+} activate both activities (18).
This activation by CO_2 or by effectors, which is necessary for catalysis, is to be avoided at other times in the presence of limiting amounts of CO_2 because of the oxygenase reaction. Upon activation, the O_2 attack may be an essential part of the overall photosynthetic process, or it may be an unavoidable reaction of an
intermediary enolate form of the RuBP in the active site of the
enzyme. But exactly how the oxygenase reaction occurs is not known.

SUMMARY

The catalytic region of each large subunit of RuBP carboxylase/
oxygenase from spinach must contain two anionic binding sites for
RuBP, of which one or both appear to be arginine. This site may

serve for storage of labile RuBP and as a catalytic site upon activation. The CO_2 substrate site includes a lysyl residue which
might take part in orienting the CO_2 adjacent to carbon 2 of the
RuBP. The lack of evidence for an O_2 binding site suggests that
the oxygenase activity may proceed by direct oxidation of the activated RuBP-enzyme complex. The CO_2/Mg^{2+} activator site is also
in the large subunit and involves another lysyl residue for the
formation of a carbonate-Mg^{2+} complex. The conformational change
induced during the formation of this complex is conceptually an
orientation of the base opposite C3 of RuBP for catalyzing the
rate-limiting enolization step.

REFERENCES

1. Bowes, G., Ogren, W. L., and Hageman, R. H., Biochem. Biophys.
    Res. Commun. 45, 716-22 (1971).
2. Lorimer, G. H., Andres, T. J. and Tolbert, N. E., Biochemistry
    12, 18-23 (1973).
3. Tolbert, N. E. and Ryan, F. J., in CO₂ Metabolism and Plant
    Productivity, pp. 141-59, R. H. Burris and C. C. Black,
    Editors, University Park Press, 1976.
4. McFadden, B. A., Bacteriol. Rev. 37, 289-319 (1973).
5. Calvin, M., Fed. Proc. 13, 697-711 (1954).
6. Fiedler, F., Müllhofer, G., Trebst, A., and Rose, I. A.,
    Eur. J. Biochem. 1, 395-9 (1967).
7. Cooper, T. G., Filmer, D., Wishnick, M., and Lane, M. D.,
    J. Biol. Chem. 244, 1081-3 (1969).
8. Müllhofer, G. and Rose, I. A., J. Biol. Chem. 240, 1341-6 (1965).
9. Rabin, B. R. and Trown, P. W., Nature London 202, 1290-3 (1964).
10. Sjödin, B. and Vestermark, A., Biochim. Biophys. Acta 297,
    165-73 (1973).
11. Siegel, M. I. and Lane, M. D., J. Biol. Chem. 248, 5486-98 (1973).
12. Wishnick, M., Lane, M. D., and Scrutton, M. C., J. Biol. Chem.
    245, 4939-47 (1970).
13. Chollet, R., Anderson, L. L., and Hovsepian, L. C., Biochem.
    Biophys. Res. Commun. 64, 97-107 (1975).
14. Pon, N. G., Rabin, B. R., and Calvin, M., Biochem. Z. 338,
    7-19 (1963).
15. Sugiyama, T., Nakayama, N., and Akazawa, T., Arch. Biochem.
    Biophys. 126, 737-45 (1968).
16. Murai, T., and Akazawa, T., Biochem. Biophys. Res. Commun.
    46, 2121-6 (1972).
17. Lorimer, G. H., Badger, M. R., and Andrews, T. J., Biochemistry
    15, 529-36 (1976).
18. Badger, M. R. and Lorimer, G. H., Arch. Biochem. Biophys. 175,
    723-9 (1976).
19. Paech, C., Ryan, F. J., McCurry, S. D., and Tolbert, N. E.,
    Plant Physiol. Suppl. 57, 54 (1976).

20. Paech, C., Ryan, F. J., and Tolbert, N. E., Arch Biochem. Biophys. 179, 279-88 (1977).
21. Paech, C. and Tolbert, N. E., J. Biol. Chem., in press (1978).
22. Whitman, W. and Tabita, F. R., Biochem. Biophys. Res. Commun. 71, 1034-9 (1976).
23. Whitman, W. and Tabita, F. R., Biochemistry 17, 1282-7 (1978).
24. Whitman, W. and Tabita, F. R., Biochemistry 17, 1288-93 (1978).
25. Norton, I. L., Welch, M. H., and Hartman, F. C., J. Biol. Chem. 250, 8062-8 (1975).
26. Schloss, J. V. and Hartman, F. C., Biochem. Biophys. Res. Commun. 75, 320-8 (1977).
27. Schloss, J. V. and Hartman, F. C., Biochem. Biophys. Res. Commun. 77, 230-6 (1977).
28. Hartman, F. C., Welch, M. H., and Norton, I. L., Proc. Natl. Acad. Sci. USA 70, 3721-4 (1973).
29. Stringer, C. D. and Hartman, F. C., Biochem. Biophys. Res. Commun. 80, 1043-8 (1978).
30. Blackburn, M. N. and Schachman, H. K., Biochemistry 15, 1316-22 (1976).
31. Riordan, J. F., McElvany, K. D., and Borders, C. L. Jr., Science 195, 884-6 (1977).
32. Lawlis, V. B. and McFadden, B. A., Biochem. Biophys. Res. Commun. 80, 580-5 (1978).
33. Schloss, J. V., Norton, I. L., Stringer, C. D., and Hartman, F. C., Fed. Proc. 37, 1310 (1978).
34. Vater, J., Salnikow, J., and Kleinkauf, H., Biochem. Biophys. Res. Commun. 74, 1618-25 (1977).
35. Ryan, F. J., Barker, J. R., and Tolbert, N. E., Biochem. Biophys. Res. Commun. 65, 39-46 (1975).
36. McCurry, S. and Tolbert, N. E., J. Biol. Chem. 252, 8344-6 (1977).
37. Chu, D. K. and Bassham. J. A., Plant Physiol. 55, 720-6 (1975).
38. Jensen, R. G. and Bahr, J. T., Annu. Rev. Plant Physiol. 28, 379-400 (1977).
39. Paech, C., Pierce, J., McCurry, S. D., and Tolbert, N. E., Biochem. Biophys. Res. Commun., submitted (1978).
40. Hitz, W. D. and Stewart, C. R., Plant Physiol. Suppl. 61, 100 (1978).
41. Bahr, J. T. and Jensen, R. G., Arch Biochem. Biophys. 185, 39-48 (1978).
42. Bowes, G. and Ogren, W. L., J. Biol. Chem. 247, 2171-6 (1972).
43. Chollet, R. and Anderson, L. L., Arch. Biochem. Biophys. 176, 344-51 (1976).
44. Branden, R., Biochem. Biophys. Res. Commun. 81, 539-46 (1978).
45. Wishnick, M., Lane, M. D., Scrutton, M. C., and Mildvan, A. S., J. Biol. Chem. 244, 5761-3 (1969).
46. Wildner, G. F., Ber. Dtsch. Bot. Ges. 89, 349-60 (1976).

DISCUSSION

McFADDEN: There are about 80 to 100 sulfhydryl groups per
molecule of the higher plant enzyme.   How extensively modified by
mercurial reagent (PCMB) was your RuBP carboxylase when tested for
RuBP binding?   If extensive modification had occurred, it would be
important to probe directly for RuBP binding (i.e., by gel filtration or equilibrium dialysis).   Was this done?

PAECH: The enzyme was incubated with a 50-fold molar excess
of p-chloromercuribenzoic acid for 20 min.   This resulted in modification of ~30 SH groups, according to T. Sugiyama et al. (Arch.
Biochem. Biophys. 125, 98-106), and complete loss of enzyme activity.
RuBP binding was determined directly by gel filtration (see ref. 21).

MIZIORKO: You mentioned that a very tight complex is formed
with both XuBP and xylitol BP.   Is this type of complex stable to
treatment by passage over a Sephadex column?

McCURRY: Yes, very stable.

MIZIORKO: Is there a divalent ion requirement for formation
of the complex?

McCURRY: No.

MIZIORKO: Is a covalent adduct formed in either or both cases?

McCURRY: We cannot envision one.   We tried to reduce the
complex, as it had been suggested that RuBP might form a Schiff
base in the active site with carbonyl, and we thought of applying
the same hypothesis to XuBP.   We could not show any reduction, but
that is no guarantee it will not occur.

MIZIORKO: Do you have to age the incubation mixture in order
to get a tight complex that is stable to Sephadex chromatography
or can you just do an initial mixing?

McCURRY: Do you mean running over the column within 1 or
2 min?

MIZIORKO: Yes.   You mentioned that it was a slow process.

McCURRY: Yes, ~90% inhibition occurs within 2 min but
complete inhibition takes 20 min.

MIZIORKO: According to a model which requires two CO_2 sites
per subunit of enzyme, modification of either the substrate CO_2
site or the activator CO_2 site should result in loss of catalytic
activity.   The modification data clearly show two types of site,
with substantial enzyme activity remaining in the region of pyridoxal phosphate titration where virtually complete modification of
one type of site is expected.   Is the residual activity consistent
with the two-site model?

TOLBERT: According to our proposal, binding of the first
group of pyridoxal phosphate molecules is facilitated because
pyridoxal phosphate is structurally similar to RuBP.   At higher
pyridoxal phosphate concentrations, significant amounts also begin
to form Schiff base at the second site.   Because of the increasing
competition between the two groups of pyridoxal phosphate binding
sites, which becomes significant after 6 pyridoxal phosphate molecules

are bound to the enzyme, the reaction with the first set of amino groups would not go to completion when 8 pyridoxal phosphate molecules were bound, nor would full inactivation occur until saturation of all 16 amino groups with pyridoxal phosphate.

HARTMAN: The lack of protection by bicarbonate against modification by pyridoxal phosphate appears inconsistent with your hypothesis that the site labeled is involved in binding of CO_2.

PAECH: The data from kinetic experiments in Figure 2 of ref. 20 clearly demonstrate protection by bicarbonate against pyridoxal phosphate inhibition. In order to measure spectrophotometrically Schiff base formation between the enzyme and pyridoxal phosphate, the enzyme was equilibrated at 1 mg/ml with 0.5 mM pyridoxal phosphate and 0 to 300 mM bicarbonate. Although dissociation constants are not available for the enzyme-CO_2 complex and the enzyme-pyridoxal phosphate complex, it is clear from the kinetic experiments in ref. 20 that the two competing ligands were present in an unfavorable ratio, i.e., too little bicarbonate. Therefore, this second research approach indicated incorrectly a lack of protection by CO_2.

WILDNER: What is the chemical evidence that the activation of RuBP carboxylase is based on carbamate formation at the lysine of the activation site?

TOLBERT: Carbamate formation with a basic amino group, in this case ε-amino group of lysine, has been the mechanism generally cited for CO_2 modification of enzyme activity. As pointed out in the text, carbamate formation is exothermic and pH sensitive, as required for RuBP carboxylase activation. Further chemical and physical proof for this carbamate formation will be forthcoming.

CHOLLET: Are the products of RuBP degradation irreversible inhibitors? If you assay the enzyme after passage over a Sephadex G25 column, or after dilution, is it still inhibited?

TOLBERT: The xylitol BP reaction is essentially irreversible.

CHOLLET: In doing chemical modification studies, we, and perhaps the Oak Ridge people and Bob Divita, have used substrate protection, as you have. If we use control enzymes with no modifier, plus or minus 5 mM RuBP, preincubated for 1 to 1½ hr, then dilute out or pass over a Sephadex G25 column, the two enzymes (plus or minus RuBP) have identical activity. Since the time is sufficient for degradation products to form, either they are readily reversible or we do not see them; the data of Fred Hartman and Bob Divita indicate that they do not see them either.

PIERCE: It has not been shown whether or not the diketo compounds are reversible. Enzyme incubated with what we have proposed to be inhibitors arising from RuBP have not been passed over columns.

HARTMAN: I am not aware of any evidence suggesting that ionized sulfhydryl is the essential base obstructing the C3 proton.

TOLBERT: We do not know what base is there. The only thing we do know is the sulfhydryl protection. It has been proposed that there are two sulfhydryl groups.

ATTEMPTS TO APPLY AFFINITY LABELING TECHNIQUES TO

RIBULOSE BISPHOSPHATE CARBOXYLASE/OXYGENASE[*]

Fred C. Hartman, I. Lucile Norton, Claude D. Stringer,
and John V. Schloss

Biology Division, Oak Ridge National Laboratory, and
University of Tennessee-Oak Ridge Graduate School of
Biomedical Sciences, Oak Ridge, Tennessee 37830

INTRODUCTION

Although its results are subject to uncertainties in interpretation, chemical modification is a proven method for defining structure-function relationships in enzymes. Innumerable times, tentative conclusions based on chemical modification studies regarding identities of active-site residues, or even their precise catalytic function, have been substantiated by x-ray crystallography.

The application of general (group-specific) protein reagents to RuBP carboxylase/oxygenase has not provided detailed information about the active site, probably because of the enzyme's molecular complexity. A number of sulfhydryl reagents inactivate the enzyme, and substrate protection is observed. However, the data do not allow differentiation between the potential roles of certain sulfhydryl groups in maintenance of tertiary and quaternary structure and in catalysis. Akazawa's group (1, 2) has thoroughly studied modification of the enzyme with p-chloromercuribenzoate. At slightly alkaline pH, disruption of quaternary structure results, leaving no doubt that modification of sulfhydryls can lead to conformational changes.

The shortcoming of studies with sulfhydryl reagents is that direct correlations between inactivation and modification of

---

[*] By acceptance of this article, the publisher or recipient acknowledges the right of the U.S. Government to retain a nonexclusive, royalty-free license in and to any copyright covering the article.

Abbreviations: RuBP, D-ribulose 1,5-bisphosphate; bromodihydroxybutanone BP, 3-bromo-1,4-dihydroxy-2-butanone 1,4-bisphosphate; Bicine, N,N'-bis(2-hydroxyethyl)glycine.

specific residues were not (or could not be) demonstrated because
of the large number of sulfhydryls being modified. For example,
even though RuBP protects against inactivation by ^{14}C-iodoacetic
acid, comparative analysis of the tryptic peptides derived from
inactivated and substrate-protected enzyme revealed extensive,
random modification of sulfhydryls and no clear-cut difference
in labeling pattern between the two samples (3).

More recently, the lysyl reagent pyridoxal phosphate and two
arginyl reagents, 2,3-butanedione and phenylglyoxal, have been
used to determine whether the carboxylase/oxygenase contains such
essential residues. The encouraging results (4-6) obtained with
pyridoxal phosphate are discussed below, since this reagent may
be an affinity label for a binding site (probably the active site)
that accomodates phosphate esters. The results (7) provided by
the reaction of butanedione with carboxylase/oxygenase from barley
and Pseudomonas oxalaticus were consistent with the presence of
active-site arginyl residues. We interpret our data (8) from modification of spinach and Rhodospirillum rubrum enzymes as showing
that 2 to 3 arginyl residues per protomer are essential to the
stabilization of native conformation. The two sets of data are
not necessarily contradictory, since enzymes from different species
were used, and since equivalent arginyl residues may not be accessible to both reagents.

Because general protein reagents have the inherent disadvantage
of lacking specificity, we have attempted systematic development of
affinity labels for use in characterizing the active site of RuBP
carboxylase/oxygenase. The major advantages of affinity labels over
general protein reagents are their potentially absolute specificity
for the catalytic site of a given enzyme and the mechanistic information they can provide (Methods in Enzymology, Vol. 47, W. B.
Jakoby and M. Wilchek, Editors, is an excellent source of references
to successful affinity labeling). The selectivity of an affinity-
labeling reagent is due to its structural similarity to substrate
and it results in the formation of a dissociable enzyme-reagent
complex comparable to that formed by a competitive inhibitor.
Complex formation causes a localized high concentration of reagent
at the active site and thus increases the likelihood of modification of a given residue within this region rather than elsewhere
in the protein molecule. In contrast to ideal general protein
reagents, which, because of their chemical nature, are selective
for a given functional group, affinity labels are selective for
a particular type of binding site and often are reactive toward
several functional groups found in proteins. The specificity with
respect to enzyme is determined by the substratelike features of
the reagent, and the specificity with respect to the kind of residue
modified depends on which reactive side-chain within the active site
is in proper juxtaposition to the leaving group of the reagent.

Reagents that we have synthesized as potential affinity labels
for RuBP carboxylase/oxygenase include bromodihydroxybutanone BP, N-
bromoacetylethanolamine phosphate, cis- and trans-2,3-epoxybutane-

1,4-diol 1,4-bisphosphate, N-bromoacetyldiethanolamine bisphosphate, 2-phosphoglycolic acid azide, and N-bromoacetylphosphoserine (Figures 1 to 4).

CRITERIA OF AFFINITY LABELING

It should be emphasized that no single experiment can prove that enzyme inactivation reflects an active-site-directed modification. Rather, a variety of criteria must be met before a strong argument can be made with respect to active-site modification. Note also that proof of the presence of a given residue at the active site does not prove that the residue plays a catalytic role. In fact, unequivocal proof of catalytic functionality cannot be provided by chemical modification alone.

Criteria that are useful in verifying affinity labeling include the following:

1. Complete Inactivation. If the residue subject to modification is absolutely essential to catalysis or substrate binding, total loss of enzymic activity should occur.

2. Pseudo-First-Order Loss of Activity. At high molar ratios of reagent/enzyme, first-order inactivation kinetics will be observed provided that inactivation correlates with the modification of one residue.

3. Rate Saturation. True affinity labels bind reversibly to the active site as an obligatory step preceding inactivation:

$$E + I \underset{k_2}{\overset{k_1}{\rightleftharpoons}} E \cdot I \overset{k_3}{\longrightarrow} E_I \quad . \tag{1}$$

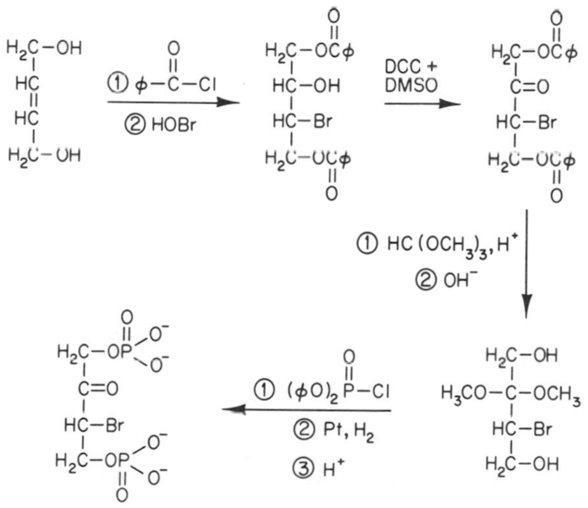

Figure 1. Synthesis of bromodihydroxybutanone BP.

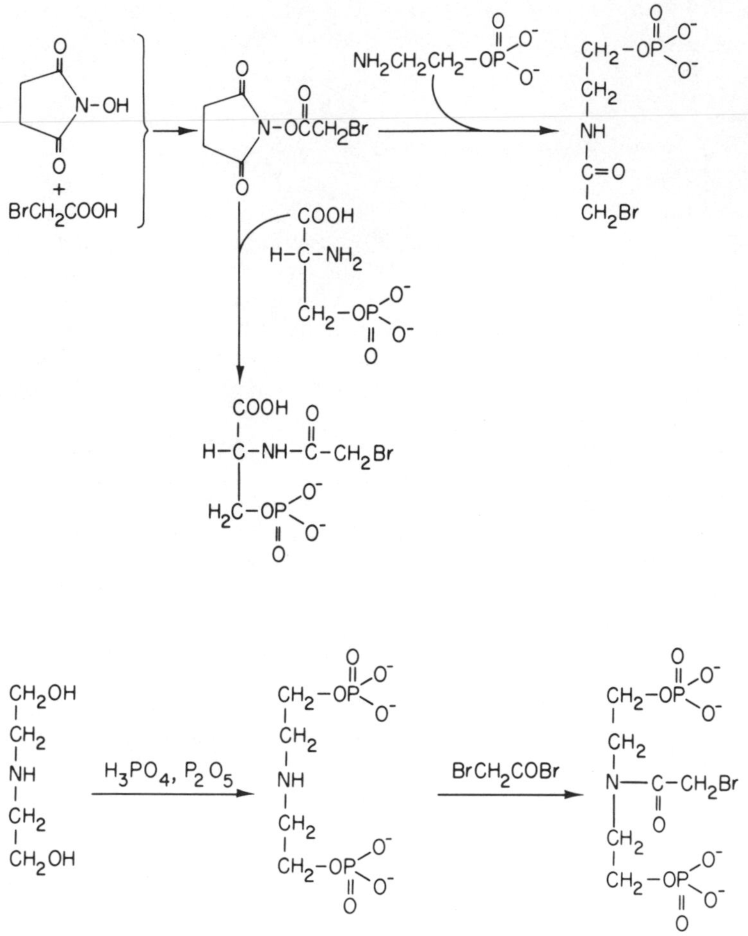

Figure 2. Synthesis of N-bromoacetylethanolamine phosphate, N-bromoacetylphosphoserine, and N-bromoacetyldiethanolamine bisphosphate.

Thus the rate of inactivation is proportional to the concentration of enzyme-reagent complex (E·I), and as the reagent concentration is increased the rate of inactivation approaches a limiting, finite value (rate saturation). A linear expression for the rate of inactivation has been derived (9, 10),

$$\tau = (TK_{inact} \cdot 1/[I]) + T , \qquad (2)$$

in which K_{inact} is $(k_2 + k_3)/k_1$ and is comparable with K_M in the Michaelis-Menten expression, T is the minimal inactivation half-

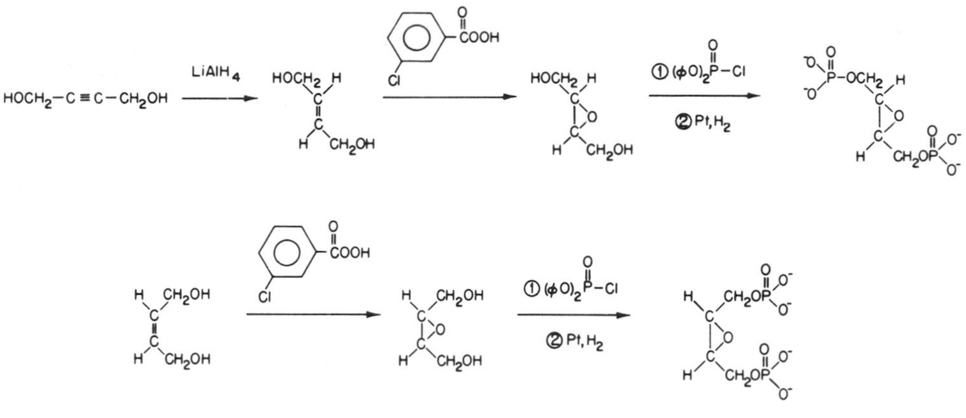

Figure 3. Synthesis of <u>cis</u>- and <u>trans</u>-2,3-epoxybutane-
1,4-diol 1,4-bisphosphate.

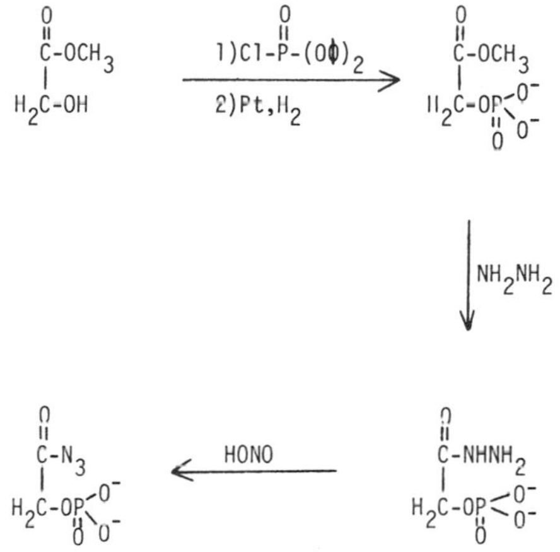

Figure 4. Synthesis of 2-phosphoglycolic acid azide.

time, and τ is the observed inactivation half-time at inhibitor
concentration $[I]$. Plots of τ vs. $1/[I]$ are used to calculate
K_{inact} and T.

4. Protection by Substrate (or Competitive Inhibitor). If
the presumed affinity label binds reversibly to the active site
before covalent modification, substrates should protect against
inactivation in a competitive fashion (9, 10),

$$\tau = (TK_{inact}/[I])(1 + [S]/K_S) + T \quad , \tag{3}$$

i.e., the maximal rate of inactivation should be unaltered by a decrease in apparent affinity (increased K_{inact}) of reagent for enzyme. K_S is the apparent dissociation constant for E·S and is calculated from plots of τ vs. $1/[I]$. Similarity of the K_S for protector to its K_M would provide a strong argument for the reagent initially binding reversibly to the active site.

5. Competitive Inhibition. Any reagent that inactivates an enzyme at a rate slow compared with the time needed for enzyme assays can be tested as a classical competitive inhibitor. If competitive inhibition is observed and the K_I proves to be equal to K_{inact} calculated from inactivation kinetics, the enzyme-reagent complex that leads to inactivation must be the same complex that is visualized in the direct assays for inhibition.

6. Site-Specific Modification Requires Native Three-Dimensional Structure. Since initial complexing with affinity label requires a functional substrate-binding site, disruption of tertiary structure with a protein denaturant can eliminate the specificity in modification.

7. Stoichiometry. An ideal affinity label is specific for a single site, and only one mole of reagent is incorporated per mole of catalytic subunit inactivated.

8. Interspecies Homology Around Modified Residue. In general, active-site structure has been conserved during evolution. Thus, comparisons (at the level of amino acid sequences) of the reactions of a suspected affinity label with analogous enzymes from different species can demonstrate whether the site of modification represents a species-invariant feature. The validity of this approach requires that the enzymes compared be derived from a common ancestral gene.

9. Pseudo Substrate. In some cases the affinity label so closely resembles the normal substrate that catalytic turnover of reagent occurs; this provides compelling documentation of active-site modification. For example, Meloche et al. (11) showed that 3-bromopyruvate, an affinity label for the aldolase that catalyzes condensation between pyruvate and D-glyceraldehyde 3-phosphate, undergoes enzyme-catalyzed proton exchange at C3 as does the substrate pyruvate. The constant ratio (regardless of bromopyruvate concentration) of moles of bromopyruvate undergoing proton exchange per mole of enzyme inactivated, and the observation that the rates of both processes are half-saturated at identical reagent concentration, strongly suggest that catalysis and alkylation occur at the same site.

AFFINITY LABELING STUDIES

Affinity labeling studies on RuBP carboxylase/oxygenase are complicated by the existence of distinct allosteric sites that bind phosphate esters (12) and by the conformational changes induced by Mg^{2+} and CO_2. Since the enzyme is in an inactive conformation unless both Mg^{2+} and CO_2 are present (13, 14), apparent inactivation by a chemical reagent could be caused by prevention

of the activation by CO_2 and Mg^{2+} that normally takes place upon
introduction of the inactive conformer into the assay medium.   We
have therefore compared results obtained in the presence and absence of CO_2 and Mg^{2+}.   Such a comparison also provides an approach
to distinguishing the catalytic site from the allosteric site,
because the inactive conformer that exists in the absence of CO_2
and Mg^{2+} is still functional in binding phosphate esters (12).

Because of these complexities, studies on carboxylase/
oxygenase from different species may be necessary to confirm that
a residue implicated as essential is indeed an active-site component.   To provide an especially stringent test case for the
identification of species-invariant structural features, we chose
for comparison the enzymes from two phylogenetically distant species,
spinach and Rhodospirillum rubrum.   The latter enzyme is the structurally simplest carboxylase/oxygenase known, and its structure and
properties differ dramatically from those of the higher plant
enzymes (15).

To date, the reactions of bromodihydroxybutanone BP and N-
bromoacetylethanolamine phosphate with the spinach enzyme have
been fairly well characterized (16-19), but only preliminary
experiments have been completed with the R. rubrum enzyme (20).
Both enzymes were isolated by slight modification of published
procedures (21, 22), and they were assayed for carboxylase activity,
spectrophotometrically (23) or by $^{14}CO_2$-fixation (24), and for oxygenase activity (24).   Syntheses for bromodihydroxybutanone BP and
N-bromoacetylethanolamine phosphate have been reported (25, 26).

Reaction of Bromodihydroxybutanone BP

1.   Inactivation.   Under conditions frequently used to assay
RuBP carboxylase/oxygenase (pH 8.0, room temperature), the spinach
enzyme is inactivated by bromodihydroxybutanone BP (Figure 5).
RuBP protects against inactivation, and CO_2 greatly stimulates
the rate of activity loss. Mg^{2+} does not influence the inactivation
rate, and we assume therefore that it does not alter the sites of
modification.   Within the limits of the assay method, inactivation
proceeds to completion.   Activity is not regained upon exhaustive
dialysis of the treated enzyme, which is suggestive of covalent
modification.

2.   Competitive Inhibition.   Reagent instability (25) precludes
the detailed kinetic study needed to demonstrate a rate saturation
effect for inactivation of the carboxylase/oxygenase by bromodihydroxybutanone BP.   However, the reagent is a competitive inhibitor
(with respect to RuBP) of the enzyme both from spinach and from
R. rubrum, with a K_I of 1.0 mM and 1.2 mM, respectively (Figure 6).

3.   Extent of Incorporation.   The incorporation into spinach
enzyme measured by reduction of the carbonyl group of the protein-
bound reagent moiety with sodium ^3H-borohydride subsequent to protein derivatization with bromodihydroxybutanone BP is shown in
Table 1.   Also shown is the decrease in free sulfhydryl content.

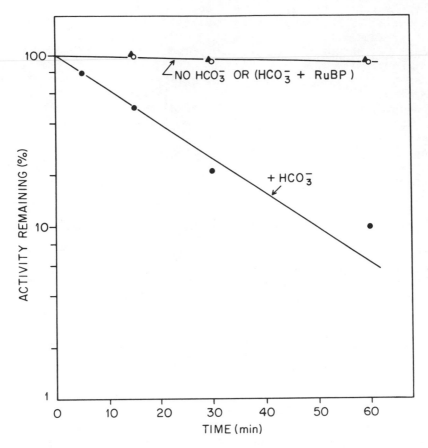

Figure 5. Loss of carboxylase activity upon incubation of RuBP
carboxylase/oxygenase (1 mg/ml) with bromodihydroxybutanone BP
(0.1 mM) in 0.1 M Bicine/60 mM KHCO₃/1 mM EDTA (pH 8.0) (●). In
other experiments, RuBP (1 mM) was added to the buffer (▲) or bicarbonate was omitted (o).

Note that sulfhydryl alkylation is more extensive in the presence
of substrate (without activity loss) than in its absence. Also,
in the presence of substrate the number of free sulfhydryls lost
is about the same as the total reagent incorporation, whereas
in the inactivated sample it is clearly smaller; thus, modification
of residues other than cysteinyl must account for inactivation.
Simple subtraction of the amount of reagent incorporated into the
sample protected with RuBP from the amount incorporated during
inactivation indicates that inactivation correlates with modification of only 1.6 residues per native molecule (8 protomeric
units). This value is misleadingly low because of the greater

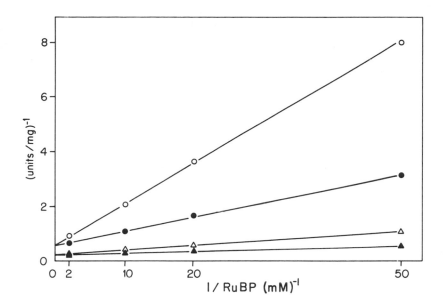

Figure 6. Lineweaver-Burk plots for the inhibition of spinach (●, o)
and R. rubrum (▲, Δ) carboxylases by bromodihydroxybutanone BP.
Prior to assay, enzyme was incubated in 10 mM MgCl₂/66 mM NaHCO₃/
50 mM Bicine (pH 8.0). Assays were carried out during a 60-sec
interval by the $^{14}CO_2$-fixation method (24) in the absence (▲, ●) or
presence (Δ, o) of 2 mM inhibitor.

Table 1

Extent of Incorporation of Bromodihydroxybutanone BP Into RuBP
Carboxylase and Number of Sulfhydryl Groups Modified

Sample	Enzymatic activity (% remaining)	^3H (mol reagent per mol enzyme)	No. of residues of carboxymethyl cysteine*	No. of sulfhydryl groups modified
Inactivated	1	13.8	90	8
Substrate-protected	95	12.2	87	11
Control	100		98	

*Determined by amino acid analysis after carboxymethylation of the
protein sample with iodoacetate.

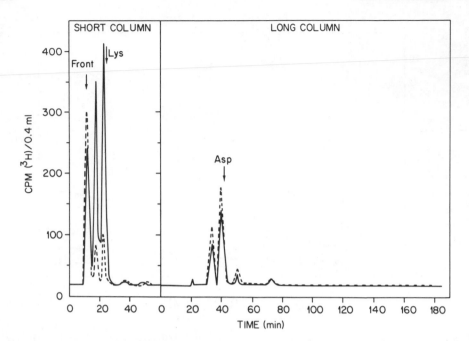

Figure 7. Radioactivity in hydrolysates of spinach carboxylase/
oxygenase after treatment, in the absence (——) and presence (- - -)
of RuBP, with bromodihydroxybutanone BP followed by reduction with
sodium ^{3}H-borohydride. Hydrolysates were chromatographed on the
amino acid analyzer, and 1-min fractions of the effluents, after
their emergence from the flow cell, were collected and counted.
The absorbance at 570 nm is not shown, but the elution positions
of some amino acids are indicated.

degree of modification of nonessential sulfhydryl groups in the
presence of RuBP. If, in the absence of RuBP, all residues modified
other than sulfhydryls contribute to the inactivation process, loss
of enzymic activity is due to derivatization of 5.8 residues (^{3}H
incorporation minus sulfhydryl residues modified).

        4. Kinds of Residues Modified. To determine what kinds of
residues other than cysteinyl are modified by bromodihydroxybutanone
BP, total acid hydrolysates of derivatized spinach protein labeled
with ^{3}H-borohydride were chromatographed on the amino acid analyzer
(Figure 7). The elution positions of tritiated compounds were compared with those of authentic standards prepared by the reactions
of glutathione and N-α-acetyllysine with bromodihydroxybutanone BP
followed by reduction with ^{3}H-borohydride and finally acid hydrolysis. Most of the radioactivity from the substrate-protected sample
elutes from the long column in the same positions (36 min and 42 min)

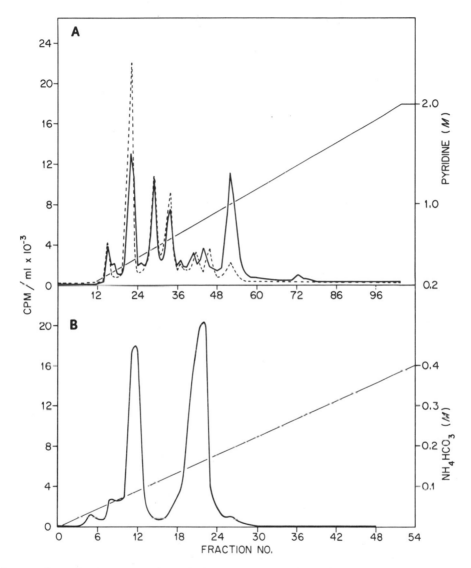

Figure 8. Chromatography on BioRad AG 50W-X2 of tryptic digests
of spinach carboxylase/oxygenase after treatment with bromodihydroxybutanone BP in the absence (———) and presence (---) of RuBP
(A). The major peak unique to the inactivated enzyme (fractions
50 to 57 in A) was chromatographed on DEAE-cellulose (B).

Table 2

Amino Acid Compositions of Purified Peptides Containing
Essential Residues of RuBP Carboxylase That Are
Modified by Bromodihydroxybutanone BP

| | Number of residues | | | |
| | | | After periodate oxidation | |
Amino acid	I	II	I	II
Derivative (18 min)*	0.5	0.6		
Derivative (24 min)*	0.7	0.7		
Lysine			2.1	1.0
Histidine		2.0		1.7
Arginine	1.1	1.0	1.0	1.0
Carboxymethylcysteine	1.0		0.6	
Aspartic acid		1.0		1.1
Threonine	1.0	1.0		
Serine		2.0		
Glutamic acid		2.1		2.2
Proline	1.9		1.8	
Glycine	1.9	5.2	2.2	4.9
Alanine				
Valine		2.1		2.1
Methionine				
Isoleucine**	1.0	1.0	1.0	1.0
Leucine	2.0	2.0	2.0	2.0
Phenylalanine				
Tyrosine	1.0		0.4	

*Time refers to elution position of the derivative from the short
column of the amino acid analyzer.
**The quantity of isoleucine in the sample was arbitrarily set to
1.0 residue.

as the cysteinyl derivatives prepared from glutathione (25). In
contrast, two radioactive peaks (emerging ahead of lysine at 18 min
and at 24 min) are seen in the short column runs on hydrolysates
of the inactivated enzyme which are not prominent in the protected
samples. The first (at 18 min) is coincident with the product
obtained from alkylation of acetyllysine by bromodihydroxybutanone
BP (25). The second peak has not been completely characterized
chemically, but it too represents a lysyl derivative, as shown below.
     The modification of lysyl residues must account for most of the
inactivation, since this represents the major difference between

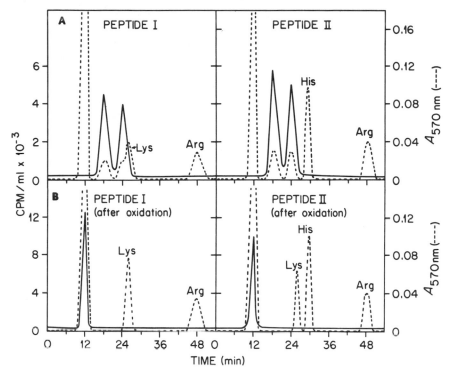

Figure 9. Analyses, on the short column of the amino acid analyzer, of hydrolysates of peptides I and II before (A) and after (B) oxidation with sodium metaperiodate.

inactivated and protected samples. This conclusion is supported by results of a differential labeling experiment (data not shown). The carboxylase/oxygenase was first treated with bromodihydroxybutanone BP under protective conditions, reduced with unlabeled borohydride, and dialyzed; the enzyme was then inactivated by retreatment with bromodihydroxybutanone BP and reduced with ^{3}H-borohydride. Hydrolysates of this material contained the radioactive lysyl derivatives, but the cysteinyl derivatives were virtually absent.

RuBP carboxylase/oxygenase from R. rubrum is also inactivated by bromodihydroxybutanone BP, and the major difference between substrate-protected and inactivated enzyme is again in the level of lysyl derivatization (20).

    5. Purification of Peptides Unique to Inactivated Carboxylase/ Oxygenase and Identity of Modified Residues. Samples of inactivated and substrate-protected spinach enzyme were digested with trypsin, and the digests were chromatographed on a cation-exchange resin (Figure 8A). One of the major radioactive peaks in the profile of the inactivated enzyme digest is virtually absent in that of the

substrate-protected enzyme digest. This is due to a peptide com-
prising 3.3 residues per molecule of carboxylase. Since the profiles from the two digests are otherwise quite similar, this peak
must represent the residues whose modifications result in inactivation. It is resolved into two radioactive peptides upon chromatography on DEAE-cellulose (Figure 8B); the peptide eluting first
is designated I and the other II. Peptides I and II are present in
a ratio of about 1:2 and thus represent, respectively, 1.1 and 2.2
modified residues per molecule. After successive chromatography
on phosphocellulose, Bio-Rad Aminex Ag 1-X4, and Sephadex G-25,
peptides I and II appear pure by peptide mapping and amino acid
composition (Table 2). The hydrolysate of each peptide contains
both radioactive derivatives eluting from the short column just
ahead of lysine (Figure 9A). The sum of the two derivatives approxi-
mates one residue. Both derivatives represent lysyl residues, as
shown by treatment of the peptide hydrolysates with sodium metaperiodate followed by chromatography on the amino acid analyzer

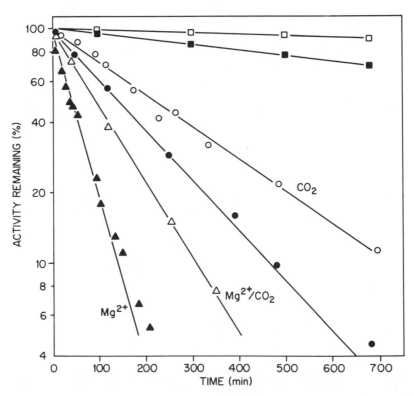

Figure 10. Time course of inactivation of RuBP carboxylase/oxygenase
(10 mg/ml) by 5 mM N-bromoacetylethanolamine phosphate in 50 mM
Bicine buffer (pH 8.0) with no additions (●), 5 mM $MgCl_2$ (▲), 66 mM
$NaHCO_3$ (o), 5 mM $MgCl_2$/66 mM $NaHCO_3$ (Δ).

(Table 2 and Figure 9B).   After oxidation, the hydrolysates contain
one additional residue of lysine and lack both labeled derivatives.
The radioactivity that was associated with the derivatives elutes
with the front.

        We believe that the presence of two distinct lysyl derivatives
in seemingly pure peptides is a consequence of bromodihydroxybutanone BP reacting with lysyl residues to form two different products.
Thus, peptides I and II could be identical in structure except for
the chemical nature of a derivatized lysyl residue.   This hypothesis
is unproven.

Reaction of N-Bromoacetylethanolamine Phosphate

        1.   Kinetics of Inactivation and Demonstration of Substrate
Protection.   Incubation of spinach RuBP carboxylase/oxygenase with
N-bromoacetylethanolamine phosphate results in parallel loss of
both activities in a pseudo-first-order fashion (Figure 10).   There

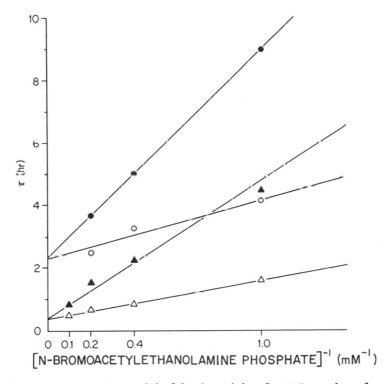

Figure 11.   Inactivation of half-time (τ) of RuBP carboxylase as a
function of the reciprocal of N-bromoacetylethanolamine phosphate
concentration.   The following conditions were used: 5mM $MgCl_2$/50 mM
Bicine ($\triangle$); 5 mM $MgCl_2$/66 mM $NaHCO_3$/50 mM Bicine ($\blacktriangle$); Mg^{2+}-free 50
mM bicine (o); Mg^{2+}-free 66 mM $NaHCO_3$/50 mM Bicine ($\bullet$).   All buffers
were pH 8.0.

is apparent inactivation of the conformer that lacks enzymic activity (modifications carried out in the absence of CO_2 and Mg^{2+}), the rate of which is stimulated by Mg^{2+} alone and decreased by CO_2 alone. This apparent effect of CO_2 is due to the increased ionic strength from added sodium bicarbonate, as the same effect is observed with sodium chloride. With a combination of CO_2 and Mg^{2+} (conditions under which the enzyme is fully activated), the rate of inactivation by N-bromoacetylethanolamine phosphate is slightly lower than in the presence of Mg^{2+} alone. RuBP and the competitive inhibitor butane-1,4-diol 1,4-bisphosphate (17) protect against inactivation.

Plotting the half-time of inactivation (τ) against the reciprocal reagent concentration (Figure 11) shows rate saturation. The data make it clear that Mg^{2+} increases the maximal velocity of inactivation (decreases T from 138 min to 24 min) and that $NaHCO_3$ merely alters the apparent affinity of N-bromoacetylethanolamine phosphate for the enzyme. Without CO_2, the apparent dissociation constant for the reagent-enzyme complex (K_{inact}) is 0.8 mM in the presence of Mg^{2+} and 3.0 mM in its absence. Thus, Mg^{2+} alters both T and K_{inact}. In the presence of 66 mM $NaHCO_3$, K_{inact} increases to 2.9 mM without Mg^{2+} and 11 mM with it.

2. Extent of Incorporation. ^{14}C-N-bromoacetylethanolamine phosphate is used to determine the degree of protein modification. The enzyme inactivated in the absence of Mg^{2+} contains about 2.7 moles reagent per mole protomer; a corresponding sample protected by RuBP contains about 0.9. The presence of Mg^{2+} in the modification reaction mixture reduces the level of incorporation to about 1.4 moles reagent per mole protomer, and the corresponding sample protected by butanediol bisphosphate contains about 0.5. By determining the level of incorporation during the time course of the incubation with reagent, a direct proportionality is seen between loss of enzymic activity and incorporation (expressed as the difference between incorporation in unprotected and protected samples) (Figure 12). Extrapolation of these data to complete inactivation gives values of 1.8 moles reagent per mole protomer in the absence of Mg^{2+} and 1.2 in its presence.

3. Kinds of Residues Modified. The sites of reaction of ^{14}C-N-bromoacetylethanolamine phosphate are readily determined by amino acid analysis because the reagent is an N-substituted carboxamidomethyl compound and therefore all derivatized residues will appear in acid hydrolysates as carboxymethyl (Cm) amino acids. Radioactivity profiles from the amino acid analyzer for hydrolysates of carboxylase/oxygenase modified in the presence and absence of Mg^{2+} are compared in Figure 13. Since CO_2 was present during all modifications, the differences observed must reflect the Mg^{2+}-induced conversion of enzymically inactive conformer to the active one. Carboxymethylcysteine is the major radioactive compound found in hydrolysates of enzyme modified in the absence of Mg^{2+}. The corresponding substrate-protected sample contains one less residue of carboxymethylcysteine per protomer. In addition, carboxymethyl-

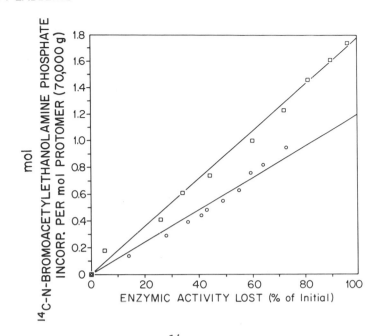

Figure 12.   Incorporation of ^{14}C-N-bromoacetylethanolamine phosphate
as a function of enzymic activity lost in the absence (□) and presence (o) of Mg^{2+}, expressed as the difference between incorporation
by unprotected and protected enzyme (the protector without Mg^{2+} is
RuBP; with Mg^{2+} it is butanediol).

lysine is found in hydrolysates of carboxylase/oxygenase after
modification in the presence of Mg^{2+}.   Butanediol bisphosphate gives
slight protection of cysteinyl residues (0.2 per protomer) but
essentially complete protection of lysyl residues (0.9 per protomer).
Thus, inactivation correlates with cysteinyl or lysyl alkylation
depending on the absence of presence of Mg^{2+}, respectively.
        4.   Purification of Peptides Unique to Inactivated Carboxylase/
Oxygenase.   Tryptic digests of modified enzyme were subjected to
ion-exchange chromatography on BioRad Aminex AG 50W-X2.   Two
major radioactive peptides (designated C1 and C2, which on the
basis of radioactivity represent 0.75 and 0.46 moles peptide per
mole inactive protomer, respectively) are resolved from tryptic
digests of enzyme inactivated in the absence of Mg^{2+}; these peptides are missing from digests of substrate-protected enzyme (Figure 14A).   In contrast, enzyme inactivated in the presence of Mg^{2+}
yields a single major radioactive tryptic peptide (designated L1,
and representing 0.73 mole peptide per mole inactive protomer) that
is not seen in samples protected by butanediol bisphosphate or the
transition-state analog carboxyribitol bisphosphate (27) (Figure
14B).   To determine the type of residue modified in the labeled peptides, an aliquot of each peak was subjected to amino acid analysis;

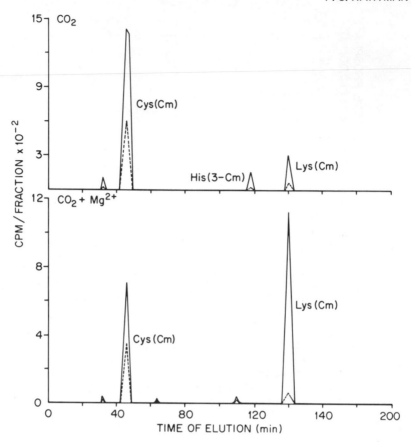

Figure 13.   Chromatographic profiles of hydrolysates of carboxylase/
oxygenase after modification with ^{14}C-N-bromoacetylethanolamine
phosphate.   Chromatography was done on an amino acid analyzer without use of its ninhydrin system; 2-min fractions were collected and
counted. Cm = carboxymethyl.   Upper panel: profiles for enzyme
samples modified in the absence of Mg^{2+} with (---) or without (——)
protector (RuBP).   Lower panel: profiles for enzyme samples modified
in the presence of Mg^{2+} with (---) or without (——) protector
(butanediol bisphosphate).

these identifications are indicated in the figure.   Both of the two
major substrate-protected radioactive peptides in Figure 14A (from
enzyme inactivated in the absence of Mg^{2+}) are derivatized at
cysteinyl residues, whereas the major radioactive peptide in
Figure 14B (from enzyme inactivated in the presence of Mg^{2+}) is
derivatized at a lysyl residue.   Peptides C1, C2, and L1 were
purified to homogeneity by successive chromatography on DEAE

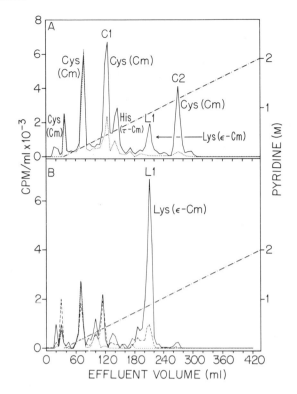

Figure 14.   Chromatography on BioRad AG 50 in a pyridine gradient
(—·—) of tryptic digests of enzyme modified by ^{14}C-N-bromoacetyl-
ethanolamine phosphate.   The kind of labeled carboxymethyl (Cm)
amino acid found in hydrolysates of each major radioactive peptide
is indicated.   (A) The enzyme was modified in the absence of Mg^{2+}
with (···) or without (——) protector (1 mM RuBP).   (B) The enzyme
was modified in the presence of Mg^{2+} without protector (——), with
20 mM butanediol bisphosphate (---), or with 0.2 mM carboxyribitol
bisphosphate (···).

cellulose and phosphocellulose, and their amino acid compositions
are given in Table 3.   As expected, hydrolysates of peptides C1
and C2 contain one radioactive Cys(cm) residue, and the hydrolysate
of peptide L1 contains one radioactive Lys(ϵ-Cm) residue.   The
latter also contains one Cys(Cm) residue; this, however, is not
radioactive and is therefore not reagent-derived (preceding tryptic
digestion, all free -SH groups in the enzyme were carboxymethylated
with iodoacetate).   Peptides C2 and L1 have identical compositions
(except for the residue that bears the reagent moiety), which are

Table 3

Amino Acid Compositions of Peptides Unique to RuBP Carboxylase/
Oxygenase Inactivated With N-Bromoacetylethanolamine Phosphate

Amino acid	Number of residues		
	C1	C2	C3
Lysine	0.9	2.0	1.0
ε-Carboxymethyllysine			1.0*
Arginine		1.1	1.0
Tryptophan**	2.2		
Carboxymethylcysteine	0.9*	1.0*	0.9
Threonine		0.9	1.0
Serine	0.9		
Glutamic acid	2.1		
Proline	0.9	1.9	1.9
Glycine		2.0	2.2
Alanine	2.9		
Valine	0.9		
Isoleucine		1.0***	1.0***
Leucine	1.0***	2.1	2.1
Tyrosine		0.9	1.0

*On the basis of assays for radioactivity in the effluent
 from the amino acid analyzer, this residue is labeled and
 therefore bears the reagent moiety in the intact peptide.
**Trp was determined from the A_{280} nm.
***This amino acid was arbitrarily assigned a value of 1.0.

the same as the composition of peptide I obtained from the enzyme
derivatized by bromodihydroxybutanone BP (see Table 2).

Sequence Determinations

Peptides I and II obtained from carboxylase/oxygenase modified
by bromodihydroxybutanone BP and peptides C1, C2, and L1 obtained
from the enzyme modified by N-bromoacetylethanolamine phosphate
were sequenced by automated Edman degradations in a Beckman 890C
sequencer. The established sequences follow (the residues underlined carry the reagent moiety):

    II   Leu-Ser-Gly-Gly-Asp-His-Ile-His-Ser-Gly-Thr-Val-Gly-Lys-
         Leu-Glu-Gly-Glu-Arg
    I   Tyr-Gly-Arg-Pro-Leu-Leu-Gly-Cys-Thr-Ile-Lys-Pro-Lys
    L1   Tyr-Gly-Arg-Pro-Leu-Leu-Gly-Cys-Thr-Ile-Lys-Pro-Lys

```
C2   Tyr-Gly-Arg-Pro-Leu-Leu-Gly-Cys-Thr-Ile-Lys-Pro-Lys
C1   Trp-Ser-Pro-Glu-Leu-Ala-Ala-Ala-Cys-Glu-Val-Trp-Lys
```

As anticipated from their amino acid compositions, peptides I, L1, and C2 are derived from the same region of the polypeptide chain and differ only in the nature and site of derivatization.

Other Potential Affinity Labels

The cis- and trans-epoxybutanediol bisphosphates were synthesized in an effort to find a reagent with binding specificity similar to that of bromodihydroxybutanone BP but with decreased chemical reactivity so as to minimize the modification of non-essential sulfhydryl groups. The rationale for preparing N-bromoacetyldiethanolamine bisphosphate was that extension of the distance between the two phosphate groups as found in bromodihydroxybutanone BP, to approximate RuBP more closely, would result in increased affinity for the active site. N-bromoacetylphosphoserine and phosphoglycolic acid azide represent attempts to utilize reactive analogs of 3-phosphoglycerate and phosphoglycolate as affinity labels. None of these reagents inactivates RuBP carboxylase/oxygenase at a sufficiently rapid rate to merit further experimentation.

CONCLUSIONS

Despite the alkylation of nonessential sulfhydryls and the incomplete characterization of the lysyl derivatives, it is clear that inactivation of spinach RuBP carboxylase/oxygenase by bromodihydroxybutanone BP is a consequence of modification of two different lysyl residues. On the basis of the stoichiometry of lysyl modification (the combined yield of peptides I and II is <1 mole per mole protomer) and the constant ratio of peptide I/peptide II regardless of bromodihydroxybutanone BP concentration used to inactivate (data not shown), we believe that within a given subunit the two lysyl residues are mutually exclusive with respect to modification, and that modification of either is sufficient for inactivation. The affinity of bromodihydroxybutanone BP for RuBP carboxylase/oxygenase, as demonstrated by the observed competitive inhibition and the protection afforded by RuBP against inactivation, suggest that the two labeled lysyl residues are in the active-site region. Other observations that indirectly support this conclusion are the unusual reactivity of the lysyl residues modified [in model systems the reactivity of bromodihydroxybutanone BP toward sulfhydryls exceeds its reactivity toward amino groups by at least 100-fold (25)], the lack of reactivity of these lysyl residues in denatured enzyme (28) (i.e., the selective modification of lysyl residues requires a catalytically functional binding site), the presence of these lysyl residues within the large subunit as is the catalytic site (29), and the inactivation of R. rubrum RuBP carboxylase/oxygenase as a consequence of lysyl alkylation.

As regards the inactivation of RuBP carboxylase/oxygenase by
N-bromoacetylethanolamine phosphate, the lysyl residue susceptible
to alkylation in the presence of CO_2 and Mg^{2+} almost certainly
occupies a position within the active-site region. The degree of
inactivation is directly proportional to the extent of modification;
the lysyl residue is protected by a competitive inhibitor and by
the transition-state analog against alkylation; and inactivation
exhibits rate saturation, which suggests specific binding of reagent
as a prerequisite to covalent reaction. Alkylation of a single
lysyl residue represents a high degree of specificity [given the
fact that each promoter contains thirty lysines (30)], and, as with
bromodihydroxybutanone BP, represents an unusual reactivity in
comparison with model compounds in which sulfhydryl groups react
far more rapidly than amino groups. Indirect evidence that N-bromoacetylethanolamine phosphate reacts at a site for RuBP is provided
by the finding that the reagent binds more tightly to the inactive
conformer (K_{inact} of 2.9 mM in the absence of Mg^{2+}) than to the
active conformer (K_{inact} of 11 mM in the presence of Mg^{2+}), as was
observed for substrate (12, 31).

The inactivation that occurs in the absence of Mg^{2+} and that
correlates with modification of two different sulfhydryl groups
might be a consequence of prevention of the activation induced by
CO_2 and Mg^{2+}. However, this inactivation also exhibits rate saturation and is prevented by RuBP as a consequence of protection
against modification.

Sequence determinations of the tryptic peptides containing the
essential residues have proven quite informative. We find that
bromodihydroxybutanone BP and N-bromoacetylethanolamine phosphate
have one common site of reaction (see sequences of peptides I and
L1). Thus, two chemically different, reactive phosphate esters with
demonstrated affinities for the carboxylase/oxygenase alkylate the
same lysyl residue. This observation, taken together with data
already discussed, provides a rather compelling argument that
the lysyl residue in question is within a binding site for
phosphate esters.

The sequence data also suggest that the catalytic site and
the previously proposed (12) allosteric site for RuBP are equivalent, overlapping, or contiguous. The presence of the allosteric
site is inferred primarily from the knowledge that RuBP binds
tenaciously to the inactive conformer and inhibits activation by
Mg^{2+} and CO_2 (12-14). These observations, however, do not rule
out the existence of only a single site with nonproductive binding
to the inactive conformer, as suggested earlier (4, 14). In agreement with this possibility, we find that in the absence of Mg^{2+}
(inactive conformer) inactivation correlates with the modification
of two cysteinyl residues, one of which is only three residues
removed from the lysyl residue that is selectively modified in
the presence of Mg^{2+} (see sequences of peptides L1 and C2). Thus,

it seems plausible that Mg^{2+} alters slightly the topology of a single binding site for RuBP (or N-bromoacetylethanolamine phosphate) so that in the reagent-enzyme complex formed from inactive conformer a sulfhydryl is accessible for alkylation, whereas in the reagent-enzyme complex formed from active conformer an ϵ-amino group is accessible. Also consistent with equivalence of the presumed allosteric site for RuBP and the catalytic site is the fact that the lysyl residue accessible to both reagents is alkylated by N-bromoacetylethanolamine phosphate in the presence of Mg^{2+} but is alkylated by bromodihydroxybutanone BP in the absence of Mg^{2+}.

The finding of at least one cysteinyl residue in the region of RuBP binding may explain, in part, the previously recognized sensitivity of the enzyme to sulfhydryl reagents.

Studies on the inactivation of RuBP carboxylase/oxygenase by pyridoxal phosphate are also consistent with the presence of lysyl residues at the active site. The reagent is highly selective in that inactivation correlates with the modification of a small number of lysyl residues (0.5 to 1.0 per protomer) (4-6). Inhibition studies (4) indicate that pyridoxal phosphate has a high affinity for the spinach enzyme; complex formation between the R. rubrum enzyme and reagent is indicated by saturation kinetics of inactivation (6). Thus, several types of data suggest that pyridoxal phosphate is an active-site-directed reagent for carboxylase/oxygenase. It will be of interest to learn whether the lysyl residue(s) modified is the same as one of those modified by bromodihydroxybutanone BP or N-bromoacetylethanolamine phosphate.

The amino group of spinach carboxylase/oxygenase that is modified by pyridoxal phosphate has been suggested as the site for binding of CO_2 as substrate (4). This suggestion was prompted by the finding of inhibition (with respect to CO_2) of the enzyme by pyridoxal phosphate. However, whether the inhibition is competitive or noncompetitive is unclear. When pyridoxal phosphate was tested by identical methodologies under the same conditions with either CO_2 or RuBP as the variable substrate, inhibition appeared noncompetitive in both cases. Furthermore, the lack of protection against inactivation afforded by saturating levels of bicarbonate seems inconsistent with an involvement of the target lysyl residue in CO_2 binding. Another perplexing result with the spinach enzyme is the absence of rate saturation for inactivation despite a K_I of 1 μM for pyridoxal phosphate in the inhibition experiments.

Since bicarbonate stimulates the rate of inactivation of spinach carboxylase/oxygenase by bromodihydroxybutanone BP, the two lysyl residues that are sites of modification (one of these residues is also modified by N-bromoacetylethanolamine phosphate) do not appear to be involved in binding of CO_2 either as effector or as substrate. We believe that interpretations of existing chemical modification data in terms of precise catalytic roles of the residues labeled are premature.

Acknowledgments: Research sponsored by the Division of Biological and Environmental Research, U.S. Department of Energy, under contract W-7405-eng-26 with the Union Carbide Corporation. JVS is a predoctoral fellow supported by Grant GM 1974 from the National Institute of General Medical Sciences, NIH.

REFERENCES

1. Nishimura, M. and Akazawa, T., J. Biochem. 76, 169-76 (1974).
2. Takabe, T. and Akazawa, T., Arch. Biochem. Biophys. 169, 686-94 (1975).
3. Sugiyama, T., Akazawa, T., Nakayama, N., and Tanaka, Y., Arch. Biochem. Biophys. 125, 107-13 (1968).
4. Paech, C., Ryan, F. J., and Tolbert, N. E., Arch. Biochem. Biophys. 179, 279-88 (1977).
5. Whitman, W. B. and Tabita, F. R., Biochemistry 17, 1282-7 (1978).
6. Whitman, W. B. and Tabita, F. R., Biochemistry 17, 1288-93 (1978).
7. Lawlis, V. B. and McFadden, B. A., Biochem. Biophys. Res. Commun. 80, 580-5 (1978).
8. Schloss, J. V., Norton, I. L., Stringer, C. D., and Hartman, F. C., Federation Proc. 37, 1310 (1978).
9. Kitz, R. and Wilson, I. B., J. Biol. Chem. 237, 3245-9 (1962).
10. Meloche, H. P., Biochemistry 6, 2273-80 (1967).
11. Meloche, H. P., Luczak, M. A., and Wurster, J. M., J. Biol. Chem. 247, 4186-91 (1972).
12. Chu, D. K. and Bassham, J. A., Plant Physiol. 55, 720-6 (1975).
13. Lorimer, G. H., Badger, M. R., and Andrews, T. J., Biochemistry 15, 529-36 (1976).
14. Laing, W. A. and Christeller, J. T., Biochem. J. 159, 563-70 (1976).
15. McFadden, B. A. and Tabita, F. R., BioSystems 6, 93-112 (1974).
16. Norton, I. L., Welch, M. H., and Hartman, F. C., J. Biol. Chem. 250, 8062-8 (1975).
17. Schloss, J. V. and Hartman, F. C., Biochem. Biophys. Res. Commun. 77, 230-5 (1977).
18. Stringer, C. D. and Hartman, F. C., Biochem. Biophys. Res. Commun. 80, 1043-8 (1978).
19. Schloss, J. V., Stringer, C. D., and Hartman, F. C., J. Biol. Chem., in press (1978).
20. Schloss, J. V. and Hartman, F. C., Biochem. Biophys. Res. Commun. 75, 320-8 (1978).
21. Wishnick, M. and Lane, M. D., Methods Enzymol. 23, 570-7 (1971).
22. Tabita, F. R. and McFadden, B. A., J. Biol. Chem. 249, 3453-8 (1974).
23. Racker, E., in Methods of Enzymatic Analysis, pp. 188-90, H. U. Bergmeyer, Editor, Academic Press, New York, 1963.

24. Lorimer, G. H., Badger, M. R., and Andrews, T. J., Anal.
    Biochem. 78, 66-75 (1977).
25. Hartman, F. C., J. Org. Chem. 40, 2638-42 (1975).
26. Hartman, F. C., Suh, B., Welch, M. H., and Barker, R.,
    J. Biol. Chem. 248, 8233-9 (1973).
27. Siegel, M. I. and Lane, M. D., Biochem. Biophys. Res. Commun.
    48, 508-16 (1972).
28. Hartman, F. C., Welch, M. H., and Norton, I. L., Proc. Natl.
    Acad. Sci. USA 70, 3721-4 (1973).
29. Nishimura, M. and Akazawa, T., Biochem. Biophys. Res. Commun.
    54, 842-8 (1973).
30. Siegel, M. I., Wishnick, M., and Lane, M. D., in The Enzymes,
    3rd ed., Vol. 6, p. 177, P. D. Boyer, Editor, Academic Press,
    New York, 1972.
31. Wishnick, M., Lane, M. D., and Scrutton, M. C., J. Biol. Chem.
    245, 4939-47 (1970).

DISCUSSION

AKAZAWA: Might your finding on the substrate-binding site of
the enzyme molecule in the presence of Mg^{2+} be related to our observations on the Mg^{2+}-induced optimal pH-shifting phenomenon?

HARTMAN: The alteration in the specificity of N-bromoacetylethanolamine phosphate and the shift in enzymic pH optimum as induced by Mg^{2+} may both reflect a conformational change at the
active site.

CHOLLET: In your original paper (Proc. Natl. Acad. Sci. USA,
70, 3721-4, 1973) on the butanone derivative, you reported that magnesium had no effect. How do you explain the difference between
this and the results with bromoacetyl derivatives?

HARTMAN: We still claim that magnesium has no influence on the
bromodihydroxybutanone BP reagent. I can state unequivocally that
it has no influence on the rate of inactivation. We have not compared the products obtained with and without magnesium because we
have concluded that the rate does not change and neither does the
site of reaction. Why does magnesium influence the site of reaction in one of the reagents and not in the other? I think that
is explained by the mobility of the reagent within its binding
site. I would further argue that the binding site, in the presence
and absence of magnesium, is one and the same with respect to
binding RuBP. The bromodihydroxybutanone BP sees the same lysine
residues in the absence of magnesium as the bromoacetyl reagent
sees in the active conformation.

STRUCTURAL STUDIES OF RIBULOSE 1,5-BISPHOSPHATE

CARBOXYLASE/OXYGENASE

David Eisenberg, T. S. Baker, S. W. Suh, and W. W. Smith

Molecular Biology Institute and Department of Chemistry, University of California, Los Angeles, California 90024

INTRODUCTION

From studies of three crystal forms of ribulose bisphosphate (RuBP) carboxylase from <u>Nicotiana</u> <u>tabacum</u> (called I, II, and III), we have determined the subunit organization of RuBP carboxylase in increasing detail. Combined x-ray diffraction and electron microscope data from these crystals show that there must be some multiple of eight polypeptide chains in the molecule and that the polypeptides are arranged around a fourfold axis of symmetry. At low resolution the eight copies of each polypeptide are equivalent. In more formal terms, the RuBP carboxylase molecule is characterized by point group symmetry D_4 (422, see Figure 1 below). The molecule has a square cross section, about 11 nm on an edge, and a cylindrical channel about 2 nm in diameter which runs along the fourfold axis perpendicular to the square cross section. Four large subunits are arranged in a ring perpendicular to the fourfold axis, and two such rings are eclipsed, forming a two-level structure that extends about 10 nm along the fourfold axis.

Three earlier models for the subunit organization of RuBP carboxylase (summarized in ref. 2) were based on relatively incomplete crystallographic or electron microscope results and on older subunit stoichiometries. All three early models are incompatible with newer, more complete data.

The packing of molecules within crystal form III is similar to the structure of crystallites, believed to be composed of RuBP carboxylase, found in chloroplasts. This suggests that form III may be an <u>in</u> <u>vivo</u> crystal as well as an <u>in</u> <u>vitro</u> crystal form. Our <u>in</u> <u>vitro</u> crystals appear to be suitable for x-ray diffraction studies at least to a resolution of 5 Å, and diffraction data extend to 2.7 Å.

SUBUNIT ORGANIZATION OF RuBP CARBOXYLASE

Conditions for growing form I crystals as large as 1 mm on an edge were developed by Kawashima and Wildman (8). Because of the extreme fragility of the crystals (which collapse to form a drop of water over a film of protein upon the slightest provocation), crystallizing conditions had to be modified to produce the crystals directly within x-ray capillaries (2). X-ray diffraction patterns could then be recorded, but extended only to 50-$\overset{\circ}{A}$ resolution because of the unusually great solvent content of the crystals (79% by weight) and the consequent weak bonding of adjacent molecules. The diffraction pattern was sufficient, however, to narrow the space group to two possibilities, I432 and I$4_1$32. Electron micrographs of microcrystalline fragments and of sectioned crystals were found to be compatible with space group I$4_1$32. A careful measurement of the density (9), when combined with estimates of the molecular weight, led to the conclusion that there are only 12 molecules in the very large unit cell (Table 1). The minimum molecular symmetry of D_2 (222) follows from these observations (2), since with 12 molecules per unit cell in space group I$4_1$32, each must be located on a special position of symmetry D_2 (222).

The 222 symmetry restricts the possible subunit stoichiometries of the RuBP carboxylase molecule. Each subunit must be present in a multiple of four copies. This can be expressed as $L_{4n}S_{4m}$, where L represents the large subunit, S represents the small subunit, and n and m are integers.

Further information on subunit structure was gained from form II crystals (1). X-ray diffraction patterns to a resolution of 15 $\overset{\circ}{A}$ and measurements of crystal density led to the conclusion that RuBP carboxylase molecules must contain at least one fourfold axis, as well as two twofold axes. The method of analysis was similar to that with the form I crystals: x-ray diffraction patterns established that the space group is P$42_1$2, and density measurements showed that there are most likely 6 molecules per unit cell. This is possible only if molecules have a fourfold axis of symmetry.

The fourfold molecular axis required by form II is compatible with the three twofold axes demanded by form I if the actual molecular symmetry is D_4 (422). This symmetry requires a multiple of eight copies of each type of subunit, $L_{8n}S_{8m}$. Only a structure of the type L_8S_8 is compatible with the generally accepted molecular weights of the subunits and of the entire molecule. A highly schematic model of RuBP carboxylase is shown in Figure 1, having symmetry D_4 (422) and stoichiometry L_8S_8.

This model is based in part on an electron micrograph (Figure 2a) of a platelet of a form II crystal. This view down the fourfold molecular axis reveals that the molecules are square with a cross section of about 11 x 11 nm. The subunits are arranged about a cylindrical hole about 2 nm in diameter. X-ray diffraction patterns reveal that the square molecule has two layers of subunits along the fourfold axis. Our interpretation of the molecular images in Figure 2 is that one layer of the molecule is formed from four

Table 1

Crystals of RuBP Carboxylase From Tobacco

	Form I	Form II	Form III
Crystallizing: pH	7.4 to 8.8	6.0 to 6.2[*]	5.2
salts	25 mM Tris	Tris or phosphate	200 mM phosphate 300 mM $(NH_4)_2SO_4$
Crystal morphology	Rhombic dodecahedrons	Square plates & triangular prisms	Pseudo-rhombic dodecahedrons
Density, g cm^{-3}	1.058+0.005	1.096+0.006	1.184+0.007
Space Group	14_132	$P42_12$	$P4_22_12$ or $P4_222$
Dimensions, Å: a=b	383+3	230+2	148.7+0.2
c	383+3	315+3	137.5+0.2
Molecules per cell	12	6	2
Minimum molecular symmetry	D_2	C_2, C_4	D_2 (>5 Å) D_4 (<5 Å)
Finest x-ray resolution, Å	50	14	2.7
Reference	(2)	(1)	(3)

[*]Form II crystals were subsequently grown at pH 7.8 (Tris buffer) with 70 mM NaCl and 4% polyethylene glycol (MW = 6000). The diffraction pattern of the hk0 zone was identical to that from the crystals grown at pH 6.

elliptical units bonded in a square planar ring. These are almost certainly the large subunits. We have no information on the location of the small subunits; their hypothetical placement in Figure 1 is intended only to represent the observed mass distribution of the molecules with a model built from spherical subunits.

The molecular symmetry D_4 (422) is confirmed at low resolution by the form III crystals, which were discovered (3) shortly after form II. To a very good approximation the form III crystals are characterized to about 5-Å resolution by space group I422, with two molecules per unit cell. This requires that each molecule have D_4 symmetry. When crystals are examined at higher resolution, it is found that the site symmetry of molecules is lower, with molecules required only to have D_2 (222) symmetry. This lower symmetry at higher resolution could conceivably reflect the minor difference in amino acid sequences of two types of small subunits, S and S' (7,14). This would require, however, an unusual pattern of assembly, in which S and S' are distributed to form molecules of symmetry D_2. Any other factor causing LS protomers to pair would also cause a reduction in symmetry to D_2.

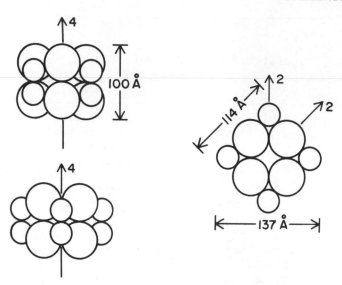

Figure 1. A schematic model of the RuBP carboxylase from tobacco, based on x-ray diffraction and electron microscope studies. The eight large and eight small polypeptide chains are arranged with symmetry D_4 (422). A channel about 20 Å in diameter runs along the fourfold axis, perpendicular to two eclipsed rings of four large subunits. The two rings of four are related by twofold axes. The positions of the small subunits are unknown, and those shown were chosen only to give a molecule with about the same mass distribution as that observed in micrographs, with no compelling reason to suppose that they are accurate.

RELIABILITY OF THE MODEL

To facilitate assessment of the reliability of our model, we should emphasize the assumptions on which it is based. The conclusions on molecular symmetry follow rigorously from the determination of the space group and from the number of molecules in the unit cell. Determination of the space group is unambiguous for form II, and for form III at low resolution. For form I, two space groups are compatible with x-ray diffraction data, but only one is compatible with the electron micrographs of sectioned crystals and of microcrystals.

Correct determination of the number of molecules per unit cell, as explained previously (1, 2), depends on several measurements. The greatest errors are introduced by uncertainties in the molecular weights and the partial specific volumes of protein and liquid of crystallization. In forms I and II, the joint uncertainties in these measurements produce a probable error that barely rules out

Figure 2. An electron micrograph of a thin platelet crystal of form II. White squares with holes are molecules, viewed down their fourfold axes, and are separated by dark strips of negative stain; nearest neighbors visible are separated by 16 nm.

other conceivable values. Because of the tighter packing in form III, the uncertainty about the number of molecules per unit cell is much less, and the conclusion of D_4 (422) symmetry at low resolution could be wrong only if the molecular weight were greatly in error.

Other conclusions (1) on subunit organization come mainly from form II. The idea of a two-layered structure is based on the measured unit cell size and the micrograph of Figure 2, though the intensity distribution of the diffraction pattern of form II is also consistent with two layers. The suggestion that four elliptical large subunits are bonded in a ring is derived almost entirely from Figure 2. We have no information whatsoever on the positions of small subunits, and the model in Figure 1 is only one of several possibilities consistent with the molecular symmetry and the mass distribution observed in micrographs.

ELECTRON MICROSCOPY OF DRIED, NEGATIVELY-STAINED SOLUTIONS

Molecular images of RuBP carboxylase from several sources have been examined in electron micrographs of dried, negatively-stained solutions. A limitation of this method is that, as the solutions

are dried, the heavy atom stain offers little support, and surface
forces between molecules and substrate tend to distort the molecules. Also, damage from the electron beam is severe and tends
further to degrade native molecular features.

Baker et al. (2) studied RuBP carboxylase from tobacco by this
method and observed poorly defined molecular images, often circular
with diameter about 11 nm. Numerous molecules, however, showed a
central, stain-filled region about 2 nm in diameter, and several
molecules showed a stain-filled line across the molecule. This
led Baker et al. to suggest that the RuBP carboxylase molecule contains a unique channel passing through the molecule, which they
identified with the fourfold molecular axis. The channel along
the axis projects either as a circle or a line, depending on the
orientation of the molecule on the substrate.

McFadden et al. (11) studied RuBP carboxylase from Euglena
gracilis, and found that it consists of two types of subunits,
with molecular and subunit masses similar to those of the tobacco
enzyme. Molecular images of the enzyme in micrographs include
projections of cubes, and also rosettes with what the authors describe as eightfold symmetry.

Purohit et al. (12) examined micrographs of RuBP carboxylase
from Thiobacillus intermedius. This molecule consists of eight
copies of a single type of subunit with a molecular weight of about
55,000. The molecular images are circular, many having a central
stain-filled hole. Some images are consistent with the eight subunits being arranged in two eclipsed rings of four.

A detailed model was put forward by Bowien et al. (4) for the
subunit organization of RuBP carboxylase from Alcaligenes eutrophus.
This organism is a facultative, chemolithotrophic, hydrogen bacterium. Its RuBP carboxylase contains eight subunits of molecular
weight about 52,000 and eight of molecular weight about 13,000.
Molecular images stained with uranyl acetate appear relatively well
preserved. Many images display a stain-filled central region, and
some molecules seem to have an elongated or double stain-filled
region. The molecular diameter is about 13 nm, and the periphery
often displays V-shaped projections. Other views of molecules
seem to contain three, and in some cases, four parallel layers.
Urea-treated samples showed some smaller rings, of diameter 8.5 nm,
and also some prominent U- or V-shaped structures.

Bowien et al. (4) suggest that the molecule is characterized
by 422 symmetry and is organized in four layers perpendicular to
the fourfold axis. Their model is based on two central layers,
each made up of four U- or V-shaped larged subunits. Each of the
two outer layers is composed of a ring of four spherical small subunits. The rings can be separated from the central layers by urea
to produce the smaller ring structures.

The model of Bowien et al. is nearly the same as that of Figure
1 for the positions of the large subunits, which together make up
75% of the mass of the molecule. In fact, the model of Figure 1 is
not based on any observations inconsistent with the Bowien four-layer

suggestion. However, the micrograph of Figure 2, and filtered images prepared from it, contain no hint of V- or U-shaped large subunits (2).

In assessing the positions of the small subunits, the cross-linking data of Roy et al. (13) are of interest. These investigators cross-linked the small subunits of RuBP carboxylase from pea leaves with four reagents. All four reagents caused formation of dimers of the small subunit. In addition, small amounts of trimers and tetramers of small subunits were detected with one reagent. These data suggest that small subunits are closely paired in the RuBP carboxylase molecule, and that three or four may be close together. Both the models of Figure 1 and of Bowien et al. are consistent with cross-linked dimers, and the latter is consistent with cross-linked tetramers.

ARE CHLOROPLAST CRYSTALLITES FORM III OF RuBP CARBOXYLASE?

Crystalline inclusions have been observed in chloroplast stroma by electron microscopy (5, 6, 10, 15-17) in both sectioned and freeze-etched chloroplasts, and in both tobacco and spinach chloroplasts. In some cases the crystallites may be induced by the chemical treatments preceding microscopy, but in other cases the crystals are believed to exist in situ. Lattice plane spacings from about 6 to 12 nm have been observed in the crystallites, and several investigators have suggested that such spacings are consistent with the crystals being composed of RuBP carboxylase.

We may ask whether any of the in vitro crystal forms I, II, or III have lattice spacings compatible with those observed in the inclusion crystallites. Comparison is not entirely straightforward because of several effects. Preparation of specimens for electron microscopy involves steps that can affect lattice dimensions, such as fixation, dehydration, infiltration of polymers, sectioning, staining, and beam damage. Any of these might distort crystallite dimensions. Also, in studies of sectioned crystallites, since chloroplasts are usually sectioned at random, there is no reason why the micrographs could present views along rational crystal directions. In freeze-etch experiments, the replica of the cleaved surface may not lie normal to the direction of view, and this would introduce a distortion in dimensions.

Even with these uncertainties, it is immediately evident that the lattice spacings of forms I and II are too great to match those of chloroplast inclusions. The spacings for form III, however, are consistent with many spacings observed in chloroplast inclusions. A detailed comparison with form III is given in Table 2 for two electron microscope studies of inclusions, and the probable packing of molecules in form III crystals is shown in Figure 3.

In making such comparisons, one must bear in mind that studies of stained sections reflect different dimensions than do studies of

Table 2

Comparison of Lattice Parameters of Crystal Form III With Two
Electron Microscope Studies of Chloroplast Inclusion Crystallites

A: Study of Larsson et al. (10) with spinach;
B: study of Willison and Davey (16) with tobacco.

Method	Inclusion parameters	Form III parameters
A. Negatively stained, sectioned inclusions	Parallel lines spaced at 7.5–8.5 nm	d_{111} = 8.4 nm
	Parallel lines spaced at 9.8–11.7 nm	d_{110} = 10.5 nm*
	Two sets of parallel lines with spacings at 8.9–11.0 nm crossing at an angle of about 90°	d_{110} = 10.5 nm* $d_{1\bar{1}0}$ = 10.5 nm Intersection angle = 90°
	Three sets of parallel lines with spacings at 8.5–9.5 nm forming a hexagonal pattern (angles 54–66°)	View along 111 direction d_{101} = 10.1 nm d_{110} = 10.5 nm Intersection angle 59°
B. Negatively stained, sectioned inclusions	Two sets of parallel lines spaced at about 10 nm crossing at an angle of about 90°	View along 001 direction d_{110} = 10.5 nm* $d_{1\bar{1}0}$ = 10.5 nm
Freeze-etched inclusions	Square arrays of molecules with spacings of about 12 nm	View along 001: square* array with spacings 14.9 x 14.9 nm
	Hexagonal arrays with spacings of about 10.4 nm	View along 110: quasi-hexagonal array with spacings of 12.6 and 13.8 nm

*See Figure 3 for an illustration of these spacings.

shadowed replicas of freeze-etched samples. Transmission micrographs of stained sections are projections. Thus, for a centered
tetragonal lattice such as form III, the projection along the 001
direction shows the primitive cell dimension $a/\sqrt{2}$, whereas a

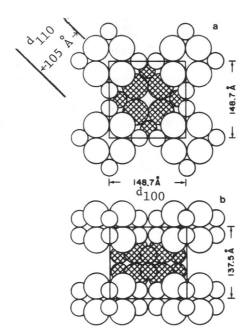

Figure 3. The probable packing of RuBP carboxylase molecules in form III, with the shaded molecule one-half unit cell below the unshaded molecules. (a) View along 001 direction. Note that the intermolecular spacing of 149 Å would be shown by the freeze-etch method, which reveals spacings between molecules in a single plane of the structure. In contrast, an intermolecular spacing of 105 Å would be indicated by transmission electron microscopy, which projects the structure and reveals spacings between molecules regardless of depth into the crystal. (b) View along 100 direction.

shadowed replica of the crystal surface shows the centered cell dimension a. This is why, in Table 2, the view along the 001 direction yields different dimensions for these two types of observations.

Evidence against the chloroplast inclusion crystallites being form III is that form III crystals have so far been grown only at pH around 5.2. When transferred to solutions at pH 7, they dissolve unless sulfate or phosphate is present at high concentration. In chloroplasts the pH is well above the range of stability of form III crystals, but there may be compensating factors. Dr. Fenna in our laboratory has recently found conditions (footnote to Table 1) under which form II is stable at a pH value 1.6 units above the original crystallizing conditions if polyethylene glycol is present.

In summary, the lattice spacings of form III are consistent with many observations on chloroplast inclusions, but identity of form III with chloroplast structures has not been definitely established.

X-RAY DIFFRACTION STUDIES OF FORM III CRYSTALS

We have initiated x-ray diffraction analysis of the structure of form III, with the preliminary goal of determining the molecular structure to 5-Å resolution. Reflections have been recorded in the hk0 and h0ℓ zones to 3.4 Å and 3.7 Å respectively.

For determining phases by the standard method of isomorphous replacement, we have undertaken a survey of the reactions of various heavy metal ions with form III crystals. Preliminary results are summarized in Table 3, which is not meant to be comprehensive but is intended to illustrate the problems of finding suitable heavy metal adducts with the form III molecule, which contains 96 cysteinyl residues. Uncontrolled sulfhydryl adducts are probably not promising for phase determination because the chances are small for locating a large number of heavy ion binding sites within the crystal with Patterson or other methods. We think heavy ions that do not attract sulfhydryl groups offer better possibilities of usefulness, but unfortunately some of these induce disorder in the crystals.

Table 3

Effects of Heavy Metal Ions on Form III RuBP Carboxylase Crystals

Compound	Concentration of heavy metal relative to LS pair	Crystal disorder	Good diffraction pattern with intensity changes
CH_3HgCl	1	No	Yes
o-Chloromercuri-phenol	1	Yes	---
PCMBS	1	Yes	---
Mercury-cyclo-hexane-butyrate	5-10	Yes	---
Mercurochrome	5-40	Yes	---
Baker mercurial	2-4	Yes	---
Uranyl nitrate	1-2	No	Yes
	>2	Yes	---
$PtCl_4$	5-10	Yes	---
K_2PtCl_6	5	Yes	---
$K_2Pt(CN)_4$	400-1000	No	Yes
	>1500	Yes	---
$KAu(CN)_2$	200-1500	Yes	---
Thallium acetate	5-30	No	Yes
	>40	Yes	---
Nb_6Cl_{14}	20	No	No (disordering along c*)

Acknowledgments: We are grateful for the enthusiasm, encourage-
ment, and gifts of RuBP carboxylase that we have received from Dr.
S. Wildman and his research group, including S. Kwok, K. Chen, and
P. Kwayuen. This work was supported by NIH grant GM 16925.

REFERENCES

1. Baker, T. S., Eisenberg, D., and Eiserling, F., Science 196,
   293 (1977).
2. Baker, T. S., Eisenberg, D., Eiserling, F. A., and Weissman,
   L., J. Mol. Biol. 91, 391 (1975).
3. Baker, T. S., Suh, S. W., and Eisenberg, D., Proc. Natl. Acad.
   Sci. USA 74, 1037 (1977).
4. Bowien, B., Mayer, F., Codd, G. A., and Schlegel, H. G., Arch.
   Microbiol. 110, 157 (1976).
5. Esau, K., J. Ultrastruct. Res. 53, 235 (1975).
6. Frederick, S. E., Gruber, P. J., and Newcomb, E. H., Proto-
   plasma 84, 1 (1975).
7. Iwai, S., Tanabe, Y., and Kawashima, N., Biochem. Biophys. Res.
   Commun. 73, 993 (1976).
8. Kawashima, H. and Wildman, S. G., Biochim. Biophys. Acta 229,
   240 (1971).
9. Kwok, S. Y., Ph.D. Thesis, University of California, Los
   Angeles, 1972.
10. Larsson, C., Collin, C., and Albertsson, P.-Å., J. Ultrastruct.
    Res. 45, 50 (1973).
11. McFadden, B. A., Lord, J. M., Rowe, A., and Dilks, S., Eur. J.
    Biochem. 34, 195 (1975).
12. Purohit, K., McFadden, B. A., and Cohen, A. L., J. Bacteriol.
    127, 505 (1976).
13. Roy, H., Valeri, A., Pope, D. H., Rueckert, L., and Costa,
    K. A., Biochemistry 17, 115 (1978).
14. Strøbaek, S., Gibbons, G. C., Haslett, B., Boulter, D., and
    Wildman, S. G., Carlberg Res. Commun. 41, 335 (1976).
15. Takabe, I., Otsuki, Y., Honda, Y., Nishio, T., and Matsui, C.,
    Planta 113, 21 (1973).
16. Willison, J. H. M. and Davey, M. R., J. Ultrastruct. Res. 55,
    303 (1976).
17. Wrischer, M. Planta 75, 309 (1967).

DISCUSSION

PAECH: Do you know whether the crystalline enzyme is able to
catalyze the carboxylase reaction?

EISENBERG: No. We have not looked into that.

WILDMAN: The remarkable property of the crystals is that, the
instant they are put into RuBP, they dissolve. They are stable in
solutions containing CO_2 and Mg^{2+}, however. The crystal dissolved
in RuBP can be dialyzed for a prolonged period without recrystalliz-
ing, but upon addition of Mg^{2+} and CO_2 it crystallizes immediately.

THE ACTIVATION OF RIBULOSE 1,5-BISPHOSPHATE CARBOXYLASE/OXYGENASE

George H. Lorimer

Abteilung für Zellchemie, Institut für Toxikologie und Biochemie
Gesellschaft für Strahlen- und Umweltforschung München mbH
8000 München 2, West Germany

Murray R. Badger

Department of Plant Biology, Carnegie Institution of Washington
Stanford, California 94305

and

Hans W. Heldt

Institut für Physiologische Chemie, Physikalische Biochemie und
Zellbiologie der Universität München, 8000 München 2, West Germany

A long-standing and continuing problem concerning ribulose bis-phosphate carboxylase/oxygenase is the discrepancy between its activity in vivo and in vitro (1, 2). The apparently low affinity of the enzyme for CO_2 was one of the better reasons Warburg had for dismissing the C3 photosynthetic carbon reduction cycle (3). When assayed in vitro with the naturally occurring concentrations of CO_2 (10 μM) and O_2 (250 μM), the purified enzyme is incapable of fixing CO_2 for sustained periods (>90 sec) at rates equal to or greater than those of photosynthesis. Similarly, sustained synthesis of phosphoglycolate under natural conditions for >2 to 3 min at rates equal to or greater than the in vivo rates has not yet been observed. Nevertheless, beginning with the observations of Bahr and Jensen (4, 5), some progress towards resolution of these discrepancies has been made, and, provided that both reactions are initiated with fully activated enzyme, adequate rates of carboxylation or oxygenation under natural conditions can be achieved or even exceeded, if only for a minute or two (6-12). Thereafter, progressive inactivation of the enzyme becomes apparent. At best this represents only a partial solution to the problem, for clearly there are no such restrictions upon the enzyme in vivo. What then

are the factors which maintain the enzyme in the activated state
in vivo, and what physiological significance can be attached to the
transformations between the activated and inactivated states?

ACTIVATION BY CO_2 AND Mg^{2+}

It has been known for a long time that the order of addition
of the reagents to the carboxylase assay profoundly affects the
time course of the subsequent reaction (13). If the reaction is
initiated with enzyme fully activated by preincubation with CO_2
and Mg^{2+}, fixation begins without a discernible lag (Figure 1a).
But if the reaction is initiated with inactivated enzyme, prepared
by the removal of CO_2 and Mg^{2+} by passage through a gel filtration
column, fixation shows a distinct lag (Figure 1a) and the enzyme
becomes progressively more active as the assay proceeds (9). Although CO_2 and O_2 compete with one another during catalysis (6, 11),
O_2 is not capable of substituting for CO_2 in the activation reaction. This can clearly be seen when the oxygenase reaction is
initiated with inactivated enzyme (Figure 1b): no O_2 uptake whatever occurs because this assay, unlike the carboxylase assay, is
done with no CO_2 in the reaction mixture. When the oxygenase
reaction is initiated with fully activated enzyme, O_2 uptake begins
without any discernible lag, but the rate of the reaction progressively declines and the enzyme seems to be reverting to the inactivated state (9).

That an enzyme should be activated by one of its substrates is
neither novel nor surprising--many examples of this have been reported. What is perhaps noteworthy about RuBP carboxylase/oxygenase
is the relative slowness of the transformations between activated
and inactivated states. Indeed the ability to detect O_2 uptake
and to demonstrate that the activated enzyme has a high affinity
or low K_m for CO_2 is to some extent dependent on the longevity of
the activated state.

The relative slowness of the activation reaction has permitted
us to examine its kinetics (7), but it was necessary to restrict
the assay time to 30 sec so that there would be essentially no
change in the activation state of the enzyme during the assay
itself. In assays conducted at 10° with 1 mM HCO_3^-, pH 8.2, and
15 mM Mg^{2+}, enzyme that was initially fully inactivated fixed only
1% as much CO_2 as did enzyme that was initially fully activated
(Figure 1a and Table 1).

Both CO_2 and Mg^{2+} are required for full activation. The
reaction is reversible, the enzyme reverting to the inactivated
state when CO_2 and Mg^{2+} are removed by gel filtration. These data
alone do not establish the order in which CO_2 and Mg^{2+} react with
the enzyme. The time course of activation was therefore followed
with use of either varying concentrations of CO_2 and a fixed concentration of Mg^{2+} (Figure 2a) or varying concentrations of Mg^{2+}
and a fixed concentration of CO_2 (Figure 2b). The initial rate of

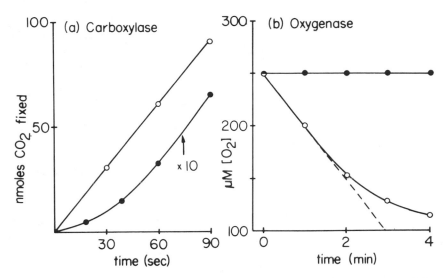

Figure 1.   Different time courses of (a) carboxylation and (b) oxygenation depending on whether the reaction was initiated with fully
activated enzyme (o) or fully inactivated enzyme (●).

Table 1

Requirements for the Activation of RuBP Carboxylase (7)

Preincubation treatment	Activity* (nmole/min)	%
Complete**	5.14	100
Mg^{2+} alone, no CO_2 added	0.90	18
CO_2 alone, no Mg^{2+} added	1.32	26
No CO_2 added, no Mg^{2+} added	0.06	1

*Activity was determined with 1 mM $NaH^{14}CO_3$, 15 mM Mg^{2+}, and 0.4
mM RuBP, pH 8.2, for 30 sec at 10°.
**The complete preincubation treatment consisted of incubating the
carboxylase for 30 min at 10° in 0.05 M Tris-HCl, pH 8.2, 1 mM
dithiothreitol, 17 mM $NaHCO_3$, and 20 mM Mg^{2+}.

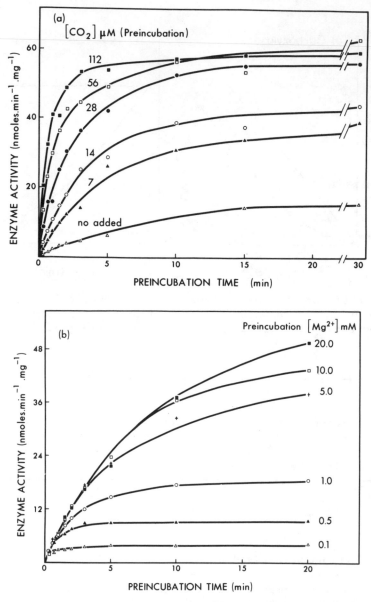

Figure 2. Time course for the activation of RuBP carboxylase (a) by varying concentrations of CO_2 and constant (20 mM) Mg^{2+} concentration, and (b) varying concentrations of Mg^{2+} and constant (30 μM) CO_2 concentration. At the times indicated the enzyme activity was determined as in Table 1 (7).

activation is proportional to the CO2 concentration and independent of the Mg^{2+} concentration. These results are consistent with an ordered equilibrium mechanism,

$$ E \quad + CO_2 \rightleftharpoons E'-CO_2 + Mg^{2+} \rightleftharpoons E'-CO_2-Mg \quad , $$
$$ \text{(inactive)} \qquad\qquad\qquad\qquad\qquad\qquad \text{(active)} $$

in which the reaction of the enzyme (E) with CO_2 or, more probably, a conformational change taking place directly thereafter, is rate determining. The reactions involving Mg^{2+} appear to be very rapid so that the system $E'-CO_2 + Mg^{2+} \rightleftharpoons E'-CO_2-Mg$ is always at or close to equilibrium.

This order of addition of CO_2 and Mg^{2+} is consistent with the EPR study of Miziorko and Mildvan (14). These authors reported that Mn^{2+} (substituting for Mg^{2+}) was tightly bound to the enzyme only in the presence of HCO_3^- (or CO_2). One Mn^{2+} per 70,000-dalton subunit of the enzyme was tightly bound under these conditions. Furthermore the Mn^{2+} was thought to be bound very closely to the HCO_3^- (or CO_2).

It is evident from Figure 2 that the final equilibrium activation state of the enzyme depends on the concentrations of both CO_2 and Mg^{2+}. If one sets

$$ K_c = \frac{[E][CO_2]}{[E'-CO_2]} \quad \text{and} \quad K_{Mg} = \frac{[E'-CO_2][Mg^{2+}]}{[E'-CO_2-Mg]} \quad , $$

it follows that at equilibrium the activity of the enzyme will be $[E][CO_2][Mg^{2+}]/K_c K_{Mg}$. If one assumes that any $E'-CO_2$ present at the end of the preincubation period is "immediately" converted to $E'-CO_2-Mg$ upon addition of the assay mixture containing 15 mM Mg^{2+}, it follows that $[E] = [E_{total}] - [E'-CO_2-Mg]$. Substituting for E and rearranging yields

$$ \frac{1}{[E'-CO_2-Mg]} = \frac{K_c K_{Mg}}{[Mg^{2+}][CO_2][E_{total}]} + \frac{1}{[E_{total}]} \quad . $$

This expression has the same form as that described by Lineweaver and Burk and indicates that a double reciprocal plot of the activity after preincubation to equilibrium, $[E-CO_2-Mg]$, against either $[CO_2]$ or $[Mg^{2+}]$ should be linear. The results (Figure 3) of such an analysis are consistent with this kinetic model.

No great significance should be attached to the values of K_c and K_{Mg} obtained from such plots because (a) the activation reaction is markedly influenced by pH (Figure 5) and the above expressions should therefore contain some proton term(s) to accommodate the changes in the equilibrium activation state with pH, and (b) K_c most probably includes a contribution due to a conformational

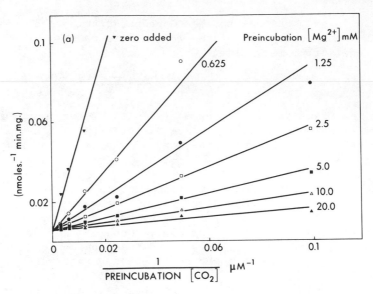

Figure 3.   Dependence of the equilibrium activity of RuBP carboxylase on the preincubation concentrations of CO_2 and Mg^{2+} (7).

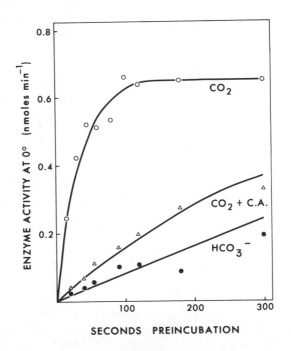

Figure 4. CO_2 as the species involved in the activation of RuBP
carboxylase (7).   The theory underlying such an analysis is described in ref. 15.   C. A.:   carbonic anhydrase.

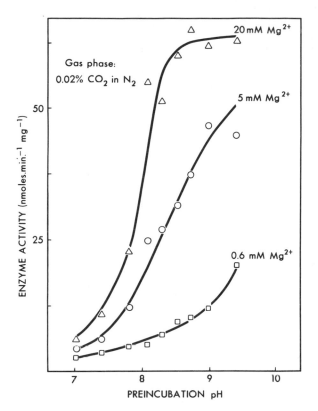

Figure 5. Dependence of the equilibrium activity of RuBP carboxylase on the preincubation pH at constant (10 μM) CO_2 concentration and Mg^{2+} concentrations indicated (7).

change and is not, strictly speaking, the equilibrium constant for the reaction of enzyme and CO_2.

Before determining the influence of pH on the activation reaction, it was necessary to find out whether CO_2 or HCO_3^- was the active species involved. Cooper's kinetic method (15) yielded the results shown in Figure 4. CO_2 rather than HCO_3^- induces the faster initial rate of activation, and from this it can be concluded that CO_2 is the species involved in the activation reaction.

After preincubation of the enzyme to equilibrium at varying pH's but under constant CO_2 partial pressure, the enzyme activities shown in Figure 5 were observed. The response to pH suggests that CO_2 reacts with a group on the enzyme which has a distinctly alkaline pK. We have suggested that this group might be an amino group and that the activation reaction might involve the formation of a carbamate (16) (Figure 6). We had a number of reasons for this. (i) Our kinetic data, although by no means constituting proof, were

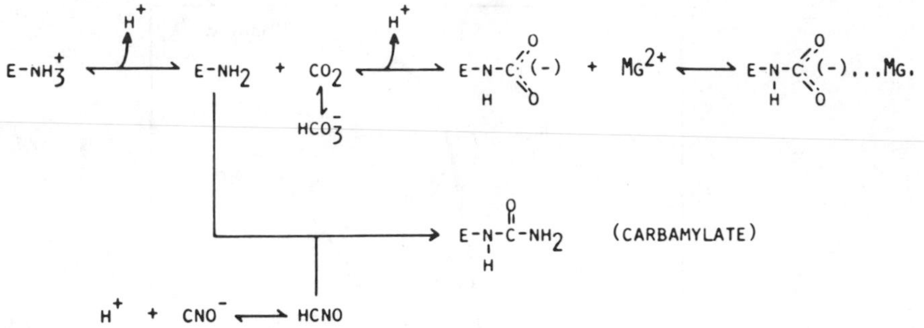

Figure 6. The reversible reaction of an enzyme-bound amino group with CO_2 to form a carbamate, to which Mg^{2+} could conceivably bind, and the irreversible reaction with HCNO to form a carbamylate. These reactions are particularly well documented in the case of hemoglobin (see refs. 16-22).

consistent with the formation of a carbamate. (ii) The well documented formation of carbamate at the N-terminal amino group of the
α-chain of hemoglobin and its relation to the alkaline Bohr effect
provide a concrete example of a carbamate functioning in a regulatory
manner (17-20). (iii) Miziorko and Mildvan (14) showed that the
metal ion was bound very closely to the carbon atom of HCO_3^- (or
CO_2) and suggested that binding of the metal ion might depend on
the formation of new anionic sites on the enzyme. The formation
of a carbamate essentially converts a cationic group into an
anionic group capable of binding Mg^{2+} or Mn^{2+}. (iv) Carbamate
formation appeared to offer an explanation for the report (21) that
diazomethane was able to stabilize an enzyme-$^{14}CO_2$ complex, and
that one radioactive peptide was obtained upon tryptic digestion.
Diazomethane would readily esterify any such carbamate and thus
stabilize the radioactivity. Unfortunately, this reaction is not
specific to the native enzyme, nor is there any relation between
the amount of radioactive material that can be trapped this way and
the activation state of the enzyme. In fact, just as much acid-
stable radioactivity can be trapped with diazomethane by boiled or
SDS-denatured enzyme as by native enzyme (Lorimer, unpublished
results). This does not disprove the involvement of a carbamate in
the activation reaction; it simply means that stabilization of the
enzyme-$^{14}CO_2$ complex with diazomethane cannot be cited as evidence
for carbamate formation. If carbamate formation is involved, one
might be able to react the responsible amino group with cyanuric
acid (22).

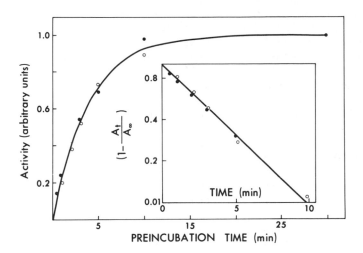

Figure 7.   Coactivation of RuBP carboxylase (●) and RuBP oxygenase
   (o) by CO_2 and Mg^{2+}.   Inset: pseudo-first-order replot (8).

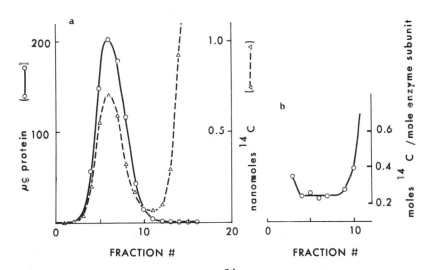

Figure 8.   Isolation of an enzyme-$^{14}CO_2$ complex by chromatography
on a Sephadex G-25 column equilibrated with 40 mM Mg^{2+} (10).   (a)
Elution profile; (b) constancy of the specific radioactivity recovered across the peak of protein.

Although CO_2 and O_2 compete with one another during catalysis in a linearly competitive manner (6, 10), which suggests that they interact at the same site, the oxygenase is activated in exactly the same manner as the carboxylase (8). For example, the time course for activation of the carboxylase is identical to that of the oxygenase (Figure 7). Thus it was difficult to see how one could observe oxygenase activity, if the activating CO_2 was also located at the catalytic site. Also, we demonstrated qualitatively the existence of an enzyme-$^{14}CO_2$ complex by gel filtration experiments (Figure 8). When this complex was applied to a Sephadex column equilibrated with CO_2-free buffer and Mg^{2+}, a distinct peak of radioactivity co-chromatographed with the enzyme. About 3 moles of $^{14}CO_2$ per mole of enzyme were recovered rather than the 8 expected for 100% recovery (Lorimer and Badger, unpublished results). The ability to isolate an enzyme-$^{14}CO_2$ complex in this way suggested that dissociation of the complex was slow, yet attempts (Badger and Lorimer, unpublished results) to trap this complex by the isotope trapping technique of Rose et al. (23) failed. This indicated either (a) that the $^{14}CO_2$ was not at the catalytic site and was therefore unavailable for reaction with RuBP, or (b) that the $^{14}CO_2$ was indeed at the catalytic site but simply exchanged with $^{12}CO_2$ faster than it underwent reaction with RuBP. We therefore suggested that the molecule of CO_2 which activates the carboxylase does not necessarily become fixed. Implicit in this, but by no means proven, was the notion that the activation site was distinct from the catalytic site. Other evidence, however, is inconsistent with the existence of a distinct activation site. For example, 2-carboxyribitol 1,5-bisphosphate, which presumably binds at the catalytic site, was found to displace the CO_2 completely from the enzyme (14), which implies that the CO_2 was also located at the active site. However, 2-carboxyribitol 1,5-bisphosphate is also thought to induce a conformational change in the structure of the enzyme (24), and this might be responsible for the displacement of the activating CO_2. The question of whether the activation site is distinct from the catalytic site is unresolved.

Whatever its mechanism, the activation reaction has interfered with the determination of the Michaelis constant for CO_2. In such a determination it is assumed that the concentration of the catalyst remains constant. If this condition is not fulfilled by pretreatment of the enzyme, the shape and slope of the resultant Lineweaver-Burk plot will deviate more or less from a true linear relationship (Figure 9). If each reaction of the concentration series is initiated with fully activated enzyme, and the duration of the assay is restricted (9) so that the activation state of the enzyme does not appreciably alter, then the kinetics will be linear with a Michaelis constant for CO_2 (determined under N_2) in the range 10 to 20 μM. By using enzyme from intact isolated chloroplasts, where it is substantially activated, and by restricting the assay time, Bahr and Jensen (4) were able to demonstrate that a form of the enzyme existed with an appropriately high affinity for CO_2.

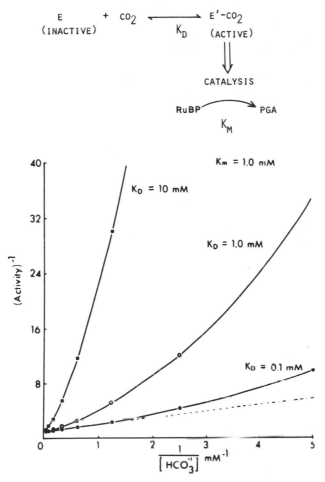

Figure 9. Deviations from linearity in the determination of the $K_m(CO_2)$ of RuBP carboxylase and the dependence of the shape and slope of the Lineweaver-Burk plot on the extent of activation.

(IN)ACTIVATION OF THE CARBOXYLASE IN RELATION TO EVENTS IN VIVO

The exact concentration of free Mg^{2+} in chloroplast stroma is not known precisely but is unlikely to exceed 5 mM. Illumination of intact isolated chloroplasts causes an increase in the stromal Mg^{2+} concentration of about 3 mM (26, 27) and a shift in the stromal pH from 7.0 to about 8.0 (28). It can be seen from Figure 5 that these increases should tend to activate the carboxylase. However, closer inspection reveals that full activation of the carboxylase at pH 8.2 and with 10 μM CO_2 (conditions thought to be similar to those in vivo) requires unrealistically high concentrations (20 mM) of Mg^{2+}. With a more realistic Mg^{2+} concentration (5 mM) the enzyme

Table 2

Effects of Various Metabolites on the Activity of RuBP Carboxylase

(+, activation; -, inhibition or inactivation; ±, activation at low
concentrations but inhibition at higher concentrations; 0, little
effect if any.)

Metabolite	Reference								
	30	35	34	10	44	36	12	25	37
Sedoheptulose 7-phosphate	+								
Fructose 6-phosphate	+	0	0						
Fructose 1,6-bisphosphate	-	+	±	+					
Glucose 6-phosphate	+	-							
Glucose 1,6-bisphosphate	-								
6-Phosphogluconate	+	+	±	+		-			
Ribulose 5-phosphate	+								
Ribulose 1,5-bisphosphate			-				-	-	-
Xylulose 5-phosphate	+								
Xylulose 1,5-bisphosphate								-	
Ribose 5-phosphate	+	-							
Erythrose 4-phosphate	+								
Dihydroxyacetone 3-phosphate	+				+				
3-Phosphoglycerate	0	+	+		+				-
ADP	+								
ADP-glucose	+								
ATP	+								
NADPH		+	+	+					
P_i						+			-

would be only about one-third activated, a value that is probably
too low. One is therefore forced to the conclusion that additional
components are required.

In the last five years a number of laboratories have reported
that various sugar phosphates, some of them intermediates of the
photosynthetic carbon reduction cycle, stimulate or inhibit the
activity of RuBP carboxylase (Table 2) (29-37). Unfortunately the
findings have been inconsistent. For example, fructose 1,6-bisphos-
phate has been reported to inhibit or inactivate the enzyme by one
group (29, 30) but to stimulate or activate the enzyme by other
groups (34, 35). These discrepancies can be attributed partly to
differences in methods. There is, however, general agreement that
these effectors do not elicit differential responses between the
carboxylase and the oxygenase (10, 35). The results of Chollet

Table 3

Comparison of the Effects of Chloroplast Metabolites on
RuBP Carboxylase and Oxygenase (35)

Metabolite	Conc. (mM)	Relative Activity (%)		Ratio
		Carboxylase	Oxygenase	
None		100	100	1.00
NADPH	0.1	121	125	0.97
	0.5	155	165	0.94
6-Phosphogluconate	0.1	161	176	0.91
Fructose 1,6-bisphosphate	0.1	136	138	0.99
3-Phosphoglycerate	0.5	123	127	0.97
Glucose 6-phosphate	0.5	86	94	0.91
Fructose 6-phosphate	0.1	100	97	1.03
	0.5	92	91	1.01
Ribose 5-phosphate	0.1	81	93	0.87
	0.5	44	47	0.94

and Anderson (Table 3) (35) and our own unpublished results indicate
that the carboxylase and the oxygenase respond in parallel. A
large element of nonspecificity is apparent in Table 2: it seems
that practically any phosphate ester is capable of influencing
the reactivity of the carboxylase. An exception is that NADPH
(33-35) induces activation but the oxidized form $NADP^+$ does not
(33,34). It is not known whether the activation involves the
oxidation of NADPH.

An additional element of uncertainty arises because the
in vitro studies were all performed with very favorable effector:
enzyme molar ratios, which in no case apply in vivo. Of the
various metabolites known to activate the carboxylase in vitro,
only hexose monophosphates, 3-phosphoglycerate, and inorganic
phosphate are present in sufficient quantities within the chloroplast to approach or exceed a molar ratio of unity (Table 4) (38,
39). For example, the value for NADPH is much too low for it to
be permanently bound to the carboxylase in vivo. Whether or not
these molar ratios are significant depends on how these various
effectors function. If full activation of the enzyme requires
that 8 molecules of effector be bound, then these values are
highly significant, and only those compounds present in molar excess can be considered as physiological effectors. On the other
hand, if all 8 catalytic sites are activated upon the binding of
only one effector molecule, then the number of potentially active
compounds increases.

Table 4

Molar Ratios of Metabolites to RuBP Carboxylase Active
Sites Within the Chloroplast Stroma

Spinach leaves contain about 5.6 mg RuBP carboxylase per
mg Chl (38), corresponding to about 80 nmoles carboxylase
active sites per mg Chl and confined to the stroma of the
chloroplasts, whose volume is 23 μl per mg Chl. The concentration of active sites within the stroma is therefore
about 3.4 mM. The quantities of the various metabolites
were determined during steady-state photosynthesis of
intact spinach chloroplasts with 5 mM HCO_3^-, pH 7.6, and
0.5 mM P_i (after ref. 39).

Metabolite	Molar ratio metabolite:carboxylase active site
Hexose monophosphates	0.67
3-Phosphoglycerate	1.10
P_i	1.70
Ribulose 1,5-bisphosphate	0.24
Pentose monophosphates	0.04
Sedoheptulose 1,7-bisphosphate	0.10
Fructose 1,6-bisphosphate	0.05
Triose 3-phosphates	0.10
ADP	0.30
ATP	0.07
NADPH	0.04
$NADP^+$	0.08

     Our recent experiments (40) have been aimed at finding out
which factors are important in activation of the carboxylase in
situ, i.e., within the chloroplast itself. We were motivated by the
report from Bahr and Jensen (41) that activation of the carboxylase
in situ depended at least in part on the same factors we have shown
to be important for activation of the purified enzyme, namely CO_2
and a pH shift from about 7.0 towards 8.5 (7). We had also observed (Table 5) that in isolated chloroplasts under conditions of
phosphate limitation, contrary to expectation, the stromal concentration of RuBP increased even though the rate of CO_2 fixation
was reduced. This suggested that RuBP carboxylase might be inactivated or inhibited by phosphate deficiency.
     Our initial experiments qualitatively confirmed the results

Table 5

Effect of P_i on CO_2 Fixation and Metabolite Levels
in Stroma of Spinach Chloroplasts (40)

Metabolite	Phosphate in medium (mM): Rate of CO_2 fixation*:	0.5 108	0.1 60
P_i		7.0[†]	2.2
3-Phosphoglycerate		6.0	8.9
Hexose monophosphates		3.7	4.9
Triose phosphates		0.33	0.25
Fructose 1,6-bisphosphate		0.29	0.31
Sedoheptulose 1,7-bisphosphate		0.29	0.31
Ribulose 1,5-bisphosphate		0.28	0.57

*μmol per hr per mg Chl.

[†]Stromal metabolite concentration (mM).

of Bahr and Jensen (41). Upon illumination of the chloroplast
suspension, the carboxylase underwent three-fold activation
(Figure 10), which was reversed upon darkening. At all times there
was sufficient carboxylase activity (even when measured with 14 μM
CO_2 and 250 μM O_2) to account for the rates of photosynthesis
by these chloroplasts. The carboxylase obtained from darkened
chloroplasts could be further activated upon lysing the chloroplasts in a hypotonic medium containing 5 mM HCO_3^- and 20 mM Mg^{2+},
conditions known to achieve full activation of the enzyme (7).
By this criterion the enzyme from illuminated chloroplasts was
completely activated. No change occurred in the activity of the
carboxylase when the chloroplasts were kept in the dark. Possibly
the relatively high activity already present in the dark may be
an artefact of isolated chloroplasts, and in vivo the enzyme may
become fully inactivated in the dark.
    The results in Figure 11 confirm that activation of the enzyme
in situ, like that of the purified enzyme, requires CO_2. Figure 12
presents data from an experiment on activation of the enzyme by
stromal Mg^{2+}. Illumination of intact chloroplasts leads to an
increase in the stromal Mg^{2+} concentration of about 3 mM (26, 27)
as a result of Mg^{2+} transport across the thylakoid membrane. This
increase can be inhibited by the addition of ionophore A23187,
which renders the chloroplast envelope membrane permeable to Mg^{2+}
(26) (Figure 13). The loss of Mg^{2+} from the stroma leads to
rapid inhibition of CO_2 fixation. This inhibition can be overcome

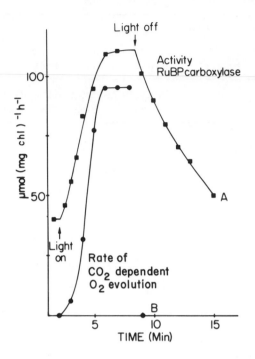

Figure 10.   Light activation of RuBP carboxylase _in situ_ (i.e.,
within intact isolated spinach chloroplasts) and the rate of CO_2-
dependent O_2 evolution.   The carboxylase activity was measured at
the times indicated by rupturing the chloroplasts in hypotonic
assay medium containing 14 μM CO_2, 250 μM O_2, 5 mM Mg^{2+}, 0.5 mM
RuBP, pH 8.2.   The chloroplasts were illuminated in a medium containing 0.33 M sorbitol, 0.05 M Hepes-NaOH, pH 7.6, 1 mM $MnCl_2$,
1 mM $MgCl_2$, 1 mM P_i, and 5 mM $NaHCO_3$ (40).

by subsequent addition of excess Mg^{2+}.   As shown in Figure 14,
addition of the ionophore to illuminated chloroplasts caused rapid
inactivation of RuBP carboxylase, and as expected, subsequent
addition of excess Mg^{2+} led to reactivation.   _In situ_ activation
of the carboxylase clearly requires Mg^{2+}.

     The influence of stromal pH on the activation process _in situ_
was investigated by using nitrite to modify the internal pH.   Illumination of intact chloroplasts increases the stromal pH by about
one unit via light-driven proton transport from the stroma into

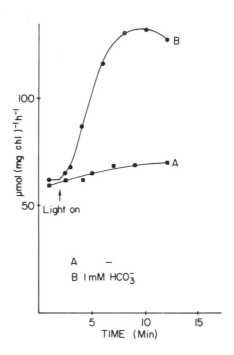

Figure 11. Light activation of RuBP carboxylase *in situ* and its
dependence on CO_2. Conditions for incubation and assay as for
Figure 10 except that the chloroplast suspension contained either
no added HCO_3^- (A) or 1 mM HCO_3^-, and the pH was 7.2 (40).

the thylakoid space (28), but the alkalization can be reduced from
about pH 7.8 to about pH 7.2 by adding 6 mM nitrite to the external
medium (42). Nitrite facilitates indirect proton transport across
the envelope membrane via nitrous acid (Figure 13). Under these
conditions CO_2 fixation is inhibited by 96%, even though the level
of RuBP remains more or less unaltered (42). The data in Table 6
show that the addition of nitrite (6 mM) diminished the light-
dependent activation of the carboxylase, an effect we attribute to
the ability of nitrite to reverse partially the light-driven alkalization of the stroma. Table 6 also shows that light activation
of RuBP carboxylase depends on the presence of inorganic phosphate.
Neither 3-phosphoglycerate nor dihydroxyacetone 3-phosphate could
substitute for P_i, although both are known to activate the purified
enzyme (29, 34, 35).

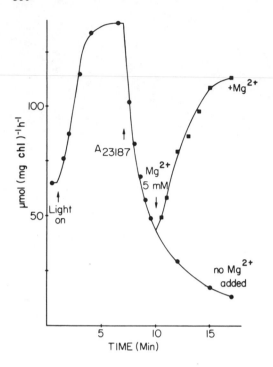

Figure 12

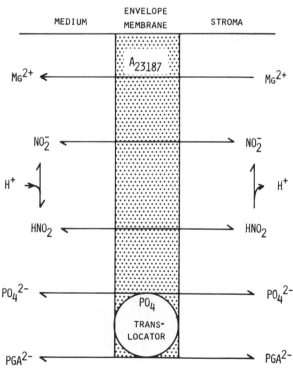

Figure 13

← Figure 12. Light activation of RuBP carboxylase in situ and its dependence on Mg^{2+}. The chloroplast medium was as for Figure 10 except that Mg^{2+} and Mn^{2+} were omitted. At the times indicated 2 μM A23187 and 5 mM Mg^{2+} were added (40).

Table 6

RuBP Carboxylase Activity in Chloroplasts Dependent
on Conditions of Preincubation

(Activity of enzyme fully activated with CO_2 and
Mg^{2+} was 102 μmol per hr per mg Chl.)

Additions to medium	RuBP carboxylase activity (μmol per hr per mg Chl)	
	Dark	Light
P_i (1 mM)	28	119
+ NO_2^- (6 mM)	27	62
+ A23187 (2 μM)	24	18
3-Phosphoglycerate (1 mM)	38	36
+ NO_2^- (6 mM)	36	31
+ A23187 (2 μM)	27	25
Dihydroxyacetone 3-phosphate (1 mM)	25	29
+ NO_2^- (6 mM)	19	26
+ A23187 (2 μM)	22	23

← Figure 13. The mechanism of action of (i) A23187, which renders the envelope membrane permeable to Mg^{2+} (26), (ii) nitrite, which facilitates proton transport across the envelope membrane via nitrous acid (42), and (iii) the phosphate translocator, located in the envelope membrane, which catalyzes a counter-exchange of P_i for either 3-phosphoglycerate or dihydroxyacetone 3-phosphate (43).

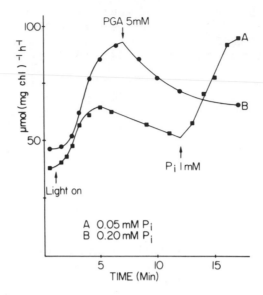

Figure 14. Light activation of RuBP carboxylase <u>in situ</u> and its dependence on P_i. Independent measurements showed that, starting with 0.05 mM P_i in the external medium, the supply of P_i was exhausted (through conversion to 3-phosphoglycerate and dihydroxyacetone 3-phosphate) after 4 to 5 min of illumination. The addition of 5 mM 3-phosphoglycerate to the medium depletes the stroma of P_i (40).

     In the experiment of Figure 14, the chloroplasts were illuminated first with a limiting concentration of P_i (0.05 mM). Initially the carboxylase activity rose almost as rapidly as in the control with 0.20 mM P_i, but after 3 to 4 min of illumination, when the P_i had been depleted by CO_2 fixation (to 3-phosphoglycerate and dihydroxyacetone 3-phosphate), it declined. When excess P_i was again provided, the enzyme underwent reactivation. Another means of depleting the stromal concentration of P_i is to supply excess 3-phosphoglycerate or dihydroxyacetone 3-phosphate to the chloroplasts; then the phosphate translocator in the envelope membrane brings about counterexchange of internal P_i for external 3-phosphoglycerate or dihydroxyacetone 3-phosphate (43) (Figure 13). When excess 3-phosphoglycerate was added to illuminated chloroplasts containing fully activated enzyme, immediate inactivation of the carboxylase was observed (Figure 14), an effect which we attribute to depletion of the stromal P_i through counterexchange.

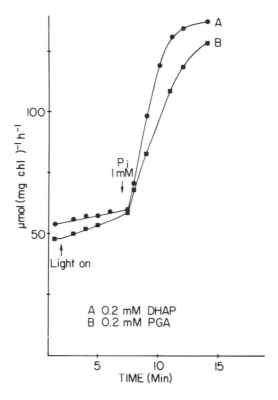

Figure 15.   Inhibition of the light activation of RuBP carboxylase
in situ by 3-phosphoglycerate and dihydroxyacetone 3-phosphate and
the reversal of this inhibition by the addition of P_i (40).

The requirement for both CO_2 and P_i for activation of the
carboxylase in situ could be attributed to the need to synthesize
some sugar phosphate effector compound such as fructose 1,6-bis-
phosphate, which would then activate the carboxylase. Although
we have by no means disproved this possibility, it strikes us as
unlikely because such a putative activator could equally well be
synthesized from 3-phosphoglycerate or dihydroxyacetone 3-phosphate,
and yet these compounds fail to activate the enzyme in situ
(Figure 15); on the contrary, they inhibit or reverse its in situ
activation.
       Simultaneous measurements of the stromal concentrations of
3-phosphoglycerate, triose phosphates, H^+, P_i, and adenine and
pyridine nucleotides have been reported (39). The results indicated

that 3-phosphoglycerate and the triose phosphates were not far from
equilibrium with their corresponding adenine and pyridine nucleotides. Both 3-phosphoglycerate and dihydroxyacetone 3-phosphate
are transported into the stroma in counterexchange with P_i via the
phosphate translocator. Addition of 3-phosphoglycerate to the
medium would therefore cause a reduction in the NADPH/NADP$^+$ ratio
within the stroma whereas dihydroxyacetone 3-phosphate would cause
an increase. Thus, if NADPH were involved in activation of the
carboxylase, one would expect the responses to 3-phosphoglycerate
and dihydroxyacetone 3-phosphate to be different. However, both
compounds induce a common response, which can be attributed to the
ability of both to deplete the stroma of P_i by counterexchange
through the phosphate translocator.

   We currently believe that the effect of P_i is much more direct,
and preliminary experiments with purified enzyme indicate that P_i
is also an effective activator of the enzyme (44).

   Acknowledgments: Some of this work was performed while GHL
and MRB were at the Research School of Biological Sciences, The
Australian National University. The work was supported in part
by the Deutche Forschungsgemeinschaft. We appreciate the skillful
assistance of Frau Chon and Fräulein Freisl.

REFERENCES

1. Walker, D. A., New Phytol. 72, 209-35 (1973).
2. Walker, D. A., Curr. Top. Cell. Regul. 11, 203-41 (1976).
3. Warburg, O. and Kripphal, G., Hoppe Seyler's Z. Physiol. Chem.
    322, 225-37 (1963).
4. Bahr, J. T. and Jensen, R. G., Plant Physiol. 53, 39-44 (1974).
5. Bahr, J. T. and Jensen, R. G., Arch. Biochem. Biophys. 164,
    408-13 (1974).
6. Badger, M. R. and Andrews, T. J., Biochem. Biophys. Res.
    Commun. 60, 204-10 (1974).
7. Lorimer, G. H., Badger, M. R., and Andrews, T. J., Biochemistry
    15, 529-36 (1976).
8. Badger, M. R. and Lorimer, G. H., Arch. Biochem. Biophys. 175,
    723-9 (1976).
9. Lorimer, G. H., Badger, M. R., and Andrews, T. J., Anal.
    Biochem. 78, 66-75 (1977).
10. Badger, M. R., Ph.D. Thesis, The Australian National University,
    Canberra, 1976.
11. Laing, W. A., Ogren, W. L., and Hageman, R. H., Biochemistry
    14, 2269-75 (1975).
12. Laing, W. A. and Christeller, J. T., Biochem. J. 159, 563-70
    (1976).
13. Pon, N. G., Rabin, B. R., and Calvin, M., Biochem. Z. 338,
    7-19 (1963).

14. Miziorko, H. M. and Mildvan, A. S., J. Biol. Chem. 249, 2743-50 (1974).
15. Filmer, D. L. and Cooper, T. G., J. Theor. Biol. 29, 131-45 (1970).
16. Faurholt, C., J. Chim. Phys. 22, 1-44 (1925).
17. Morrow, J. S., Keim, P., and Gurd, F. R. N., J. Biol. Chem. 249, 7484-94 (1974).
18. Kilmartin, J. V., Fogg, J., Luzzana, M., and Rossi-Bernardi, L., J. Biol. Chem. 248, 7039-48 (1973).
19. Arnone, A., Nature London 247, 143-5 (1974).
20. Bauer, C., Baumann, R., Engels, V., and Pacyna, B., J. Biol. Chem. 250, 2173-6 (1975).
21. Akoyunoglou, G., Argyrondi-Akoyunoglou, J. H., and Methenitou, H., Biochem. Biophys. Acta 132, 481-91 (1967).
22. Manning, J. M., Cerami, A., Gillette, P. N., de Furia, F. G., and Miller, D. R., Adv. Enzymol. 40, 1-27 (1974).
23. Rose, I. A., O'Connell, E. L., Litwin, S., and Tarra, J. B., J. Biol. Chem. 249, 5163-6 (1974).
24. Siegel, M. and Lane, M. D., Biochem. Biophys. Res. Commun. 48, 508-16 (1972).
25. McCurry, S. D. and Tolbert, N. E., J. Biol. Chem. 252, 8344-6 (1977).
26. Portis, A. R. and Heldt, H. W., Biochim. Biophys. Acta 449, 434-46 (1976).
27. Krause, G. H., Biochim. Biophys. Acta 460, 500-10 (1977).
28. Werdan, K., Heldt, H. W., and Milovancev, M., Biochim. Biophys. Acta 396, 276-92 (1975).
29. Buchanan, B. B. and Schürmann, P., J. Biol. Chem. 248, 4956-64 (1973).
30. Buchanan, B. B. and Schürmann, P., Curr. Top. Cell. Regul. 7, 1-20 (1973).
31. Chu, D. K. and Bassham, J. A., Plant Physiol. 50, 224-7 (1972).
32. Chu, D. K. and Bassham, J. A., Plant Physiol. 52, 373-9 (1973).
33. Chu, D. K. and Bassham, J. A., Plant Physiol. 54, 556-9 (1974).
34. Chu, D. K. and Bassham, J. A., Plant Physiol. 55, 720-6 (1975).
35. Chollet, R. and Anderson, L. L., Arch. Biochem. Biophys. 176, 344-51 (1976).
36. Tabita, F. R. and McFadden, B. A., Biochem. Biophys. Res. Commun. 48, 1153-8 (1972).
37. Paulsen, J. M. and Lane, M. D., Biochemistry 5, 2350-7 (1966).
38. Andrews, T. J. and Hatch, M. D., Phytochemistry 10, 9-15 (1971).
39. Portis, A. R., Chon, C. J., Mosbach, A., and Heldt, H. W., Biochim. Biophys. Acta 461, 313-25 (1977).
40. Heldt, H. W., Chon, C. J., and Lorimer, G. H., FEBS Lett., submitted (1978).
41. Bahr, J. T. and Jensen, R. G., Arch. Biochem. Biophys. 185, 39-48 (1978).
42. Purczeld, P., Chon, C. J. Portis, A. R., Heldt, H. W., and Heber, U., Biochim. Biophys. Acta 501, 488-98 (1978).

43. Fliege, R., Flügge, U. I., Werdan, K., and Heldt, H. W.,
    Biochim. Biophys. Acta, submitted (1978).
44. Lorimer, G. H. and Heldt, H. W., unpublished results.

INTERACTION OF CHLOROPLAST AND NUCLEAR GENOMES

IN REGULATING RuBP CARBOXYLASE ACTIVITY[*]

S. D. Kung and P. R. Rhodes
Department of Biological Sciences
University of Maryland Baltimore County, Catonsville, Maryland 21228

INTRODUCTION

Extensive studies of ribulose 1,5-bisphosphate (RuBP) carboxyl-ase/oxygenase have provided much information on its structure, function, biosynthesis, mode of inheritance, and genetic variability (1-3). RuBP carboxylase, which in higher plants has been termed Fraction I protein, is found in all photosynthetic organisms as an oligomeric protein of high molecular weight (1-4). It catalyzes the carboxylation of RuBP, producing two molecules of glycerate 3-phosphate (PGA), the primary product of photosynthesis (3). In the presence of molecular oxygen, RuBP carboxylase can also catalyze the oxidation of RuBP, forming PGA and phosphoglycolate, which can be converted to glycolate, the substrate for photorespiration (3, 5). Since the balance between photosynthetic and photorespiratory processes determines a plant's maximum yield, the ability of RuBP carboxylase to participate in either process has generated renewed interest in this enzyme.

It is well established that RuBP carboxylase is composed of eight large subunits coded by the chloroplast genome, each of which has a molecular weight of 5.5×10^4, and eight small subunits coded by the nuclear genome, each with a molecular weight of 1.2×10^4 (1-4). Studies by Baker et al. (1) indicate that the subunits are arranged in a two-layered structure with each layer consisting of four large and four small subunit pairs. In cases where there are two or more types of small subunits, Hirai (6) has shown that they are distributed at random. Attempts to dissociate these subunits and then reassemble

_____

[*]This study was supported by USDA cooperative agreements 12-14-1001-967 and 12-14-1001-810.

the enzyme by biochemical means resulted in recovery of only 15 to
25% of the original activity (7).   Nevertheless, this as well as
other approaches provided some information on the function of the
large subunit, which is catalytically active even in the absence
of the small subunit (7-9).   The function of the small subunit may
well be regulatory (7), but the evidence is not conclusive.   Nothing
is known about any interaction between the large and small subunits
in regulating enzyme activity.

Isoelectric focusing of RuBP carboxylase from Nicotiana has resolved these subunits into many polypeptide patterns characteristic
of each species (10).   These patterns disclose four groups of large
subunits and a wide variety of small subunits within this genus.
Since the large subunit is transmitted via the maternal line, a
series of crosses and backcrosses can create almost any desired
combination of large and small subunits and circumvent the problems
associated with in vitro reconstitution of the enzyme.   By using
this approach, combinations of large and small subunits can be constructed that have either variable large and identical small subunits or identical large and variable small subunits.   Each large
and small subunit can be characterized by isoelectric focusing techniques (11) followed by amino acid analysis and tryptic peptide
mapping (12).   This system offers the unique advantage of examining
not only the regulation of RuBP carboxylase activity through the
joint expression of chloroplast and nuclear genomes but also the
functional role of each subunit in performing and modulating both
carboxylation and oxygenation reactions of RuBP.

BIOSYNTHESIS AND SUBUNIT ASSEMBLY

RuBP carboxylase has been isolated from a variety of species
representing the major plant phyla and examined by isoelectric
focusing techniques (13).   In a wide range of plant species examined
thus far RuBP carboxylases are composed of three major equally-spaced
large-subunit polypeptides (14, 15).   These three polypeptides
could have originated either from three distinct polypeptides coded
by three separate chloroplast genes or from post-translational modifications, such as deamidation of glutaminyl or asparaginyl residues (16, 17), of a single gene product producing three polypeptides
of different charge.   Since these polypeptides cannot be differentiated by amino acid analysis or by fingerprinting of trypsin or
chymotrypsin digests (14), current evidence favors the second alternative.   This is in agreement with the recent finding that a single
2000- to 2200-base-pair segment of maize chloroplast DNA (18) contains the information for the large subunit.   This segment is not
large enough to code for three distinct polypeptides of 450 amino
acids each.   Moreover, one would expect to find some unequally
spaced polypeptides arising from independent mutation of separate
genes during the course of plant evolution, but such isoelectric
focusing patterns have not been found.   Instead, the relative intensity of these polypeptide bands varies from experiment to experiment

Figure 1.   Isoelectric focusing of subunit polypeptides from
N. tabacum RuBP carboxylase in which isoelectric points were altered
through chemical modification of thiol groups with iodoacetic acid
(a) but not with iodoacetamide (b).

with the center band often being most intense (Figure 1). This is
evident in virtually all isoelectric focusing runs whether the proteins were S-carboxymethylated with iodoacetic acid or treated with
iodoacetamide (Figure 1). Such differences are more likely due to
different degrees of modification rather than to gene regulation.
     Most plant species contain a single small-subunit polypeptide
coded by the nuclear genome. The presence of multiple small-subunit
polypeptides in tobacco, tomato, and spinach (13) has been observed.
Unlike the large subunit, these have proved to be separate gene
products by sequence analysis (19).
     The synthesis of this important photosynthetic enzyme is regu-
lated by both the chloroplast and nuclear genomes. The large subunit of this enzyme is coded by the chloroplast genome; its mRNA is
present in chloroplast RNA and it is synthesized on chloroplast
ribosomes (20). The small subunit is coded by the nuclear genome;

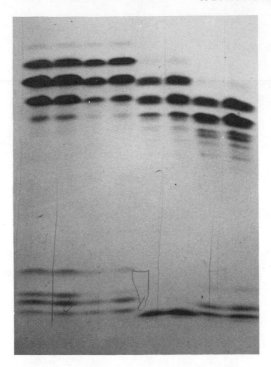

Figure 2. Isoelectric focusing of RuBP carboxylase from four
species of <u>Nicotiana</u> representing three of the four types of large
subunit and the variation from one to four polypeptides of the
small subunit. Each species is duplicated (15).

its mRNA is present in cytoplasmic RNA and it is synthesized on
cytoplasmic ribosomes (21). Recent evidence indicates that the
small subunit is synthesized as a precursor of higher molecular
weight (22, 23), which is transported across the chloroplast membrane to link up with the large subunit. Highfield and Ellis (22)
have suggested that the extra segment of the small-subunit precursor
is needed for assembly of the RuBP carboxylase molecule. If this
proves to be the case, it could explain why the attempt to recon-
stitute this enzyme from isolated subunits was not successful (7).
Under such circumstances new combinations of large and small subunits
could be assembled only <u>in vivo</u>.

GENETIC MANIPULATION OF RuBP CARBOXYLASE
SUBUNIT COMPOSITION IN <u>NICOTIANA</u>

In <u>vivo</u> modification of RuBP carboxylase subunit composition
can be readily achieved through genetic manipulation. Since cytoplasm is contributed exclusively by the female parent in a sexual

cross, chloroplasts are transmitted only via the maternal line, and
thus the large subunit will follow non-Mendelian laws of inheritance
(24).   Crosses between two species having different small subunits
would result in all small subunits being expressed in the progeny.
     Many species of <u>Nicotiana</u> have different subunit compositions,
some of which are illustrated in Figure 2.   Among these species,
examples can be found of interspecific hybridizations that have
modified the subunit composition of RuBP carboxylase.   <u>N. tabacum</u>,
which has been widely studied by cytogenetic and biochemical
approaches (25-27), arose from the interspecific hybridization of
<u>N. sylvestris</u> ♀ x <u>N. tomentosiformis</u> ♂ (25).   The RuBP carboxylase
of <u>N. tabacum</u> has the large-subunit polypeptides of <u>N. sylvestris</u>
and two small-subunit polypeptides, one from each of the parental
species.   Thus, the single small subunit of <u>N. tomentosiformis</u> has
been added to the RuBP carboxylase of <u>N. sylvestris</u> and perpetuated

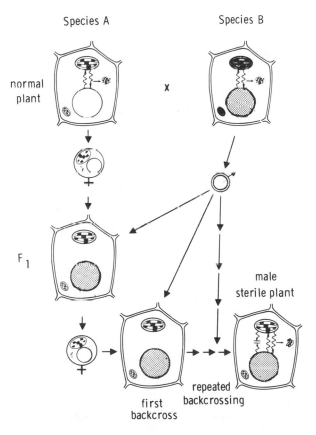

Figure 3.   A model for constructing new combinations of
    chloroplast and nuclear genomes (32).

in N. tabacum as a new RuBP carboxylase. Another example of genetic manipulation is N. digluta, which arose by chromosome doubling following the interspecific hybridization of N. glutinosa ♀ x N. tabacum ♂ (26). The large subunit of N. digluta is identical to that of the female parent, but the two small-subunit polypeptides of each parent species have been combined in the new species to create a RuBP carboxylase with four small-subunit polypeptides.

The above two cases represent new species in which a chloroplast genome has been combined with two nuclear genomes, one of which was the original counterpart of the chloroplast genome. There are many cases in Nicotiana in which a completely new combination of chloroplast and nuclear genomes can be created, the most familiar being the cytoplasmic male sterile cultivars of N. tabacum. In all such cultivars so far examined the new combination of chloroplast and nuclear genomes was achieved by making a cross between species A ♀ and species B ♂ and using the F_1 as female parent in backcrosses to species B (Figure 3). The backcross progeny (F_2) receive all of their chloroplast genome from the female parent (F_1) and 75% of their nuclear genome from the male parent (species B). After a few generations of backcrossing, virtually all the nuclear genome will be from the male parent (Figure 3). By this approach, many cytoplasmic male sterile cultivars of N. tabacum have been obtained with N. tabacum as the male parent in each case. The RuBP carboxylase subunit compositions of several cytoplasmic male sterile cultivars have been analyzed by isoelectric focusing techniques (28). The large-subunit polypeptides were characteristic of the female parent, as expected from the maternal inheritance of chloroplast genes. However, the small-subunit polypeptides were identical to those of N. tabacum. Consequently, each cytoplasmic male sterile cultivar had a new combination of chloroplast and nuclear genomes, and a group of RuBP carboxylases having variable large- but identical small-subunit polypeptides had been created (Figure 4). In a similar manner, RuBP carboxylase having identical large- but different small-subunit polypeptides can be obtained by using N. tabacum as the female parent in all crosses and any other species having small-subunit polypeptides different from N. tabacum as the male parent.

CHEMICAL PROPERTIES OF RuBP CARBOXYLASES
HAVING VARIABLE SUBUNIT COMBINATIONS

In order to investigate the interaction of chloroplast and nuclear genomes in regulating RuBP carboxylase activity, two groups of enzyme were selected. Those in the first group consisted of variable large subunits with two small subunits identical to those of N. tabacum. They were prepared from five cultivars of N. tabacum including three male sterile lines (Figure 4). The male sterile lines were obtained by interspecific hybridization of N. megalosiphon (Figure 4b), N. plumbaginifolia (d and g), and N. undulata (e) with N. tabacum (a and f) as the male parent. The F_1 progeny was

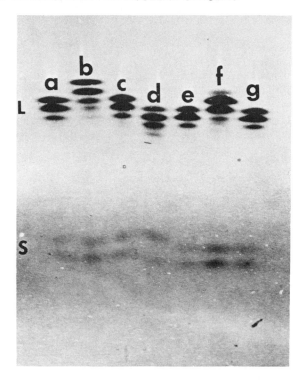

Figure 4.   Isoelectric focusing patterns of variable large (L) and
identical small (S) subunit combinations of RuBP carboxylase from
the following cultivars of N. tabacum: (a and f) Maryland 609;
(b) Burley 21 ♂ sterile from N. megalosiphon ♀;   (c) Burley 21;   (d
and g) Burley 21 ♂ sterile from N. plumbaginifolia ♀;   and (e) ♂
sterile from N. undulata ♀ (32).

repeatedly backcrossed to N. tabacum as described above (Figure 3).
Together with the enzyme from normal Burley 21 cultivar of N. tabacum
(Figure 4c), these enzymes contain three of the four types of large
subunits found in Nicotiana (10, 13).
    Since differences in amino acid composition and tryptic finger-
prints have been detected in large-subunit polypeptides having identical isoelectric points (12), several subunits from this first
group of enzymes were characterized by isoelectric focusing (Figure
4), amino acid composition analysis (Table 1), and tryptic fingerprinting (Figure 5). By these criteria, the large subunit of the
Burley 21 ♂ sterile line from N. megalosiphon (Figure 4b) was found
to differ from that of the normal Burley 21 cultivar (c), which was
identical (29) to that of N. tabacum MD609 (a and f). Since N.
plumbaginifolia and N. tabacum have large-subunit polypeptides with

Table 1

Amino Acid Compositions of the Small Subunits
of RuBP Carboxylase From Different Cultivars of
N. tabacum, N. sylvestris, and N. tomentosiformis

Results are expressed as probable numbers of
residues in a 1.2 x 10^4-dalton peptide (32).

| | Cultivars of N. tabacum | | | | | |
| | | | Burley 21* | | | |
Amino Acid	Maryland 609	Burley 21	male sterile	Kostoff hybrid	N. sylvestris	N. tomentosiformis
Aspartic						
acid	8	8	8	8	8	8
Threonine	4	4	4	4	4	4
Serine	4	3	3	4	4	4
Glutamic						
acid	16	16	18	17	17	16
Proline	7	7	6	8	8	7
Glycine	8	8	7	8	7	9
Alanine	6	6	6	6	6	6
Valine	6	6	6	6	8	6
Methionine**	1	1	1	1	1	1
Isoleucine	5	5	5	5	5	5
Leucine	9	9	9	9	9	9
Tyrosine	8	8	8	7	7	8
Phenylalanine	5	5	5	5	5	5
Lysine	8	9	9	8	7	7
Histidine	1	1	1	1	1-2	0.6
Arginine	5	5	4	4	4	5

*Male sterile line derived from N. megalosiphon ♀.
**Not corrected for methionine sulfoxone.

the same isoelectric point (30), the male sterile line derived from
N. plumbaginifolia ♀ was expected to have an enzyme with an isoelectric focusing pattern indistinguishable from that of the enzyme from
N. tabacum. The large subunit in this cultivar, however, is more
acidic than that in N. tabacum (compare f and g). Tryptic digest
fingerprints (Figure 5) and amino acid analyses (31) also reveal
differences between the large subunits and suggest that a spontaneous
mutation of the chloroplast gene for this subunit has taken place
(30). Although the male sterile line derived from N. undulata
(Figure 4e) has not been further characterized, differences in the
large-subunit primary structure are expected, with an isoelectric
focusing pattern deviating from that of N. tabacum. The small subunits of all N. tabacum cultivars in this group are indistinguishable
by any of the analyses used (Figures 4 and 5 and Table 1) (32).

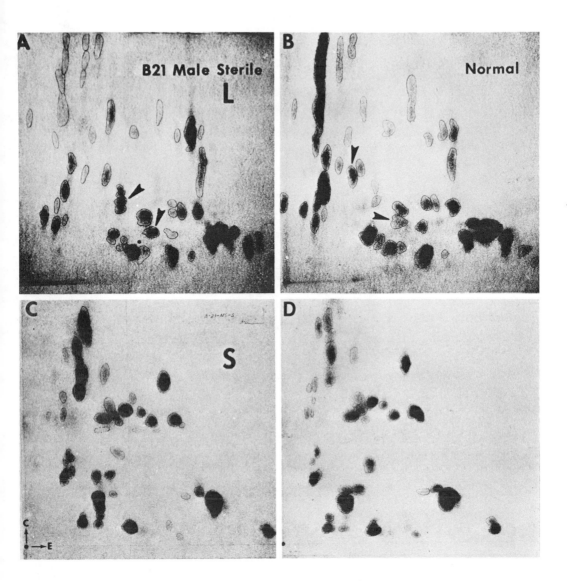

Figure 5. Peptide maps from tryptic digests of the large (L) and small (S) subunits of RuBP carboxylase from cultivars of N. tabacum. Differences between the large subunits of cv. Burley 21 ♂ sterile from N. plumbaginifolia ♀ (A) and cv. Maryland 609 (B) are marked with arrows. Differences between the small subunits of cv. Burley 21 ♂ sterile from N. megalosiphon ♀ (C) and cv. Maryland 609 (D) are not significant (32).

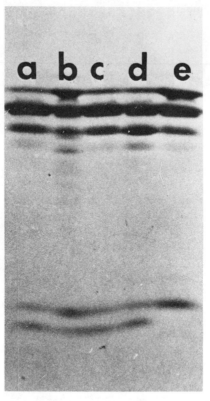

Figure 6.   Isoelectric focusing of RuBP carboxylase subunit combinations from (a) N. tabacum Maryland 609; (b) Burk's hybrid; (c)
Kostoff's hybrid; (d) Burk's hybrid x N. tabacum cv. NC95; and
(e) N. sylvestris (31).

     The RuBP carboxylases in the second group, consisting of iden-
tical large subunits but different small subunits, were prepared
from N. tabacum and its maternal progenitor N. sylvestris (25).
N. tabacum and N. sylvestris are identical with respect to the
large-subunit polypeptides and one of the small-subunit polypeptides;
the only difference between them is the single small-subunit polypeptide from N. tomentosiformis, the paternal ancestor of N. tabacum.
The interspecific hybridization between N. sylvestris ♀ x N. tomentosiformis ♂ which gave rise to the species N. tabacum must have occurred
millions of years ago (30).   Since mutations have been introduced in
these three species in the course of time, it was desirable to make
a more recent cross between N. sylvestris ♀ and N. tomentosiformis ♂.
Kostoff (33) did so in the 1930's and Burk (34) in the 1970's.
     The isoelectric focusing patterns of the enzymes from N. tabacum,
Burk's hybrid, Kostoff's hybrid, and N. sylvestris (Figure 6) show

that all have identical large and small subunits except that from
N. sylvestris, which has only a single small-subunit polypeptide.
From the amino acid compositions of the small subunits listed in
Table 1, it is evident that the small subunit from N. tabacum and
that from Kostoff's hybrid are indistinguishable. The higher valine
and histidine content of N. sylvestris and the higher glycine and
tyrosine content of N. tomentosiformis small subunits are reflected
as an equal mixture in N. tabacum (Table 1), in accord with results
obtained by Strøbaek et al. (19). Tryptic peptide mapping and amino
acid composition of all the large subunits in this group of enzymes
are currently under investigation.

The chemical analyses conducted thus far have established that
the carboxylases in the first group are composed of variable large
and identical small subunits, whereas those in the second group are
composed of identical large and variable small subunits. In con-
junction with enzymatic analyses, these subunit combinations can now
be used to relate structural and functional aspects of the enzyme.
Differences in enzymatic activity within the first group, for example,
would be attributable to variation in the large subunit, which may
involve the active site (7, 9). On the other hand, differences within the second group would be assigned to modifications in the small
subunit which affect its interaction with the large subunit.

ENZYMATIC PROPERTIES OF RuBP CARBOXYLASE
HAVING VARIABLE SUBUNIT COMBINATIONS

Before enzymatic activities of various RuBP carboxylases within
the two groups can be studied, one enzyme must be chosen for inclusion
in every experiment as a standard of comparison to allow correction

---

Table 2

Magnesium Requirements for Activation at pH 8.6 and Assay at pH 8.2
of RuBP Carboxylase From N. tabacum MD 609 (standard) and
the Male Sterile (MS) Cultivar Derived From N. plumbaginifolia ♀

Specific activities, expressed as nmoles $H^{14}CO_3^-$ incorporated
per mg protein per min, are the means of two experiments (31).

Assay $[Mg^{2+}]$(mM)	$[Mg^{2+}]$ in activation medium					
	5 mM		10 mM		20 mM	
	Std.	MS	Std.	MS	Std.	MS
4	353	700	389	802	440	884
10	364	612	343	738	408	745
20	276	380	295	437	311	435

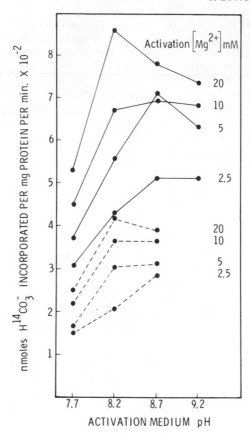

Figure 7. The relationship between magnesium concentration and pH in the CO_2 activation of RuBP carboxylase from <u>N. tabacum</u> cv. Burley 21 ♂ sterile line from <u>N. plumbaginifolia</u> ♀ (——) and cv. Maryland 609 (---)(31).

for variation in enzymatic activity among experiments. The RuBP carboxylase from N. tabacum MD 609 (Figure 4a and f) was used to standardize the CO_2 activation (35) and the assay conditions in this study. As shown in Table 2, the magnesium requirements of this enzyme for activation at pH 8.6 and assay at pH 8.2 were 20 mM and 4 mM $MgCl_2$, respectively. The pH 8.6 activation medium used in this study, though not optimal for the standard enzyme (Figure 7), was adequate for comparing a variety of enzymes. Likewise, the standard pH 8.2 assay medium was within a broad pH range suitable for measuring carboxylase activity (Table 3) in air. For measuring the oxygenase function of this enzyme, since similar activation and assay conditions have been found suitable, the only change made was to reduce the CO_2 content in the oxygenase reaction mixture (36).

Table 3

The pH Requirements for Assay of RuBP Carboxylase
From N. tabacum MD 609 (standard) and From the
Male Sterile Line Derived From N. plumbaginifolia ♀

Enzymes were CO_2 activated at pH 8.6 with 20 mM $MgCl_2$, and assays
were conducted in the presence of 4 mM $MgCl_2$. Specific activities,
expressed as nmoles $H^{14}CO_3^-$ incorporated per mg protein per min,
are the means of three replications (31).

Enzyme source	Assay medium pH				
	7.2	7.7	8.2	8.7	9.2
Standard	472	483	470	333	103
Male sterile	864	1006	966	579	307

Table 4

Specific RuBP Carboxylase and Oxygenase Activities of
Nicotiana Enzymes With Different Large and Small Subunit
Combinations and Their Ratios (32)

Species or cultivar, source of enzyme	No. of experiments		Specific activity (nmole/mg/min)*		
	Carboxylase	Oxygenase	Carboxylase	Oxygenase	Ratio
N. tabacum, cv. Md. 609	16	12	520±113 b	45.8±5.8 b	11.4±1.0
N. tabacum, cv. Burley 21 ♂ sterile from N. megalosiphon ♀	8	7	540±130 b	40.4±6.0 bc	13.4±1.4
N. tabacum, cv. Burley 21	6	7	488± 91 bc	44.4±9.2 b	11.0±0.2
N. tabacum, cv. Burley 21 ♂ sterile from N. plumbaginifolia ♀	7	8	881±148 a	66.0±7.6 a	13.3±0.7
N. tabacum, cv.♂ sterile from N. undulata ♀	9	8	477±108 bc	34.6±6.1 c	13.8±0.7
N. sylvestris	8	3	339±106 c	38.8±6.6 bc	8.7±0.8

*Data subjected to Student-Newman-Keuls test. Means followed by the
same letter are not significantly different at 0.05 level.

    The CO_2-activated enzymes (35) were assayed for carboxylase
and oxygenase activities under standard conditions, as described
by Kung et al. (32), and data from both RuBP carboxylase groups
are presented in Table 4. Three types of enzymatic activity were
found: high, shown by the enzyme from the male sterile line from
N. plumbaginifolia ♀; intermediate, shown by the standard enzyme;
and low, shown by the enzyme from N. sylvestris. Except for the
enzyme from the male sterile line from N. plumbaginifolia ♀, enzymes from all cultivars of N. tabacum in the first RuBP carboxylase
group have an intermediate level of carboxylase activity. Presumably
differences in primary structure of their large subunits do not involve the catalytic site. The decidedly high value for carboxylase
activity of the enzyme from the male sterile line from N. plumbaginifolia ♀ is an exciting result in view of the spontaneous mutation of
its large subunit (Figure 5). This unexpected finding suggests the
possibility of identifying some of the peptides associated with the
active site(s) for small-subunit interaction.
    To verify that the differences in RuBP carboxylase activity
reported here (Table 4) are real, the enzyme from the male sterile
line from N. plumbaginifolia ♀ and the standard enzyme were compared
under modified activation and assay conditions. Similar activation
and assay magnesium requirements (Table 2) as well as assay pH conditions (Table 3) were found for both enzymes; and under all conditions
tested the RuBP carboxylase from the male sterile was more active
than the standard enzyme. With variation of both [Mg^{2+}] and pH in
the activation medium (Figure 7), a high magnesium concentration
(20 mM $MgCl_2$) seems to be optimal at all pH levels tested. A shift
in optimal pH from 9.2 at 2.5 mM $MgCl_2$ to 8.2 at 20 mM $MgCl_2$ for the
enzyme from the male sterile line seems to apply also for the
standard enzyme.
    Kinetic properties of RuBP carboxylase also reflect differences
in activity (31) between the enzyme from the male sterile line from
N. plumbaginifolia ♀ and that from N. tabacum MD 609. Experiments
indicate that the affinity of RuBP carboxylase for RuBP appears to
be greater for the former than for the latter: Km(RuBP) = 60 μM
and 120 μM respectively.
    The differences described above are not likely to be due to ex-
perimental error, since a number of separate preparations were used
in these studies and the purity of each was insured by repeated crystallization (36). All samples were examined microscopically for uni-
formity of crystal size and shape; routinely, two recrystallizations
were sufficient to remove contaminants. It is evident that in all
tests, without exception, the RuBP carboxylase from the male sterile
line from N. plumbaginifolia ♀ is more active than the enzyme from
N. tabacum MD 609.
    In the second RuBP carboxylase group (Figure 6), the activity
of enzyme from N. sylvestris is decidedly lower than that of enzyme
from N. tabacum MD 609 (Table 4). Since the subunit compositions
of these enzymes are identical except for an additional polypeptide
on one small subunit of the latter (Figure 6) inherited from N. to-

mentosiformis (25), the higher activity of the N. tabacum enzyme
could be attributed to this single small-subunit polypeptide. However, minor differences in primary structure, which could affect
the catalytic site of the large subunit, may have been introduced
by spontaneous mutation since the two species diverged. In a
thorough analysis of chymotryptic peptides from N. tabacum and
N. sylvestris large subunits, Kawashima et al. (37) failed to
detect any evolutionary differences; however, not all peptides are
resolved by their techniques. For this reason, the activities of
the enzymes from N. tabacum MD 609 from Burk's and Kostoff's recent
interspecific crosses of N. sylvestris ♀ x N. tomentosiformis ♂
will be compared, to define functionally any changes that may have
occurred. Should the specific activities of these enzymes resemble
each other, differences between the standard enzyme and that from
N. sylvestris can be due only to the additional small-unit polypeptide altering the regulation of a catalytically active large subunit.

     Although the results of this study can be used to analyze the
functional roles of the large and small subunits, it is not clear
whether the ratio of carboxylase to oxygenase activity can be altered.
It is interesting that a change in carboxylase activity is always
accompanied by a similar change in oxygenase activity (Table 4).
The ratio is quite uniform at either 11 or 13 (Table 4) for all six
RuBP carboxylases studied except that from N. sylvestris, which requires additional oxygenase determinations. Those with the higher
ratio, 13, are all from male sterile cultivars, but whether this has
any biological significance is not certain. For the N. sylvestris
enzyme, the levels of both activities are low, and also their ratio.
Interestingly enough, N. sylvestris is a more compact plant and
grows much more slowly than N. tabacum. We are investigating
whether all of these characteristics are correlated.

CONCLUDING REMARKS

     The study of interaction between chloroplast and nuclear genomes
is a new and exciting development in plant biology. Much of the
published work has been limited to the effects of the nuclear genome
on chloroplast biogenesis. Only recently has intergenomic cooperation and communication between organelles (including chloroplasts)
and nuclei been firmly established (24, 38). This interaction is
truly a two-way process with input from organelles as well as from
the nuclei. Many organellar proteins, for example, are known to be
coded by nuclear genomes (24), whereas the biogenesis of some nuclear
components is regulated by organelles (39). Furthermore, evidence
is appearing that the biosynthesis of many organellar proteins requires joint expression of both genomes, as in the well documented
cases of the synthesis of RuBP carboxylase (24), coupling factor
(40), and cytochrome c oxidase (41).

     Advances in technology and the cooperation of molecular biolo-
gists have made it possible to study the joint expression of chloroplast and nuclear genomes in the biosynthesis of a single protein

and also the metabolic consequences of this process in plants. The data obtained represent the first evidence of intergenomic cooperation in regulating RuBP carboxylase activity. This important photosynthetic enzyme is composed of eight large subunits, which contain the catalytic site, and eight small subunits, which contain the regulatory site. Together, these two separate sites modulate enzyme activity. Since the large subunits are coded by the chloroplast genome and the small subunits by the nuclear genome, any changes in the coding information involving either the catalytic or the regulatory site are likely to affect enzyme activity. One may speculate that a multiple subunit structure coded by separate genomes serves to allow some means of joint regulation by those genomes. It can be postulated, therefore, that the activities of coupling factor and cytochrome c oxidase are likely to be jointly regulated by organellar and nuclear genomes through their subunit structure.

Evidence provided here suggests that either a mutation of the chloroplast genome or a combination of nuclear genomes can alter the level of RuBP carboxylase activity. This alteration involves both carboxylase and oxygenase activities. It was theorized on evolutionary grounds that carboxylation and oxygenation are inseparable reactions because a proper balance of CO_2 and O_2 in the atmosphere has to be maintained (42). This balance could be achieved enzymatically through a common catalytic site for both reactions. If this is so, attempts to alter the ratio of these two reactions by manipulating this enzyme will be futile.

It is hoped that this approach may eventually provide detailed structural and functional information on the interaction between the large and small RuBP carboxylase subunits. The use of Nicotiana provides an ideal model system for such studies. Its major advantage is that Nicotiana is one of a few closely related genera that yield RuBP carboxylase preparations in crystalline form. The extraction process is so simple and rapid and the yield so great that gram quantities of RuBP carboxylase can be readily produced. The value of this for enzymatic studies cannot be overemphasized. A second advantage is the wide variety of both large and small subunits within this genus. Other genera, such as Avena (43) or Triticum (44), contain a variety of subunits, but these do not yield gram quantities of enzyme in crystalline form. A further advantage of Nicotiana is that crosses producing the desired subunit combinations can readily be made.

REFERENCES

1. Baker, T. S., Suh, S. W., and Eisenberg, D., Proc. Natl. Acad. Sci. USA 74, 1037-41 (1977).
2. Kung, S. D., Science 191, 429-34 (1976).
3. Jensen, R. G. and Bahr, J. T., Annu. Rev. Plant Physiol. 28, 379-400 (1977).

4. Kawashima, N. and Wildman, S. G., Annu. Rev. Plant Physiol. 21, 325-58 (1970).
5. Tolbert, N. E. and Ryan, F. J., in Proc. 3rd Int. Congr. Photosynthesis, pp. 1303-19, M. Avron, Editor, Elsevier, Amsterdam, (1974).
6. Hirai, A., Proc. Natl. Acad. Sci. USA 74, 3443-5 (1977).
7. Nishimura, M. and Akazawa, T., Biochemistry 13, 2277-81 (1974).
8. McFadden, B. A., Biochem. Biophys. Res. Commun. 60, 312-17 (1974).
9. Paech, C. et al., See paper in this Symposium.
10. Chen, S., Johal, S., and Wildman, S. G., in Genetics and Biogenesis of Chloroplasts and Mitochondria, pp. 3-11, Th. Bucher et al., Editors, Elsevier, Amsterdam, 1976.
11. Kung, S. D., Sakano, K., and Wildman, S. G., Biochim. Biophys. Acta 365, 138-47 (1974).
12. Kung, S. D., Lee, C. I., Wood, D. D., and Moscarello, M. A., Plant Physiol. 60, 89-94 (1977).
13. Chen, K., Kung, S. D., Gray, J. C., and Wildman, S. G., Plant Sci. Lett. 7, 429-34 (1976).
14. Gray, J. C., Kung, S. D., and Wildman, S. G., Arch. Biochem. Biophys. 185, 272-81 (1978).
15. Sakano, K., Kung, S. D., and Wildman, S. G., Mol. Gen. Genet. 130, 91-7 (1974).
16. Robinson, A. B. and Rudd, C. J., Curr. Top. Cell. Regul. 8, 247-95 (1974).
17. Williamson, A. R., Salaman, M. R., and Kreth, H. W., Ann. N.Y. Acad. Sci. 209, 210-16 (1973).
18. Link, G. L. et al., See paper in this Symposium.
19. Strøbaek, S., Gibbons, G. C., Haslett, B., and Wildman, S. G., Carlsberg Res. Commun 41, 335-9 (1976).
20. Blair, G. E. and Ellis, R. J., Biochim. Biophys. Acta 319, 223-34 (1973).
21. Gray, J. C. and Kekwick, R. G. O., Eur. J. Biochem. 44, 491-500 (1974).
22. Highfield, P. E. and Ellis, R. J., Nature London 271, 420-4 (1978)
23. Chua, N. H. and Schmidt, G. W., See paper in this Symposium.
24. Kung, S. D., Annu. Rev. Plant Physiol. 28, 401-37 (1977).
25. Gray, J. C., Kung, S. D., Wildman, S. G., and Sheen, S. J., Nature London 252, 226-7 (1974).
26. Kung, S. D., Sakano, K., Gray, J. C., and Wildman, S. G., J. Mol. Evol. 7, 59-64 (1975).
27. Sheen, S. J., Evolution 26, 143-54 (1972).
28. Kung, S. D., Unpublished data.
29. Lee-Chang, C. I., Master's thesis, U. of Maryland Baltimore County, Catonsville, 1977.
30. Chen, K., Ph.D. Thesis, U. of California, Los Angeles, 1974.
31. Rhodes, P. R. and Kung, S. D., In preparation.
32. Kung, S. D., Rhodes, P. R., Lee-Chang, C. I., Marsho, T. V., and Wood, D. D., Plant Physiol., submitted.
33. Kostoff, D., Cytogenetics of the Genus Nicotiana, p. 1071, State Printing House, Sofia, Bulgaria, 1943.

34. Burk, L. G., J. Hered. <u>64</u>, 348-50 (1973).

35. Lorimer, G. H. and Badger, M. R., Anal. Biochem. <u>78</u>, 66-73 (1977).

36. Kung, S. D., Chollet, R., and Marsho, T. V., Methods Enzymol., in press (1978).

37. Kawashima, N., Tanabe, Y., and Iwai, S., Biochim. Biophys. Acta <u>427</u>, 70-7 (1976).

38. Bogorad, L., Science <u>188</u>, 891-8 (1975).

39. Blamire, J., Flechtner, V. R., and Sager, R., Proc. Natl. Acad. Sci. USA <u>71</u>, 2867-71 (1974).

40. Mendiola-Morgenthaler, L. R., Morgenthaler, J. J., and Price, C. A., FEBS Lett. <u>62</u>, 96-100 (1976).

41. Klein, J. L. and Edwards, D. L., J. Biol. Chem. <u>250</u>, 5852-6 (1975).

42. Lorimer, G. H. and Andrews, T. J., Nature London <u>243</u>, 359-60 (1973).

43. Steer, M. W. and Kernoghan, D., Biochem. Genet. <u>15</u>, 273-86 (1977).

44. Chen, K., Gray, J. C., and Wildman, S. G., Science 190, 1304-6 (1975).

IN VITRO SYNTHESIS, TRANSPORT, AND ASSEMBLY OF

RIBULOSE 1,5-BISPHOSPHATE CARBOXYLASE SUBUNITS

Nam-Hai Chua and Gregory W. Schmidt

The Rockefeller University, New York, New York 10021

INTRODUCTION

It is well established that the chloroplast synthesizes only a limited number of its own proteins. Many stromal enzymes and thylakoid membrane polypeptides are made in the cytosol and must be transferred subsequently from their sites of synthesis to their final locations inside the chloroplast (1). The mechanism of intracellular transport of proteins across the two chloroplast envelope membranes and the events associated with it are unknown.

In order to obtain some clues to this important problem, we investigated, in collaboration with Drs. Dobberstein and Blobel, the in vitro synthesis of the small subunit (S) of ribulose 1,5-bisphosphate (RuBP) carboxylase in the unicellular green alga Chlamydomonas reinhardi (2). This enzyme was chosen as a model system because it accounts for about 5 to 7% of the total cell protein (3, 4) and because it has been established that S is synthesized by cytoplasmic ribosomes (5, 6). We found that translation of poly(A)-containing mRNA from C. reinhardi in a wheat-germ cell-free system resulted in the synthesis predominantly of a protein with a molecular weight of 20,000. (It has an apparent MW of 21,000 in the 7.5 to 15% SDS gradient gel system used in the work described here.) This protein, which is about 4500 daltons larger than authentic S, was identified by indirect immunoprecipitation as a precursor (pS) to the latter. Cell fractionation studies showed that pS is synthesized by free polysomes and, after the complete precursor molecule has been made, can be processed to S by a specific endoprotease present in post-ribosomal supernatants. These results strongly suggest that the synthesis and transport of pS in vivo are not coupled to one another but rather are separate events. We therefore proposed that pS is an extrachloroplastic

325

form of S and that the transfer of pS across chloroplast envelope
membranes is similar, in principle, to the transfer of certain
plant and microbial toxins across plasma membranes of animal cells
(7).   In both cases, the transport of proteins across membrane
barriers is a post-translational event.

In an attempt to obtain direct evidence for the transport
model postulated above we have extended our investigations to
higher plants from which intact chloroplasts could be readily
isolated.   In this paper, we describe the in vitro synthesis and
identification of pS in Pisum sativum (pea) and Spinacia oleracea
(spinach).   We show that pS could be transported into intact chloroplasts without concomitant protein synthesis, as predicted by the
model.   Furthermore, either during or immediately after transport,
pS is converted to S, and a large proportion of the latter becomes
assembled into the holoenzyme.   While this work was in progress,
Cashmore et al. (8) and Highfield and Ellis (9) reported the in
vitro synthesis of a precursor to the small subunit of pea RuBP
carboxylase.   The latter authors also found that the precursor
could be taken up and processed to the small subunit by a crude
chloroplast preparation.

IN VITRO SYNTHESIS AND TRANSPORT OF pS INTO INTACT CHLOROPLASTS

Polyadenylated RNA was isolated from total nucleic acid extracts of spinach leaves by chromatography on poly(U)-Sepharose
(10). Addition of this RNA preparation to a wheat-germ cell-free
system (11) stimulated the incorporation of ^{35}S-Met into proteins
by 30 to 40-fold over the endogenous level (Figure 1, lane 1).
Product analysis by SDS gradient gels (12) revealed that a large
number of polypeptides were synthesized in vitro (lane 2).   Among
them, only the polypeptide band at 18,000 daltons was precipitated
specifically by antibodies against spinach S (lane 3) but not by
a pre-immune IgG fraction (results not shown).   Because the 18,000-
dalton polypeptide is immunochemically related to spinach S (MW
14,000) but larger by about 4000 daltons, it was identified as a
precursor (pS).   By similar techniques a precursor (pS) to S of
pea RuBP carboxylase was also identified among the polypeptides
synthesized by pea polyadenylated mRNA in a wheat-germ cell-free
translation system (Figure 1, lanes 5 and 6).   Comparison of the
precursors of spinach, pea, and C. reinhardi revealed that each
is larger than its corresponding authentic S by about 4000 to 5000
daltons (Figure 1).   Note that in no case was S detected among the
in vitro translation products, which suggests that the wheat-germ
cell-free extract lacks the processing enzyme necessary for converting pS to S.

The finding of pS in spinach and pea supports and extends the
previous observation with C. reinhardi that S is synthesized as
a larger precursor (2).   We suggested previously that pS is an
extrachloroplastic form of S containing an additional sequence

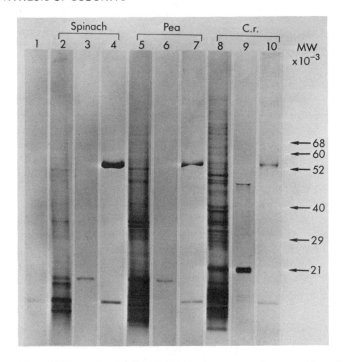

Figure 1. Identification of a precursor to the small subunit of
RuBP carboxylase in spinach, pea, and C. reinhardi. Analysis was
by SDS polyacrylamide gel electrophoresis (PAGE) (12) followed by
fluorography of dried gels (13). Lanes: 1, polypeptides synthesized
in the wheat-germ system without added mRNA; 2, polypeptides synthesized in the wheat-germ system in the presence of polyadenylated
mRNA from spinach; 3, immunoprecipitation of 2 with anti-spinach S
IgG fraction; 4, ^{35}S-RuBP carboxylase from spinach; 5, polypeptides
synthesized in the wheat-germ system in the presence of polyadenylated mRNA from pea; 6, immunoprecipitation of 5 with anti-spinach
S IgG fraction; 7, ^{35}S-RuBP carboxylase from pea; 8, polypeptides
synthesized in the wheat-germ system in the presence of polyadenylated mRNA from C. reinhardi; 9, immunoprecipitation of 8 with
anti-C. reinhardi S IgG fraction; 10, ^{35}S-RuBP carboxylase from
C. reinhardi. In vitro protein synthesis with wheat-germ cell-
free extract was carried out according to Roman et al. (11). Antibodies to spinach S were prepared as described previously (2), and
immunoprecipitation was done with S. aureus used as an adsorbant
for IgG (14).

which plays a role in the transport of pS into chloroplasts (2).
To see whether pS can be taken up by intact chloroplasts, a post-
ribosomal supernatant was first prepared from the wheat-germ translation mix containing polypeptide products of pea polyadenylated RNA.

Table 1

Protocol for In Vitro Reconstitution of
Protein Transport into Chloroplasts

Translation Mixture*   Intact Chloroplasts**
↓
Centrifuge at 140,000 x g_{max}, 45 min
↓
Adjust to 0.33 M sorbitol and 10 mM methionine

Incubate at 25° for 1 hr at chlorophyll concentration
of 0.5 mg/ml
↓
Centrifuge by acceleration to 4000 x g & Brake
↓
Discard supernatant   Pellet (intact chloroplasts)
↓
Resuspend in 0.33 M sorbitol - 50 mM tricine-KOH (pH 8.4)
↓
Add TPCK-trypsin & α-chymotrypsin to 200 µg/ml each
↓
Incubate at 25° for 30 min
↓
Centrifuge by acceleration to 4000 x g & Brake
↓
Discard supernatant   Pellet (Intact chloroplasts)
↓
Resuspend in lysis buffer: 5 mM tris-HCl (pH 7.5) -
1000 units/ml trasylol - 2 mM PMSF
↓
Vortex to lyse intact chloroplasts
↓
Adjust to 0.2 M NaCl - 25 mM tris-HCl (pH 7.5) -
0.5 mM EDTA
↓
Membrane pellet   Centrifuge at 30,000 x g for 10 min
analyzed by SDS-PAGE
↓
Layer supernatant on 10-30% sucrose gradients containing
0.2 M NaCl - 25 mM tris-HCl (pH 7.5) - 0.5 mM EDTA - 5 mM DTT
↓
Centrifuge at 170,000 x g_{av} for 17 hr

*Translation mixture contained 20 mM HEPES-KOH, pH 7.5; 130 mM KCl, 2 mM
$MgAc_2$; 2 mM DTT; 250 µM spermine; 1 mM ATP; 120 µM GTP; 8.4 mM phosphocreatine;
24 µM 19 amino acids; 40 µg/ml creatine phosphokinase; 0.15 µM methionine
(150 µCi/ml); 40 $O.D._{260}$/ml of wheat germ extract.

**Intact chloroplasts from silica sol gradients[15,16] were resuspended in
0.33 M sorbitol - 50 mM tricine-KOH (pH 8.4). One volume of the suspension
(1 mg chlorophyll/ml) was mixed with an equal volume of a post-ribosomal
supernatant (20,000 cpm/µl) derived from the translation mixture.

This supernatant, which includes pS (Figure 1, lane 2), was incubated with pea chloroplasts isolated by centrifugation on silica
sol gradient according to Morgenthaler et al. (15, 16). The procedure gave preparations containing ~70% intact chloroplasts.
The incubation conditions and subsequent processing of the incubation mix are detailed in Table 1. Briefly, after incubation,
intact chloroplasts were separated from the translation products
by centrifugation, and the pellet was resuspended in 0.33 M
sorbitol, 50 mM Tricine-HCl (pH 8.4). Trypsin and α-chymotrypsin
were then added to degrade any residual translation products
trapped in the chloroplast pellet or bound adventitiously to the
outer aspects of the chloroplast envelope. After proteolytic
treatment the suspension was diluted with one volume of the sorbitol-Tricine medium. Chloroplasts were then pelleted and lysed
in a hypotonic buffer containing proteolytic inhibitors. The
chloroplast lysate was centrifuged at 30,000 x g for 10 min to
yield a membrane pellet and a supernatant containing soluble proteins. With this procedure any ^{35}S-Met-labeled polypeptides
present in the chloroplast membrane pellet and supernatant were
considered to have become sequestered within the chloroplasts during
the previous incubation period. This conclusion is justified only
if the proteolytic treatment was effective; experiments discussed
below showed that this was indeed the case (cf. Figure 6).

    In vivo, S is normally found in association with L to form
the RuBP carboxylase holoenzyme. We consider that the transport
of pS into chloroplasts in vitro would have greater physiological
significance if the newly transported S could be shown to assemble
into the holoenzyme. To this end, we exploited the observation
that RuBP carboxylase holoenzyme has a sedimentation coefficient
of 18 S. After incubation, RuBP carboxylase in the 30,000 x g
chloroplast supernatant was separated from the rest of the stromal
proteins by sedimentation on a 10 to 30% sucrose density gradient
(3, 4). As shown in Figure 2, the supernatant proteins were in
fact resolved into two major fractions, and SDS-PAGE analysis
showed that the 18 S peak contained almost exclusively the holoenzyme.

    Figure 3 shows that incubation of intact pea chloroplasts in
the dark with translation products of pea polyadenylated mRNA
resulted in the uptake of a major polypeptide (lanes 3 and 4)
identical in electrophoretic mobility to S of pea RuBP carboxylase
(lane 9). This polypeptide was detected in the top fraction (lane
3) and in the 18 S peak (lane 4) but not in the 30,000 x g membrane
pellet (data not shown). Because of its electrophoretic mobility
and its co-sedimentation with RuBP carboxylase holoenzyme it was
tentatively identified as S of pea. Definitive identification was
provided by two-dimensional gel analysis (17). Figure 4 shows
that this polypeptide in the 18 S fraction could be resolved into
two isoelectric forms of identical molecular weight (panel b),
and that the two forms have the same pI and MW as the two
major isoelectric species of pea RuBP carboxylase small subunit
(panel a). Mixing experiments did not resolve any differences

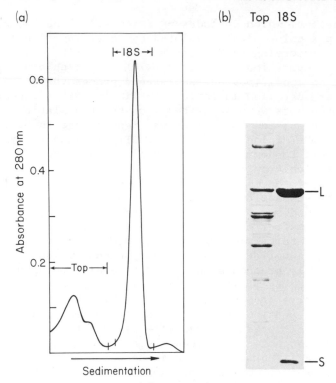

Figure 2. Fractionation of chloroplast soluble proteins by sucrose density gradient centrifugation and SDS gradient gel analysis of the top fraction and 18 S peak. (a) 30,000 x g supernatant from lysed pea chloroplasts (250 µg chlorophyll) was layered onto a 10 to 30% sucrose gradient containing 25 mM Tris-HCl (pH 7.5), 0.2 M NaCl, 5 mM DTT, and 0.5 mM EDTA. After centrifugation at 170,000 x g_{av} for 17 hr at 4° it was fractionated by an ISCO gradient frac-tionator, and the absorbance was monitored at 280 nm. (b) Poly-peptide patterns of the top fraction and 18 S peak. The fractions were collected separately as shown in (a) and precipitated by ad-justing to 10% TCA. Aliquots were analyzed by 7.5 to 15% SDS gradient gel electrophoresis (12). The loads were 45 µg for the top fraction and 22 µg for the 18 S peak. The 18 S peak, which contains almost exclusively RuBP carboxylase, comprises about 65% of the total soluble proteins in the chloroplasts.

between the isoelectric species of S synthesized in vitro and in vivo (data not shown).

   The above results show that S of pea RuBP carboxylase appeared in the soluble fraction of intact pea chloroplasts after the latter had been incubated with polypeptides synthesized in vitro by pea polyadenylated mRNA. Since pS but not S was present among the

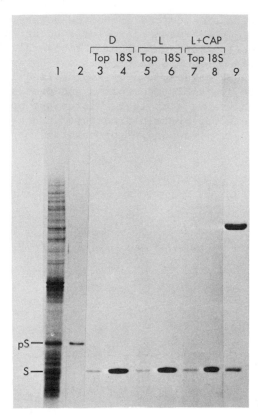

Figure 3. In vitro reconstitution of protein transport into intact pea chloroplasts does not require concomitant protein synthesis. Experiments were carried out as described in Table 1, and the samples were analyzed by SDS-PAGE followed by fluorography of dried gels. Lanes: 1, polypeptides synthesized in the wheat-germ system by polyadenylated mRNA from pea; 2, immunoprecipitation of 1 with anti-spinach S IgG fraction; 3 and 4, top fraction and 18 S peak, respectively, from pea chloroplasts incubated with translation products (lane 1) in the dark; 5 and 6, top fraction and 18 S peak, respectively, from pea chloroplasts incubated with translation products (lane 1) in the light; 7 and 8, top fraction and 18 S peak from pea chloroplasts incubated with translation products (lane 1) in the light in the presence of 100 μg chloramphenicol (CAP) per ml; 9, ^{35}S-RuBP carboxylase from pea. Dark (D), light (L), and L + CAP samples were incubated at 25° for 1 hr. Light intensity was 4000 lux.

in vitro polypeptide products, our results strongly suggest that pS was taken up by the chloroplasts and immediately processed to yield S. It should be emphasized that pS was never detected in the soluble or the membrane fraction of the chloroplasts after uptake.

In the experiment described above the uptake of pS and its
conversion to S by intact pea chloroplasts occurred in the absence
of protein synthesis either in the post-ribosomal supernatant of
the translation mix or in the chloroplasts. However, since the
large subunit of RuBP carboxylase is synthesized inside the chloroplast (16, 18, 19), possibly the uptake and processing of pS may be
enhanced under conditions such that chloroplast protein synthesis
is active. Furthermore, protein transport into chloroplasts could
conceivably be an energy-dependent process and may be stimulated

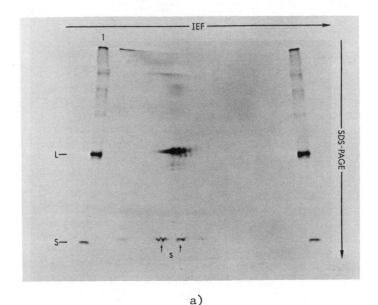

a)

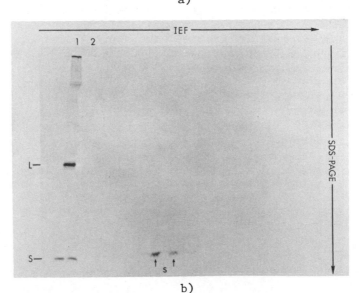

b)

by ATP produced inside the organelle. To test these possibilities, intact chloroplasts were incubated with in vitro translation products in the light with or without chloramphenicol (100 μg/ml). Under these conditions (Figure 3, lanes 5 to 8) the amounts of radioactive S in the top fraction and the 18 S peak were about the same as for the dark samples. Therefore, it can be concluded that transport of pS into chloroplasts and its conversion to S are not dependent on or enhanced by chloroplast ATP production or active chloroplast protein synthesis.

In an attempt to assess the relative distribution of newly transported S, radioactive bands were excised from gels of the top fraction and 18 S peak of the dark sample (Figure 3, lanes 2 and 3), and the radioactivity was measured directly. The finding of ~80% of S in the 18 S peak suggested that most of the newly transported S had been assembled into the holoenzyme. However, since the 18 S fraction was broad, one could argue that the radioactive S in this peak was due not to its assembly into the holoenzyme but rather to aggregation of newly transported S and sedimentation of the aggregates in the same region of the sucrose gradient as the holoenzyme. This possibility was excluded by our observation of exact coincidence between the radioactivity profile of S and the absorbance profile of the RuBP carboxylase (18 S) peak (data not shown).

Further evidence for assembly was obtained from electrophoresis of the 18 S peak in polyacrylamide gels without SDS. Under nondenaturing conditions the RuBP carboxylase holoenzyme in the 18 S peak migrated as a single band near the top of the 4 to 8% gel, and any contaminants sedimenting in the 18 S region of the sucrose gradient would have been resolved by the gel (Figure 5). All the radioactivity in the 18 S peak was confined to the RuBP carboxylase band, and no radioactive bands were detected in other gel regions (Figure 5, S and F). The RuBP carboxylase holoenzyme in the nondenaturing gel could be dissociated into its constituent subunits

← Figure 4. Two-dimensional gel analysis of the 18 S peak from intact pea chloroplasts after reconstitution of protein transport in vitro. Intact pea chloroplasts were incubated with translation products of pea polyadenylated mRNA in the dark at 25° for 1 hr as described in Table 1. Isoelectric focusing in the first dimension was carried out as described by O'Farrell (17), and SDS-PAGE in the second dimension was with 7.5 to 15% gradient gel (12). (a) ^{35}S-RuBP carboxylase from pea. Slot 1 in the second-dimension SDS gel contained pea ^{35}S-RuBP carboxylase; (b) 18 S peak from pea chloroplasts after reconstitution of protein transport in vitro. In the second-dimension SDS gel, slot 1 contained pea ^{35}S-RuBP carboxylase and slot 2 contained 18 S peak used in the two-dimensional analysis. The pH gradient from right to left was 4.5 to 8.3. Small arrows indicate the two major isoelectric species of S with pI's of ~6.0 and 6.6.

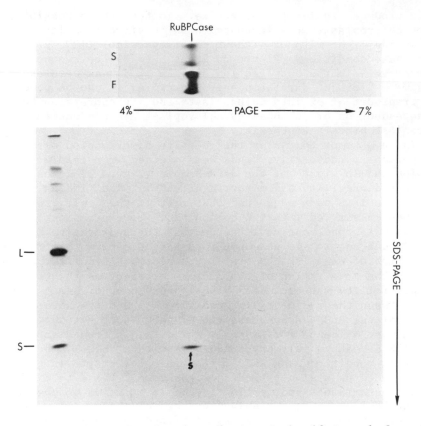

Figure 5. Two-dimensional gel analysis of the 18 S peak from intact
pea chloroplasts after reconstitution of protein transport in vitro.
The incubation conditions were the same as for Figure 4. The 18 S
peak was precipitated with 60% $(NH_4)_2SO_4$, dissolved in 25 mM Tris-
HCl (pH 7.5), 0.2 M NaCl, 5 mM DTT, and 0.5 mM EDTA and dialyzed
against the same buffer overnight. In the first dimension, the
sample was analyzed by 4 to 7% gradient gel according to Laemmli
(20) but without SDS in the gel or the buffers. The second dimension was by SDS-PAGE with 7.5 to 15% gradient gel as described
previously (12). Top: S, Coomassie Blue staining pattern of the
first-dimension gel strip; F, fluorograph of S. L and S are large
and small subunits of pea RuBP carboxylase.

upon second-dimensional electrophoresis into a 7.5 to 15% gradient
gel containing SDS (12). Upon fluorography of the two-dimensional
gel a single radioactive band was obtained coincident with the protein stain of the small subunit (Figure 5). These results demonstrate conclusively that the newly transported S in the 18 S region
was indeed assembled into the holoenzyme.

Figure 6. Homologous and heterologous reconstitution of protein
transport into intact pea chloroplasts in vitro. Intact pea chloroplasts were incubated with translation products of polyadenylated
mRNA isolated from pea (lanes 1 and 2), spinach (lanes 3 and 4), or
C. reinhardi (lanes 5 and 6) at 25° for 1 hr in the dark. Top
fractions and 18 S peaks were obtained by sucrose gradient fractionation as described for Figure 2. Proteins were precipitated
with 10% TCA and analyzed by SDS-PAGE and fluorography.

HETEROLOGOUS RECONSTITUTION OF PROTEIN TRANSPORT INTO
INTACT PEA CHLOROPLASTS

     The finding of pS in spinach and pea raises the possibility of
reconstituting the pS transport system of chloroplasts with use of
heterologous components. Furthermore, it is of interest whether pS
of C. reinhardi, a unicellular alga, could be taken up by higher
plant chloroplasts. To test these possibilities, pea chloroplasts
were incubated with polypeptides synthesized by polyadenylated mRNA
of spinach or C. reinhardi in the wheat-germ cell-free system
under the conditions outlined above (Table 1). Figure 6 shows that
spinach pS was taken up by pea chloroplasts and processed to S.

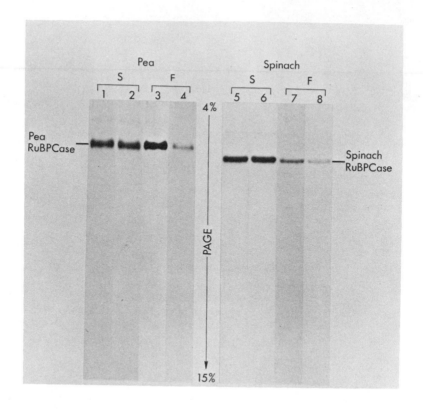

Figure 7.   Analysis of the 18 S peak from intact chloroplasts after homologous or heterologous reconstitution of protein transport in vitro.   Samples were analyzed by 4 to 15% gradient gels without SDS as described for Figure 5.   The Coomassie Blue staining patterns (S) of the dried gels are shown in lanes 1, 2, 5, and 6, and the corresponding fluorographs (F) are shown in lanes 3, 4, 7, and 8, respectively.   Lanes: 1 and 2, 18 S peak from intact pea chloroplasts after incubation with translation products of polyadenylated mRNA from either pea (lane 1) or spinach (lane 2); 5 and 6, 18 S peak from intact spinach chloroplasts after incubation with translation products of polyadenylated mRNA from either spinach (lane 5) or pea (lane 6).   Incubation was at 25° in the dark for 1 hr as outlined in Table 1.

Furthermore, the newly transported S was assembled into the holoenzyme, presumably with the large subunit of pea RuBP carboxylase (Figure 6, lane 4; Figure 7, lanes 2 and 4).   In the nondenaturing polyacrylamide gel shown in Figure 7, the RuBP carboxylase holoenzyme of spinach migrates faster than that of pea.   Since the small subunits of both enzymes are almost identical in molecular

Figure 8. Homologous and heterologous reconstitution of protein
transport into intact spinach chloroplasts in vitro. Intact
spinach chloroplasts were incubated with translation products of
polyadenylated mRNA isolated from spinach (lanes 1 and 2), pea
(lanes 3 and 4), or C. reinhardi (lanes 5 and 6) at 25° for 1 hr
in the dark. Other conditions were the same as for Figure 6.

weight the difference in electrophoretic mobility is probably due
to the difference in size of the large subunits. It is therefore
not unexpected that the hybrid RuBP carboxylase holoenzyme, which
contained radioactive S of spinach and presumably pea large subunits,
had the same electrophoretic mobility as the pea RuBP carboxylase
(Figure 7, lanes 1 to 4).

       In contrast to the case of spinach, incubation of pea chloro-
plasts with polypeptides synthesized in vitro by C. reinhardi
polyadenylated mRNA did not result in the appearance of any radioactive polypeptides in the top fraction (Figure 6, lane 5) or the
18 S peak (lane 6). Similarly, no radioactive polypeptides were
found in the chloroplast membrane fraction (30,000 x g pellet) upon
analysis by SDS-PAGE and fluorography (data not shown). Thus, pea
chloroplasts do not appear to transport pS or any other in vitro
synthesized polypeptides of C. reinhardi. It should be emphasized that the absence of radioactive polypeptides within the pea
chloroplasts provides unequivocal evidence that the proteolytic
treatment was effective in removing polypeptides that were not
transported into the organelle in vitro.

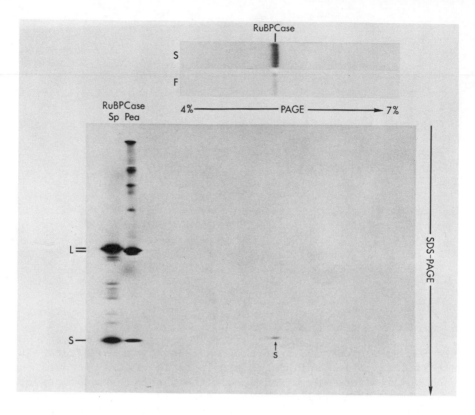

Figure 9. Two-dimensional gel analysis of the 18 S peak from intact
spinach chloroplasts after heterologous reconstitution of protein
transport in vitro. Intact spinach chloroplasts were incubated
with translation products of pea polyadenylated mRNA at 25° in the
dark for 1 hr. The 18 S peak was analyzed in the first and second
dimension as described for Figure 5. Top: S, Coomassie Blue
staining pattern of the nondenaturing gel (4 to 7% gradient) in
the first dimension; F, fluorograph of S. Pea and spinach (Sp) RuBP
carboxylase were run in the second-dimension SDS gels as standards.

HOMOLOGOUS AND HETEROLOGOUS RECONSTITUTION OF
PROTEIN TRANSPORT WITH SPINACH CHLOROPLASTS

     All the reconstitution experiments carried out with pea chloro-
plasts were repeated with spinach chloroplasts prepared according to
Morgenthaler et al. (15, 16). This procedure yielded spinach chloroplasts 90% of which were intact as monitored by phase-contrast
microscopy. Figure 8 shows that spinach chloroplasts were capable
of taking up pS of spinach and pea (lanes 1 to 4) but not of C.
reinhardi (lanes 5 and 6). As in the pea chloroplasts, the newly

transported S from pea or spinach was found to assemble into the
holoenzyme (Figure 7, lanes 5 to 8; Figure 9).

PRESENCE OF FREE LARGE SUBUNIT AMONG THE
TOP-FRACTION PROTEINS OF CHLOROPLASTS

The assembly of newly transported S into the holoenzyme strongly
suggests that in the chloroplast stroma there is a pool of free L or
assembly intermediates containing an excess of L in comparison with
S. Since these molecules are smaller than the RuBP carboxylase
holoenzyme, they would be expected to sediment in the region of the
sucrose gradient designated as top fraction (Figure 2). Double-
diffusion assay of the top fraction and the 18 S peak from pea and
spinach with anti-C. reinhardi L IgG revealed the presence of
materials immunochemically related to L in the top fractions of
both higher plants (Figure 10a). Definitive identification of
this cross-reacting material as L was provided by two-dimensional
crossed immunoelectrophoresis (22) with SDS-PAGE in the first
dimension (Figure 10b and c). Whether L exists as a free pool or
in combination with S as assembly intermediates remains to be
established by future work.

NATURE OF THE RECEPTOR ON OUTER CHLOROPLAST
ENVELOPE MEMBRANE FOR TRANSPORT OF pS

The transport of pS into chloroplasts of pea and spinach
suggests the presence of a specific receptor on the outer chloroplast envelope membrane. To probe the nature of this receptor,
intact pea chloroplasts were incubated with and without TPCK-
trypsin (200 μg/ml) and α-chymotrypsin (200 μg/ml) in 0.33 M sorbitol, 50 mM Tricine-KOH (pH 8.4). After 30 min at 25°, control
and digested chloroplasts were pelleted separately by centrifugation, washed twice in a sorbitol-Tricine medium, and resuspended
in the same medium in the presence of 100 units of Trasylol per ml.
The chloroplast suspensions were fractionated again on silica sol
gradient (15, 16), and the bands containing intact chloroplasts
were collected and used for reconstitution of protein transport
in vitro (Table 1). Intact chloroplasts pretreated with a mixture
of trypsin and α-chymotrypsin lost their ability to transport pS,
whereas control chloroplasts were active (data not shown). Because
of its sensitivity to proteases, we conclude that the receptor for
pS on the outer chloroplast envelope membrane is likely to be
a protein.

DISCUSSION

In the present study we confirm previous reports (2, 8, 9)
that the major products of in vitro protein synthesis driven by

polyadenylated RNA from algae and higher plants include precursors
to the small subunits of RuBP carboxylase. The cell-free products
of translation of mRNA from spinach, pea, and C. reinhardi which
are immunoprecipitated by antibodies to S are 4 to 5 kD larger
than the authentic polypeptides (Figure 1). pS was first identified
in C. reinhardi on the basis of its antigenic properties and its
endoproteolytic conversion to S upon incubation with post-ribosomal
supernatants from whole-cell extracts (2). A similar precursor to
the small subunit of pea RuBP carboxylase has recently been identified by immunoprecipitation of cell-free translation products with
antibodies to S (8) or to the RuBP carboxylase holoenzyme (9).
Cashmore et al. (8) reported good correspondence between tryptic
peptides from the pea precursor and those of S. In addition,

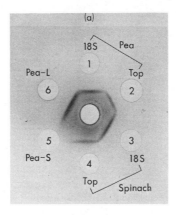

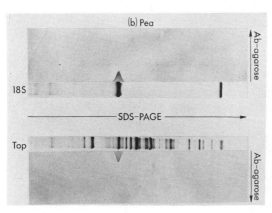 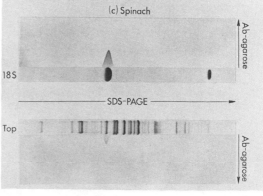

Highfield and Ellis (9) showed that pS could be processed to S by
a crude chloroplast preparation.   These results, together with
those reported in this paper, show conclusively that S of higher
plant RuBP carboxylase is also synthesized as a larger precursor
in vitro.
        The existence of pS raises the question of its physiological
function.   It has been shown previously (2) that the small subunit
of C. reinhardi is synthesized on free polysomes.   Thus, the
synthesis and transport of pS in vivo are not coupled; the two
processes are separated in time and in space.   Furthermore, pS
could be converted to S by a soluble endoprotease after completed
synthesis of the precursor in vitro (2).   These observations led
to the proposal (2) that the transport of pS into chloroplasts is
similar to the entry of certain plant and microbial toxins through
the plasma membranes of sensitive mammalian cells (7).   An important feature in both cases is the post-translational nature of the
transport process, in which the transport of protein across membrane
barrier is achieved after its complete synthesis.   This mode of
transport is to be contrasted with the co-translational transfer of
precursors of secretory proteins across microsomal membranes (23-
25).   It is well known that secretory proteins are synthesized on

← Figure 10.   Immunochemical characterization of the top-fraction
   proteins of pea and spinach chloroplasts with antibodies to the
   large subunit of RuBP carboxylase.   Antibodies were raised against
   L of C. reinhardi, and the IgG was purified according to Harboe
   and Ingild (21).   (a) Characterization by immunodiffusion.   The
   central well contained 25 μl of anti-L IgG (78 mg/ml).   The peripheral wells contained (1) 15 μl pea 18 S peak (2 μg/μl); (2) 15 μl
   pea top fraction (3 μg/μl); (3) 15 μl spinach 18 S peak (1.9
   μg/μl); (4) 20 μl spinach top fraction (2 μg/μl); (5) 20 μl
   spinach S (0.5 μg/μl); (6) 20 μl spinach L (0.7 μg/μl).   The
   immunodiffusion plate was incubated at 25° for 24 hr, washed,
   and stained with Coomassie Brilliant Blue.   (b) Characterization
   of pea top fraction and 18 S peak by crossed immunoelectrophoresis.
   Crossed immunoelectrophoresis was performed by a modification (to
   be published elsewhere) of the method described by Converse and
   Papermaster (22).   In the upper panel, 4 μg of the 18 S peak was
   used in the first-dimensional SDS gel.   However, the reference
   SDS gel shown above contained 6 times as much protein.   In the
   lower panel, 40 μg of the top fraction was used in the first dimension.   The IgG concentration in both experiments was 0.71 mg
   cm^{-2}.   (c) Characterization of the spinach top fraction and 18 S
   peak by crossed immunoelectrophoresis.   In the upper panel, 25 μg
   of the 18 S peak was used in the first dimension.   In the lower
   panel, 40 μg of the top-fraction protein was used in the first
   dimension.   The IgG concentrations were 1.065 and 0.71 mg cm^{-2}
   for the upper and lower panels, respectively.

membrane-bound ribosomes (23). According to the signal hypothesis
(24, 25), these proteins are transferred across the endoplasmic
reticulum membrane as they are synthesized, and during transport
the nascent presecretory proteins are processed by a membrane-
bound endoprotease. In vitro reconstitution studies have established that presecretory proteins are segregated into the microsomal lumen only when stripped microsomes are present during but
not after translation; completed secretory protein precursors can
no longer be segregated (24, 25).

The model for transport of pS into chloroplasts proposed previously (2), and reiterated above, receives direct experimental
support from the results presented here and also from the recent
work of Highfield and Ellis (9). Our results clearly demonstrate
that the transport of pS into chloroplasts and its processing to
S are independent of concomitant protein synthesis in the chloroplast or in the incubation mix (Figure 3). Similar results have
been reported by Highfield and Ellis (9) for the transport of pea
small subunit precursor. In reconstitution studies with heterologous components we found that pS of pea could be transported into
spinach chloroplasts and vice versa. In all these experiments, pS
was never detected in the chloroplasts, which indicated that the
precursor is processed during transport or immediately thereafter.
The apparent difference in the transport efficiency of pS between
the homologous and the heterologous systems (Figures 6 and 8)
probably reflects the difference in the amounts of pS among the
in vitro translation products of pea and spinach polyadenylated
mRNA (Figure 1, slots 2 and 5). In contrast to the higher plant
precursors, pS from C. reinhardi is not transported into either
spinach (Figure 8) or pea chloroplasts (Figure 6). C. reinhardi
S derived from in vitro processing of pS is not transported either
(unpublished data). Also, the specific endoprotease present in
the post-ribosomal supernatants of C. reinhardi whole-cell extracts does not process pS from the two higher plants (Schmidt and
Chua, unpublished data). Thus, the algal and higher plant pS have
apparently diverged during evolution so that heterologous activities
for transport and processing of pS are not demonstrable.

The observation that pS and other in vitro synthesized polypeptides of C. reinhardi are not transported into higher plant
chloroplasts demonstrates that transport of the precursor does not
result simply from a nonspecific interaction with membrane lipids.
Furthermore, the outer envelope membrane of higher plant chloroplasts must be impermeable to polypeptides of molecular weights
between 8000 and >100,000 since these polypeptides are present in
the in vitro translation products of C. reinhardi polyadenylated
mRNA. We proposed previously (2) that the additional sequence in
pS binds to a receptor in the chloroplast envelope and mediates
the transfer of the protein from the cytosol into the chloroplast
stroma. Since pre-incubation of intact chloroplasts with proteases
completely abolishes subsequent transport of pS, we conclude that
the envelope receptor is likely to be a protein.

The ability to process pS to S is an important means of establishing that the precursors are significant biologically rather than
artefacts of protein synthesis in a heterologous cell-free system.
Apparently the wheat-germ system lacks proteases that can convert
pS to S, since immunoprecipitation of the products of protein synthesis directed by mRNAs from pea, spinach, and C. reinhardi
with antibody to S results in specific recovery of pS but not S.
The findings presented here clearly show that the processing enzyme
which rapidly converts pS to S remains associated with intact chloroplasts following their purification by differential centrifugation
and fractionation on silica sol gradients (15, 16). Therefore, this
enzyme must be localized within the chloroplasts. We have presented
evidence that pS of pea is processed to an authentic form, since the
product is comprised of two species with isoelectric points and
molecular weights identical to those of authentic S found in pea
carboxylase holoenzyme (Figure 4). Similar results are mentioned
by Highfield and Ellis (9). Moreover, the precursor products are
physiologically active inasmuch as we could demonstrate specific
assembly with L within intact chloroplasts. Therefore, we are
assured of the ability of the wheat-germ system to synthesize a
protein identical to that in vivo.

Highfield and Ellis (9) have suggested that the processing
enzyme of pea is located in the chloroplast envelope because the
processing activity of their stromal fraction could be removed by
centrifugation at 30,000 x g_{av} for 40 min. Unfortunately, the
30,000 x g pellet was not checked for processing activity and therefore the possibility that the enzyme was somehow inactivated could
not be ruled out. This notwithstanding, we concur with their observation that lysed and/or fractionated chloroplasts yield little
or no activity for processing pS to S (data not shown).

In contrast to the results of our attempts to recover soluble
or membrane-associated processing activity, it is noted that polysomes isolated from wheat (6) and pea (26) seedlings yielded authentic S when allowed to complete their nascent chains in vitro. In
C. reinhardi, the processing enzyme is soluble and has been
shown to bind adventitiously to polysomes during cell fractionation
(2). Similarly, polysomes isolated from wheat (6) and pea (26)
seedlings are probably contaminated by the processing enzyme. Thus
we predict that the higher plant protease, like that of C. reinhardi, is soluble. Efficient processing of pS from pea and
spinach in vitro may require interaction of the precursors with
envelope proteins as well as a soluble chloroplast fraction.

Because in chloroplasts S is normally associated with L in
the form of the holoenzyme, the complete transport of pS into
chloroplasts could be most convincingly demonstrated if assembly
of the newly transported pS or S could be shown. Otherwise it
could be argued that pS is simply taken up into the space between
the two envelope membranes, or that transport of pS is an in vitro
artefact without physiological significance. For these reasons, we
have provided substantial evidence that 80% of S sequestered into

intact chloroplasts is recovered in association with the RuBP carboxylase holoenzyme in the 18 S peak on sucrose gradients and
following electrophoresis in nondenaturing polyacrylamide gels
(4 to 7% and 4 to 15%). For the most part, transport of pS seems
to be followed rapidly by assembly with the L of RuBP carboxylase.
The implication that a pool of free L or assembly intermediates
containing L exists in pea and spinach chloroplast is substantiated
by crossed-immunoelectrophoresis (Figure 10) with the top fractions
of chloroplast supernatants fractionated on sucrose gradients. Re-
construction experiments with ^{35}S-RuBP carboxylase from pea have
established that this pool of L does not result from dissociation
of the holoenzyme induced artefactually by our experimental pro-
tocols (unpublished data). Similar results were obtained by Roy
et al. (27), who have provided evidence for the presence of free
pools of L and S in pea seedlings.

The driving force for transport of pS across the chloroplast
envelope is unknown. We find that stimulation of chloroplast ATP
synthesis, and consequently chloroplast protein synthesis, by
incubation in the light does not enhance transport and processing
of pS and the assembly of S (Figure 3). Similarly, chloramphenicol
does not inhibit these processes. Therefore, we conclude that
chloroplast protein synthesis is not directly required for transport and processing of pS and assembly of the newly transported S
into the holoenzyme. Whether these processes require ATP, which is
present in the wheat-germ cell-free system, has not been determined.

Can assembly with L drive transport of S across the envelope?
Probably not. Although 20% of S which is sequestered by chloroplasts is not assembled into an 18 S complex (Figure 3), it might
be supposed that S and L in the top fractions of sucrose gradients
are associated with one another in the form of assembly intermediates having sedimentation coefficients lower than 18 S. On the
other hand, evidence for a free pool of S has been reported recently by Roy et al. (27). We interpret the results of Feierabend
and Wildner (28) as convincing evidence that L is not required for
transport and processing of pS. Growth of rye seedlings at elevated
temperatures induces loss of chloroplast ribosomes. Although chloroplast protein synthesis is totally inhibited, S is synthesized in
vivo and can be immunoprecipitated by antibodies to RuBP carboxylase. The S has a molecular weight identical to that of S isolated
from control plants. Moreover, S, but not L, is recovered from
the supernatant fraction of lysed plastids isolated from rye seedlings grown at high temperatures. Therefore, under these conditions transport and processing of pS occurs in the absence of L and
other products of chloroplast protein synthesis, which suggests
that in the normal plant processing of pS to S precedes assembly
into the holoenzyme.

At present, nothing is known about the envelope receptor protein or the processing enzyme for pS. We have found that wild-type
cells of C. reinhardi grown in chloramphenicol for several genera-

tions and mutant ac20cr1 (29), which is greatly deficient in chloro-plast ribosomes, retain control levels of the soluble endoprotease which converts pS to S (unpublished data). Therefore, the endo-protease and, as inferred from the results of Feierabend and Wildner (28), the envelope receptor protein are synthesized by cytoplasmic ribosome.

The post-translation transport of pS into chloroplasts, as shown in this study and also by Highfield and Ellis (9), provides support for the toxin model proposed previously (2). As yet, direct evidence that the precursor form of S is required for trans-port has not been obtained. Nevertheless, we are confident that in our _in vitro_ systems the synthesis of precursors to the small subunit of RuBP carboxylase, their transport into purified, intact chloroplasts, their processing to authentic S, and finally assembly with L accurately reconstruct the molecular events which occur _in vivo_. With these _in vitro_ systems it is now possible to examine in detail factors which regulate the synthesis, transport, and assembly of RuBP carboxylase subunits.

Acknowledgements: We thank Ms. Sally Liang for expert technical assistance. This work was supported in part by NIH grants GM-21060 and GM-25114. NHC is a recipient of NIH Research Career Development Award 1K04 GM-00223. GWS is supported by NIH National Research Service Award Postdoctoral Fellowship 4F32 GM-05390.

REFERENCES

1. Gillham, N. W., Boynton, J. E., and Chua, N.-H., Curr. Top. Bioenerg. 7 in press (1978).
2. Dobberstein, B., Blobel, G., and Chua, N.-H., Proc. Natl. Acad. Sci. USA 74, 1082-5 (1977).
3. Iwanij, V., Chua, N.-H., and Siekevitz, P., Biochim. Biophys. Acta 358, 329-40 (1974).
4. Iwanij, V., Chua, N.-H., and Siekevitz, P., J. Cell Biol. 64, 572-85 (1975).
5. Gray, T. C. and Kekwick, R. G. O., Eur. J. Biochem. 44, 491-500 (1974).
6. Roy, H., Patterson, R., and Jagendorf, A. T., Arch. Biochem. Biophys. 172, 64-73 (1976).
7. Pappenheimer, A. M. Jr., Annu. Rev. Biochem. 46, 69-94 (1977).
8. Cashmore, A. R., Broadhurst, M. K., and Gray, R. E., Proc. Natl. Acad. Sci. USA 75, 655-9 (1978).
9. Highfield, P. E. and Ellis, R. J., Nature London 271, 420-4 (1978).
10. Lindberg, U. and Persson, T., Eur. J. Biochem. 31, 246-54 (1972).
11. Roman, R., Brooker, J. D., Seal, S. N., and Marcus, A., Nature London 260, 359-60 (1976).

12. Chua, N.-H. and Bennoun, P., Proc. Natl. Acad. Sci. USA 72, 2175-9 (1975).
13. Bonner, W. M. and Laskey, R. A., Eur. J. Biochem. 46, 83-8 (1974).
14. Kessler, S. W., J. Immunol. 117, 1482-90 (1976).
15. Morgenthaler, J. J., Price, C. A., Robinson, J. M., and Gibbs, M., Plant Physiol. 54, 532-4 (1974).
16. Morgenthaler, J. J. and Mendiola-Morgenthaler, L., Arch. Biochem. Biophys. 172, 51-8 (1976).
17. O'Farrel, P. H., J. Biol. Chem. 250, 4007-21 (1975).
18. Blair, G. E. and Ellis, R. J., Biochim. Biophys. Acta 319, 223-34 (1973).
19. Bottomley, W., Spencer, D., and Whitfeld, P. R., Arch. Biochem. Biophys. 164, 106-17 (1974).
20. Laemmli, U. K., Nature New Biol. 227, 680-5 (1970).
21. Harboe, N. and Ingild, A., Scand. J. Immunol. Suppl. 1, 2, 111-64 (1973).
22. Converse, C. A. and Papermaster, D. S., Science 189, 469-72 (1975).
23. Campbell, P. N. and Blobel, G., FEBS Lett. 72, 215-26 (1976).
24. Blobel, G. and Dobberstein, B., J. Cell Biol. 67, 835-51 (1975).
25. Blobel, G. and Dobberstein, B., J. Cell Biol. 67, 852-62 (1975).
26. Roy, H., Terenna, B., and Cheong, L. C., Plant Physiol. 60, 532-7 (1977).
27. Roy, H., Costa, K. A., and Adari, H., Plant Sci. Lett. 11, 159-68 (1978).
28. Feierabend, J. and Wildner, G., Arch. Biochem. Biophys. 186, 283-91 (1978).
29. Boynton, J. E., Gillham, N. W., and Chabot, J. F., J. Cell Sci. 10, 267-305 (1972).

DISCUSSION

L. ANDERSON: Do you know anything about the specificity of the endopeptidase, pH optima, or peculiar cofactors?

CHUA: No. The endopeptidase has not been purified but we know that the preparation from Chlamydomonas containing it does not work with higher plant precursors.

S. HOWELL: Do you know whether the translational product must be in the form of a precursor to be transported into the chloroplast? Conceivably the translational product is cleaved before it is actually carried into the chloroplast. Must it really be in the form of a precursor? It is not an adequate answer to say the small subunit itself is not taken up by the chloroplast because that is obtained generally by denaturing mature enzyme.

CHUA: I agree. The important thing we have shown is that the precursor, pS, is taken up by intact chloroplasts, but the processed pS, derived by proteolytic/endoproteolytic action, is not taken up.

We hoped to show this dependence on precursor with our Chlamydomonas system and higher plant chloroplasts, but unfortunately it did not work. We also tried, unsuccessfully, to repeat this experiment with Euglena chloroplasts.

PAECH: Have you any suggestion as to how the chloroplast communicates to the protein-synthesizing system outside so that it does not overproduce small subunits?

CHUA: Perhaps Dr. Wildner can answer this question better than I can.

WILDNER: If barley is grown at 32°, it has no functional chloroplast ribosomes, and large amounts of small subunits are picked up from the cytoplasm. Actually, chloroplasts are not needed for synthesis of the small subunit. The small subunits can be detected on gels. Most of them are already in the cut form, with possibly a small band of precursor protein, and this may be cut down in the cytoplasm.

CHUA: Since the small subunit is recovered from the etioplast preparation, it could be that whatever is seen in the cytoplasm represents relocation or redistribution. This is the problem with all cell fractionation techniques.

PRICE: Could you comment on the finding by John Ellis (9) that the precursor of the small subunit from higher plants appears to be processed on the chloroplast envelope?

CHUA: His hypothesis is very attractive and logical, but we tried to check it and found the chloroplast envelope to be inactive. We can see that the precursor is processed either during transport or immediately after. The results of Wildner and co-workers suggest that the sequence of events is probably transport, processing, and then assembly, and not the other way around. One could argue that chain extension plays a dual role--that it is required for transport into the chloroplast and also for subsequent assembly with the large subunit. However, this does not appear to be the case.

THE EXPRESSION OF THE GENE FOR THE LARGE SUBUNIT

OF RIBULOSE 1,5-BISPHOSPHATE CARBOXYLASE IN MAIZE

Gerhard Link, John R. Bedbrook,[*] Lawrence Bogorad

The Biological Laboratories, Harvard University
Cambridge, Massachusetts 02138

Donald M. Coen,[**] and Alexander Rich

Department of Biology, Massachusetts Institute of Technology
Cambridge, Massachusetts 02139

During the past few years it has been established by DNA-RNA hybridization and physical mapping that the circular 85 x 10^6-dalton chloroplast chromosome of Zea **mays** contains genes for tRNAs for at least 16 amino acids (1) and two sets of genes for rRNAs (2). The chloroplast gene for ribosomal LC4 protein (3, 4) has been located on the Chlamydomonas reinhardi chloroplast chromosome by transmission genetics (5). We have now located physically the site of the gene for the large subunit (L) of the enzyme ribulose 1,5-bisphosphate (RuBP) carboxylase on the maize chloroplast genome. This has been possible because of progress in the physical mapping of this genome (6) and the availability of chloroplast (ct) DNA sequences cloned in bacterial plasmids as well as the development of linked transcription-translation systems (7).

LOCATION OF THE RuBP CARBOXYLASE LARGE SUBUNIT GENE

Wildman and co-workers (8) showed that L of RuBP carboxylase in Nicotiana species is inherited maternally and thus is likely to be coded in chloroplast DNA. The possible coding sequences for this 52,000-dalton polypeptide have been narrowed down to a 2500-base-pair maize chloroplast DNA sequence (9, 10).

---

[*]Present address: Plant Breeding Inst., Trumpington, Cambridge CB2 2LQ, England.
[**]Present address: Sidney Farber Cancer Inst., Boston, MA 02115.

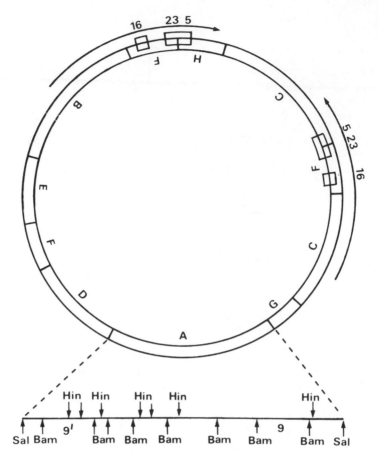

Figure 1. Physical map of Zea mays chloroplast DNA. The two
concentric circles representing the two DNA strands are inter-
sected by bars at the recognition sites for restriction endonuclease Sal 1. Letters are assigned to the Sal 1 DNA fragments
following the nomenclature of Bedbrook and Bogorad (6). The DNA
sequences coding for maize chloroplast rRNAs are marked by cross-
hatched boxes. The line at the bottom of the figure corresponding
to Sal 1 fragment A shows the locations of Bam H1 fragments 9 and
9' in relation to the recognition sequences for Hind III and Bam H1.

        The physical map of the maize chloroplast genome (Figure 1) is
based on (i) the results of double digestions of ctDNA with various
restriction endonucleases and comparison of overlapping DNA fragments, (ii) overlapping deduced from DNA-RNA hybridizations with
RNA transcribed in vitro from ctDNA fragments, and (iii) analyses
of recombinant DNA plasmids containing sequences of maize ctDNA
(6, 10). A prominent feature of the maize chloroplast genome is

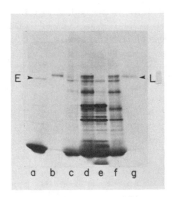

Figure 2. Polypeptides synthesized in a linked transcription-
translation system directed by chloroplast DNA sequences. Incubation for linked transcription-translation included DNA, RNA polymerase from E. coli, and nuclease-treated rabbit reticulocyte
lysate (7, 21). DNA-directed ^{35}S-methionine-labeled polypeptides
were fractionated on a 10 to 15% gradient polyacrylamide slab gel
in the presence of dodecylsulfate. The gel was then dried and
fluorographed (23). Samples a and c to f are polypeptides produced
by the linked system and directed by the indicated DNA: (a) control
without added DNA; (c) maize chloroplast DNA digested with Bam H1;
(d) pZmc 37 DNA containing Bam H1 fragment 9 digested with Bam H1;
(e) RSF 1030 isolated from pZmc 37 upon Bam H1 digestion by sucrose
gradient centrifugation. Lanes (b) and (g): purified ^{35}S-labeled
RuBP carboxylase from maize. L, position of the large subunit;
E, position of a 50,000-dalton polypeptide produced in the linked
system without addition of DNA. From Coen et al. (9).

the presence of an inverted repeated sequence containing the genes
for the ribosomal RNA species (2). These genes represent convenient
reference points when other coding sequences are to be located on
the physical map.

     In an attempt to map structural genes for proteins, linked
transcription-translation analysis (7) was performed with ctDNA
fragments that had been purified by molecular cloning. These cloned
DNA fragments served as templates for RNA synthesis in vitro with
E. coli RNA polymerase. The transcription products then directed
the cell-free synthesis of polypeptides in a reticulocyte lysate
(9, 10). It was shown that a 4200-base-pair ctDNA fragment, called
Bam H1 fragment 9 according to the nomenclature of Bedbrook and
Bogorad (6), directs the synthesis of a polypeptide with a molecular
weight of about 52,000. This is the size of the large subunit of
maize RuBP carboxylase (Figure 2). Often, a slightly larger

polypeptide with MW about 53,000 was also synthesized. Both these
[35]S-methionine-labeled in vitro products are immunoprecipitated with
anti-carboxylase serum and both, when digested with chymotrypsin or
papain in the presence of SDS (11), yield limited digestion products
resembling those of ^{35}S-methionine-labeled RuBP carboxylase L (9).
Thus, the coding sequence for RuBP carboxylase L appears to be contained within Bam H1 fragment 9. Copy RNA transcribed in vitro from
this DNA fragment with E. coli RNA polymerase does not hybridize to
other fragments generated from total ctDNA by Bam H1 digestion (6).
This suggests that there are no other regions containing the L gene
on the maize chloroplast genome.

Bam H1 fragment 9 maps ~30,000 base pairs from the 5'-end of
the closest of the two sets of rRNA genes and 70,000 base pairs from
the other set of rRNA genes (Figure 1). Fragment 9 is present in
only one copy per genome. Another Bam H1 fragment, 9' (Figure 1),
is similar in size but unrelated in sequence (10).

Bam H1 fragment 9 was joined, by DNA ligase, to the vehicle RSF
1030 (12) which had been digested with Bam H1, and chimeric plasmid
pZmc 37 was isolated (9, 10).

The structural gene for RuBP carboxylase L within the cloned
Bam H1 fragment 9 has been located more precisely by physical mapping with various restriction endonucleases with transcription-translation analysis of the fragmented DNA (10). The smallest DNA fragment known to direct the synthesis of this protein is located between recognition sites for endonucleases Bgl 11 and Sma 1 (Figure
3). This subfragment is 2500 base pairs in length. About 1600
base pairs of DNA would be sufficient to code for the amino acid
sequence in L. Thus, the structural gene coding for RuBP carboxylase L is present only once in Bam H1 fragment 9 and, accordingly,
is present in one copy per ctDNA molecule.

EXPRESSION OF THE RuBP CARBOXYLASE L GENE

In maize leaves, as in other plants, RuBP carboxylase is a
major component of the soluble protein fraction. Its synthesis is
related to the age and physiological state (etiolated versus light-
grown) of the plant (13). On the other hand, carbon assimilation
in Zea mays, unlike that in some other plants, occurs via the C_4
pathway of CO_2 fixation (Figure 4).

Huber et al. (14) and Matsumoto et al. (15) have shown that
RuBP carboxylase is present in bundle sheath (B) cells but is
absent from mesophyll (M) cells in the C_4 plants Digitaria sangui-
nealis and Zea mays. We have separated M from B cells of maize
enzymatically and, in agreement with these authors, have found 97%
of the RuBP carboxylase activity associated with B cell extracts.
From this we judge that our M cell preparations are contaminated
with 3% of the B cells, although the presence of small amounts of
the enzyme in maize M cells cannot be excluded. The differential
expression of the gene for L RuBP carboxylase in mesophyll and

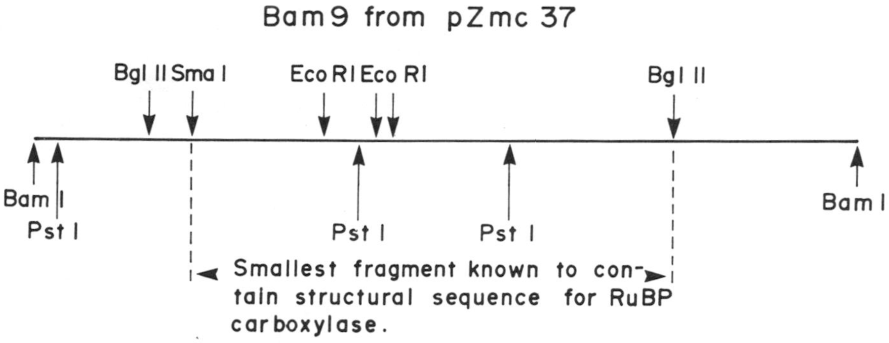

Bam 9 from pZmc 37

Bgl II Sma I   Eco RI Eco RI   Bgl II

Bam I   Bam I

Pst I   Pst I   Pst I

◄ Smallest fragment known to con-► tain structural sequence for RuBP carboxylase.

Figure 3. A map for restriction endonuclease recognition sites on Bam Hl fragment 9: The location of the structural gene for RuBP carboxylase L. The arrangement of DNA fragments is deduced from the results obtained when pZmc 37 DNA was digested with various re- striction endonucleases and resulting fragments were ordered by gel electrophoresis (10). The location of the coding sequence for L was determined by linked transcription-translation directed by various DNA fragments.

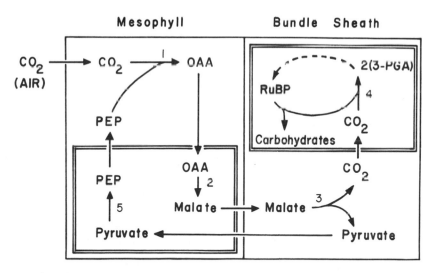

Figure 4. A schematic representation of the reactions during photo- synthetic carbon reduction by a C_4 plant, _Zea mays_. Chloroplasts are indicated within the cells by a double line. Enzymes: (1) phosphoenolpyruvate carboxylase, (2) NADP-malate dehydrogenase, (3) NADP-malic enzyme, (4) RuBP carboxylase, (5) pyruvate-P_i dikinase.

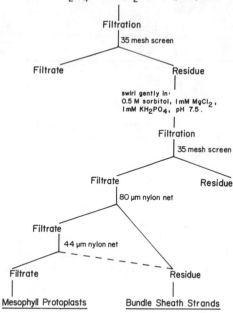

Figure 5A.   Separation of mesophyll and bundle sheath cells from
Zea mays by the procedures of Kanai and Edwards (16-18).   Leaf
segments from 16 to 20-day-old maize plants were incubated with a
mixture of cell-wall degrading enzymes.   Mesophyll (M) cells are
easily released from the tissue as protoplasts.   Bundle sheath (B)
cells are less susceptible to enzyme treatment.   "Strands" still
adhering to vascular elements can be separated from mesophyll
protoplasts by a series of filtration steps.

bundle sheath cells of maize could be controlled by any one of a
number of mechanisms, for example, (i) elimination of the coding
sequence for L RuBP carboxylase during differentiation of mesophyll
and bundle sheath cells; (ii) differential transcription of the
gene; (iii) post-transcriptional processing of the mRNA; (iv)
translational control of the transcribed gene; (v) degradation of
the translation product.   None of these addresses the question of
integration of expression of components coded and synthesized in
the nucleo-cytoplasmic and chloroplast systems.
     The first possibility, i.e., that the gene might be eliminated,
was tested by using the cloned gene for L RuBP carboxylase as a
probe for these sequences in maize M and B cell DNA.   Figure 5A
summarizes the procedure for separating B and M cells according
to Kanai and Edwards (16-18).   Upon digestion of Zea mays leaf tissue with cell-wall-degrading enzymes, M cells are released as

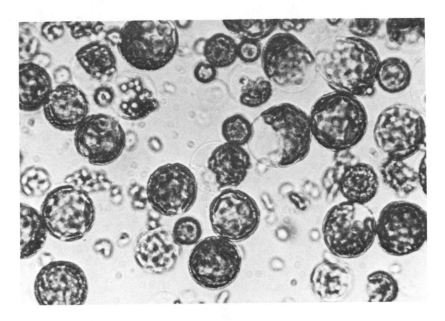

Figure 5B.   Isolated protoplasts from mesophyll cells.

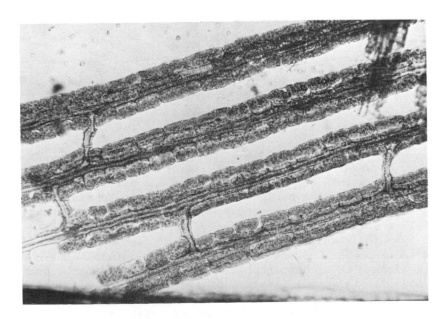

Figure 5C.   Isolated bundle sheath strands.

protoplasts and B cells as "strands" adhering to vascular elements
(Figure 5B and C). The two cell types are then separated by a
series of filtration steps. Total cellular DNA was extracted and
purified from the preparation of each of the cell types, labeled
in vitro with ^{32}P by polynucleotide kinase (19), and hybridized
with fragments of maize ctDNA that had been produced by endonuclease
digestion and separated by electrophoresis. As shown in Figure 6,
the fragments of ctDNA produced by Bam Hl digestion are similarly
hybridized by DNA from both M and B cells. Panel 1 shows that both
M and B cell DNA contain sequences that hybridize to each of the
fragments produced when the ctDNA is cut with Bam Hl. The data
would not reveal a deletion smaller than any of the fragments
generated by the endonuclease.

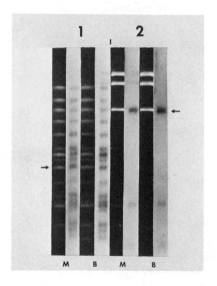

Figure 6. Hybridization of total cellular DNA from M and B cells
with chloroplast DNA fragments. Cellular DNA was isolated and
labeled with ^{32}P by polynucleotide kinase (19). Hybridizations
were with DNA fragments produced by Bam Hl digestion of ctDNA
(panel 1) and with fragments produced from pZmc 37 with Bam Hl plus
Bgl 11 (panel 2). Arrows point to the DNA fragments containing the
coding sequence for RuBP carboxylase L. Each autoradiograph is
printed along with a photograph of the corresponding gel containing 0.85% agarose (panel 1) and 1.5% agarose (panel 2). DNA bands
were photographed upon staining with ethidium bromide and illumination with long-wave UV light. DNA was transferred to nitrocellulose sheets as described by Southern (24). Hybridization was in
5x SSC and Denhardt's solution (25) for 24 hr at 65° (26).

Note that our ctDNA used here for Bam Hl digestion and also
in cloning experiments is largely, if not exclusively, derived from
M cells. The recovery of intact plastids requires that leaves be
disrupted gently. Such gentle disruption liberates primarily the
contents of M cells but may well include also some relatively un-
differentiated cells from very young tissue. The presence of the L
RuBP carboxylase gene in such ctDNA was already an indication,
though hardly proof, that this gene exists in maize M cell chloroplasts. Firmer evidence that M and B cells both contain the gene
for L RuBP carboxylase comes from hybridization of DNA from these
two types of cells to the 2500-base-pair fragments produced from
pZmc 37 by Bam Hl or Bgl 11 digestion (panel 2). This fragment has
been shown to contain the L gene (cf. Figure 3). We conclude that
this gene is present in both M and B cells. Differential expression
of the L gene must be accounted for by some other mechanism. Walbot
(20) could not detect differences in ctDNAs from M and B cells of
the C_4 grass Panicum by measuring sizes of restriction endonuclease
fragments in the electron microsope or by molecular hybridization
with total ctDNA from the two cell types. RNA preparation from M

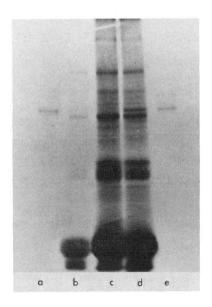

Figure 7.   Products of cell-free translation directed by M and B
cell RNA. ^{35}S-methionine polypeptides were fractionated on a 10
to 15% polyacrylamide slab gel, and the gel was fluorographed.
Samples (a) and (e): purified ^{35}S-RuBP carboxylase from maize.
Samples (b), (c), and (d): cell-free products of the reticulocyte
lysate programmed without added RNA (b), with M cell RNA (c), with
B cell RNA (d) (26).

and B cells were tested for the presence of L RuBP carboxylase mRNA by two methods: comparison of ^{35}S-methionine-labeled polypeptides produced in a reaction directed by M and B cell RNAs in a rabbit reticulocyte lysate (21), and molecular hybridization of in vitro labeled M and B cell RNAs with purified fragments of maize ctDNA and fragments of cloned ctDNA containing the gene for L RuBP carboxylase (26).

As shown in Figure 7, RNA from B cells (panel d) but not M cells (panel c) directs the synthesis of the 52,000- to 53,000- dalton polypeptide doublet comigrating with authentic maize L RuBP carboxylase (panels a and e). Both the in vitro synthesized poly- peptides are immunoprecipitated by antiserum against RuBP carboxyl- ase. In addition, RNAs from both B and M cells direct the synthesis of several other polypeptides. Although the M cell RNA fails to direct the synthesis of L, it does direct the synthesis of other proteins; this shows that it is active and does not contain a gen- eral inhibitor of translation. The results, i.e., that L is syn- thesized in the presence of B cell RNA but not in the M cell RNA directed system, indicates that L mRNA is relatively abundant in B cells but not in M cells.

Other aliquots of total M or B cell RNA were labeled in vitro with ^{32}P from α-^{32}P-ATP via polynucleotide kinase (6) and hybrid- ized with fragments of maize ctRNA known to contain the RuBP carbox- ylase L gene. As seen from Figure 8A, RNA from both M and B cell

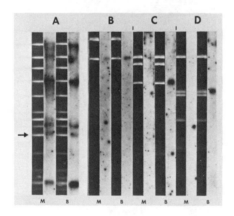

Figure 8. Hybridization of total cellular RNA from M and B cells with ctDNA fragments generated by restriction endonucleases. A. Hybridization with fragments of total ctDNA generated by Bam Hl. B to D. Hybridization with DNA fragments generated from DNA clone pZmc 37: by Bam Hl (B), by Bam Hl plus Bgi 11 (C), and by Bam Hl plus Bgl 11 plus Sma 1 (D). Arrow indicates position of Bam Hl band 9. Hybridizations were in 2x SSC for 16 hr at 65° (26).

RNA hybridizes Bam Hl-generated band 9 of a ctDNA digest. This band
contains two equal-sized DNA fragments, fragment 9 with the RuBP
carboxylase L gene and fragment 9', as outlined earlier. Hybridization with fragment 9' cannot be discriminated from hybridization
with fragment 9, which complicates the interpretation of this experiment. This difficulty, however, is eliminated when purified fragment 9 as part of plasmid pZmc 37 instead of a digest from total
ctDNA is used for hybridization.

     DNA-RNA hybridizations were carried out with ctDNA from clone
pZmc 37 digested with Bam Hl (Figure 8B), Bam Hl plus Bgl 11 (C), and
Bam Hl plus Bgl 11 plus Sma 1 (D). From the intensity of hybrid
bands on the autoradiograph it is evident that sequences hybridizing
the cloned 4200-base-pair fragment 9 are present in RNA from both M
and B cells. However, the large fragment generated from 9 by Bgl 11
and the 2500-base-pair fragment produced by Bgl 11 plus Sma 1, both
of which contain the L gene (see Figure 3), are hybridized by B
cell RNA but not by M cell RNA. These data correlate with the results of cell-free protein synthesis (Figure 7) and show that the L
gene is transcribed into RNA in B cells while transcripts of the L
gene are not detected in RNA from M cells. Hybridization of M cell
RNA to the cloned Bam Hl fragment 9 but not to the subfragment containing the L gene suggests that DNA sequences other than L within
this fragment are transcribed in both cell types.

     In an attempt to assess the relative abundance of RuBP carbox-
ylase L messenger RNA in M and B cell RNA more precisely, RNA was
"titrated" against the Bam Hl plus Bgl 11 fragment by using series
of RNA dilutions during hybridization (Figure 9). In addition to
the DNA fragments from pZmc 37, we also transferred the fragments
generated by Eco Rl from another recombinanant DNA plasmid (2),
pZmc 134, to each nitrocellulose strip used in this series of hybridizations. The Eco Rl ctDNA fragment cloned in pZmc 134, fragment a, contains the genes for maize chloroplast rRNAs (cf. Figure
1). Hybridization to Eco Rl fragment a was about equal when equal
amounts of RNA from M or B cells having same specific radioactivity
were used (Figure 9, panels 1 and 5). Electropherograms of total
RNA from M or B cells after gel electrophoresis revealed that the
16 S and 23 S chloroplast rRNA species are present in both cell
types in relative proportions similar to those in the corresponding cytoplasmic rRNAs. Thus, hybridization of RNA to Eco Rl ctDNA
fragment a containing the chloroplast rRNA genes serves as a useful internal reference for comparing the relative abundance of L
mRNA in M and B cells.

     Increasing amounts of RNA from M cells (Figure 9, panels 1 to
4) and decreasing amounts of B cell RNA (panels 5 to 8) were hybridized against constant amounts of ctDNA fragments. At the
highest concentration of M cell RNA (panel 4) hybridization to the
fragment containing the L gene (arrow) is no more prominent than
with a more than 1000-fold lower concentration of B cell RNA
(panel 8). We therefore estimate that mRNA sequences for RuBP

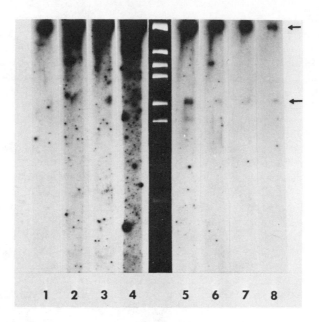

Figure 9. Hybridization of ctDNA fragments with different amounts of RNA from M and B cells. Clone pZmc 37 was digested with Bam Hl plus Bgl 11, clone pZmc 134 with Eco Rl. Both digests were applied to a 1.5% agarose gel, and the fragments were separated by electrophoresis. The stained bands, from top to bottom, are: (1) Eco Rl fragment a, (2) pMB9, (3) and (4) RSF 1030 fragments, (5), (7), and (8) ctDNA fragments generated from pZmc 37, (6) Eco fragment 1. Arrows mark Eco Rl fragment a containing the genes for chloroplast rRNAs and the fragment containing the coding sequence for L. The amounts of RNA used were 20 ng (panel 1), 50 ng (panel 2), 160 ng (panel 3), 400 ng (panel 4), 20 ng (panel 5), 5 ng (panel 6), 1.25 ng (panel 7), 0.3 ng (panel 8). Panels 1 through 4, M cell RNA; panels 5 through 8, B cell RNA (26).

carboxylase L are 200 to 2000 times as abundant in preparations from B cells as in those from M cells. Taking into account that M cell preparations are usually contaminated to about 3% by B cells (results not shown), it is conceivable that the M cell population of the leaf totally lacks L mRNA.

A more precise assessment of this mRNA in extracts from both photosynthetic cell types could be achieved by quantitative hybridization against a Bam 9 subfragment containing the structural gene. However, a plasmid containing this sequence without interfering adjacent sequences of ctDNA is not yet available. Sufficient amounts of purified L gene (a prerequisite for quantitative hybridizations) are difficult to prepare by conventional techniques.

Despite these limitations, it is clear that control mechanisms exist during M and B cell differentiation in maize which account for the described differences in the amount of RNA complementary to the L gene and in directing the cell-free synthesis of RuBP carboxylase L. Although we cannot completely rule out the possibility of post-transcriptional control, e.g., via specific ribonucleases, the differential expression of the RuBP carboxylase L gene most likely reflects control at the level of chloroplast transcription.

The small subunit of RuBP carboxylase is coded by the nuclear genome and is a product of cytoplasmic ribosomes (for a review see ref. 22). It is conceivable that much of the primary control of RuBP carboxylase production during B and M cell differentiation as well as during light-mediated chloroplast development resides in this nucleo-cytoplasmic system, although it is no less likely that control of synthesis of small-subunit mRNA in the nucleus is exercised by plastid activity or that coordinate control is exercised over both nuclear and plastid transcription.

Acknowledgments: We thank Mrs. A. R. Beaton and Mrs. B. Link for skilled technical assistance. JRB was a postdoctoral fellow of the Maria Moors Cabot Foundation; DMC was supported by an NIH Biophysics Training Grant and an Institute National Research Service Award: GL was supported by a grant from the Deutsche Forschungsgemeinschaft. This research was supported in part by grants to LB from the National Institute of General Medical Sciences (NIH) and the Maria Moors Cabot Foundation of Harvard University, and by grants to AR from the National Institutes of Health and the National Science Foundation.

REFERENCES

1. Haff, L. A. and Bogorad, L., Biochemistry 15, 4105-9 (1976).
2. Bedbrook, J. R., Kolodner, R., and Bogorad, L., Cell 11, 739-49 (1977).
3. Mets, L. J. and Bogorad, L., Science 174, 707-9 (1971).
4. Hanson, M. R., Davidson, J. N., Mets, L. J., and Bogorad, L., Mol. Gen. Genet. 132, 105-18 (1974).
5. Boynton, J. E., Gillham, N. W., Harris, E. H., Tingle, C. L., Van Winkle-Swift, K., and Adams, G. M. W., in Genetics and Biogenesis of Chloroplasts and Mitochondria, pp. 313-22, T. Bucher et al., Editors, Elsevier, Amsterdam, 1976.
6. Bedbrook, J. R. and Bogorad, L., Proc. Natl. Acad. Sci. USA 73, 4309-13 (1976).
7. Roberts, B. E., Gorecki, M., Mulligan, R. C., Danna, K. J., Rozenblatt, S., and Rich, A., Proc. Natl. Acad. Sci. USA 72, 1922-6 (1975).
8. Chan, P. H. and Wildman, S. G., Biochim. Biophys. Acta 277, 677-80 (1972).
9. Coen, D. M., Bedbrook, J. R., Bogorad, L., and Rich, A., Proc. Natl. Acad. Sci. USA 74, 5487-91 (1977).

10. Bedbrook, J. R., Coen, D. M., Beaton, A. R., Bogorad, L., and Rich, A., Unpublished (1978).
11. Cleveland, D. W., Fischer, S. G., Kirschner, M. W., and Laemmli, U. K., J. Biol. Chem. 252, 1102-6 (1977).
12. Heffron, F., Sublett, R., Hedges, R. W., Jacob, A., and Falkow, S., J. Bacteriol. 122, 250-6 (1975).
13. Bogorad, L., Dev. Biol. Suppl. 1, 1-31 (1967).
14. Huber, S. C., Hall, T. C., and Edwards, G. E., Plant Physiol. 57, 730-3 (1976).
15. Matsumoto, K., Nishimura, M., and Akazawa, T., Plant Cell Physiol. 18, 1281-90 (1977).
16. Kanai, R. and Edwards, G. E., Plant Physiol. 51, 1133-7 (1973).
17. Kanai, R. and Edwards, G. E., Plant Physiol. 52, 484-90 (1973).
18. Huber, S. C. and Edwards, G. E., Physiol. Plant 35, 203-9 (1975).
19. Hurwitz, J. and Novogradsky, A., Methods Enzymol. 12B, 207-12 (1968).
20. Walbot, V., Cell 11, 729-37 (1977).
21. Pelham, H. R. B. and Jackson, R. J., Eur. J. Biochem. 67, 247-56 (1976).
22. Kung, S. D., Annu. Rev. Plant Physiol. 28, 401-37 (1977).
23. Laskey, R. A. and Mills, A. D., Eur. J. Biochem. 56, 335-41 (1975).
24. Southern, E. M., J. Mol. Biol. 98, 503-17 (1975).
25. Denhardt, D. T., Biochem. Biophys. Res. Commun. 23, 641-6 (1966).
26. Link, G., Coen, D. M., and Bogorad, L., Cell, in press (1978).

DISCUSSION

KLESSIG: Have you tried hybridization selection of messenger RNAs using your clone probe? This would enable you to map the gene for the large subunit more finely, and then perhaps you could use the smaller fragments for $C_o t$ analysis to quantitate the level of message.

LINK: We have not yet tried that. We are following another approach at the moment. It certainly would be helpful to have the coding sequence cloned without any disturbing adjacent sequences.

THE MESSENGER RNAs AND GENES CODING FOR THE SMALL AND
LARGE SUBUNITS OF RIBULOSE 1,5-BISPHOSPHATE CARBOXYLASE/OXYGENASE
IN Chlamydomonas reinhardi*

Stephen H. Howell and Stanton Gelvin

Department of Biology, University of California at San Diego
La Jolla, California 92093

INTRODUCTION

The synthesis of ribulose 1,5-bisphosphate (RuBP) carboxylase/
oxygenase poses some interesting strategical problems for the plant
cell. The enzyme is produced in the cell in great abundance and is
sequestered within the chloroplast. It is synthesized by two different protein synthesizing systems, one in the cytoplasm and one
in the chloroplast (1-4). To produce large quantities of the enzyme, the cell must synthesize this protein preferentially over all
others. This presents gene regulation problems that are aggravated
by the enzyme being composed of two different subunits, the small
subunit (S) and the large subunit (L), which are apparently encoded
by two genomes within the cell (5-7) that may employ quite different
means for controlling the synthesis of their gene products. In this
paper we describe how Chlamydomonas reinhardi deals with these
problems of RuBP carboxylase synthesis. We are not asking why certain plant cells such as C. reinhardi produce so much of this en-
zyme--that has been the subject of other articles in this Symposium
We are concerned only with the mechanisms of how this happens.
    We have investigated the problems of RuBP carboxylase synthesis
in C. reinhardi by studying the messenger RNA (mRNA) coding for L
and S and their genes. The study of highly active genes in other
eucaryotic cells (not necessarily plant cells) has been made possible by first isolating the mRNA encoded by these genes. Such isolated messengers, when radioactively labeled, can be used as "hybridization probes" to recognize their corresponding genes. The
rationale for isolating a mRNA first, and not the gene, is that the

*This work was supported by the National Science Foundation.

mRNA from a highly "active" gene is present in higher proportion
in RNA preparations than the gene is in DNA preparations. For S
of RuBP carboxylase, this is certainly true. If S genes are present
in the nuclear genome, as suggested for tobacco by Kawashima and
Wildman (6), then they are buried among 10^5 or more other gene
sized DNA sequences (depending on the size of the plant genome) and
their outright identification or isolation is a nearly impossible
task. The argument for isolating L mRNA first to identify the L
gene is not as compelling, if the L gene is found on the chloroplast
genome as suggested by Chan and Wildman (5). Chloroplast genomes
in all plant cells (8) are far less complex than their corresponding nuclear genomes, and an L gene would be one of only 150 to 200
gene sized DNA sequences in the chloroplast genome. In a genome
of that size, one can identify and isolate single genes without using
mRNA hybridization probes. That approach was taken by Coen et al.
(9) when they identified, by using a linked transcription-translation system, a DNA fragment from the maize chloroplast genome which
codes for L. However, we have chosen the mRNA isolation approach
to study both S and L genes in C. reinhardi.

     A prerequisite to any procedure for isolating a specific mRNA
must be a way of identifying it--usually a translation assay. A
number of laboratories, including our own, have developed translation assays for both S and L mRNA. Because S and L are synthesized
in plant cells by two different translational systems (1-4), one
procaryotic-like or organellar and the other cytoplasmic, it has
been the usual practice to attempt to duplicate those conditions
in vitro, that is, to assay for L mRNA in a procaryotic translation
system and S mRNA in a eucaryotic system. Hartley et al. (10) were
the first to show that RNA isolated from pea chloroplasts could
direct L synthesis in vitro in an E. coli protein-synthesizing
system. The major translation product in that system was somewhat
smaller than bona fide L but was composed of cyanogen bromide fragments which cochromatographed with L peptides. Recently Dobberstein
et al. (11) successfully translated C. reinhardi S mRNA in a wheat-
germ protein-synthesizing system. The product (20,000 MW) was much
larger than S but could be recognized by its precipitability with
an antibody specific for RuBP carboxylase. For this and other
reasons Dobberstein et al. described the 20,000 MW protein as being
a putative precursor to S (ppS) which is processed in the cell to
yield the lower molecular weight form.

     Sagher et al. (12) demonstrated that any assumed stringency
for translation of L mRNA in a procaryotic-like protein-synthesizing system does not hold for Euglena L mRNA which can be translated in a wheat-germ system. Subsequently Rosner et al. (13)
demonstrated that chloroplast RNA from Spirodela could also spon-
sor L synthesis in wheat germ. Any apparent inability of RNA preparations to direct L synthesis in a eucaryotic protein-synthesizing system may not be a formal property of the message, but may
be simply the ability of L mRNA to compete for synthesis with other
more active mRNAs in partially purified RNA preparations. We have

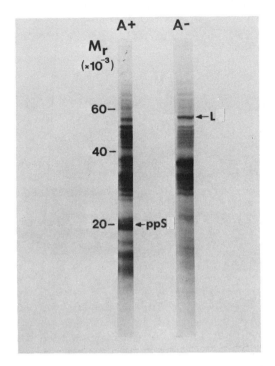

Figure 1. Synthesis of ppS and L in a wheat-germ _in vitro_ translation system. _C. reinhardi_ CW15 mt$^+$ cells were synchronized by
12 hr light--12 hr dark illumination cycles and harvested 4 hr into
the light period (14). RNA was extracted according to Weeks and
Collis (25) and fractionated into polyadenylated [poly(A)+]
nonpolyadenylated [poly(A)-] RNA by chromatography on oligo(dT)
cellulose (26). Poly(A)+ (2 µg) and poly(A)- (5 µg) were translated in a wheat-germ system according to Roman et al. (27). Poly(A)+
RNA resulted in a 20-fold greater stimulation of ^{35}S-methionine
incorporation per µg RNA than did poly(A)- RNA. Reaction products
were denatured, and aliquots containing the same amounts of cold-
acid-precipitable (hot-acid-resistant) radioactivity were subjected
to electrophoresis on 11% SDS-polyacrylamide gels. Shown is a
fluorogram of the gel which has been impregnated with 2,5-diphenyl-
oxazole, dried, and exposed to photosensitized Kodak X-0 mat R
x-ray film. Scale of molecular weights (M_r) is related to migration of standards (bovine serum albumin, 68,000; ovalbumin, 43,000;
carbonic anhydrase, 29,000; ribonuclease A, 14,700).

found, for example, that _C. reinhardi_ RNA preparations from which
polyadenylated RNA [poly(A)+ RNA] has been removed (Figure 1) will
direct the synthesis of L (a protein which co-migrates with L on
SDS-polyacrylamide gels), whereas total RNA fraction will not. In
any case, L mRNA isolated from plant cells or chloroplasts is not a

highly active mRNA in vitro, in either a procaryotic or eucaryotic
translation system. In contrast, S mRNA is highly active in vitro
(11).

ISOLATION OF S mRNA

     In the course of purifying S mRNA for use as a hybridization
probe to identify S genes, we have studied certain properties of the
message. We have confirmed the observation by Dobberstein et al.
(11) that in a wheat-germ system S mRNA codes for a protein (ppS,
20,000 MW) larger than that found in the enzyme (Figure 1), and
that only S, and not ppS, are produced when nascent chains on C.
reinhardi polyribosomes are elongated in a wheat-germ system (not
shown).
     In agreement with Dobberstein et al. (11) we have also found
that S mRNA in C. reinhardi is probably polyadenylated since it
binds efficiently to oligo(dT) cellulose (Figure 1). When poly(A)+
RNA from C. reinhardi is fractionated on a sucrose density
gradient under nondenaturing conditions, RNA which directs the
synthesis of ppS sediments sharply with a peak sedimentation coefficient of 11 S (Figure 2). The messenger activity in the 11 S
peak appears quite homogeneous in that it codes almost exclusively
for ppS (80% of the total protein coded for by RNA in fraction 17
migrates as ppS). Therefore, in two purification steps, oligo(dT)
cellulose column chromatography and sucrose density gradient sedimentation, we can obtain from whole-cell RNA a hybridization probe
of sufficient purity to detect S genes. An added step to enrich
for S mRNA in the 11 S fraction is to extract (or radioactively label)
RNA for synchronous cells during the light phase of the cell cycle.
We have found that S mRNA accumulates during the light phase (Figure 1), and therefore in all our S mRNA isolation procedures we
have used light phase cells.

ISOLATION OF L mRNA

     The isolation of L mRNA presents a greater challenge than the
isolation of S mRNA: L mRNA is more difficult to detect and to
separate from whole-cell RNAs. We assayed for L mRNA by measuring
the L-specific immunoprecipitability of translation products in an
E. coli protein-synthesizing system (14). The immunoprecipitation
assay is highly sensitive and permitted us to detect the synthesis
of small quantities of L that were not necessarily full-sized polypeptide chains (55,000 MW). In fact, with the E. coli translation
system we have never succeeded in synthesizing full-sized L chains,
only smaller immunoprecipitable polypeptides. However, we have
been able to demonstrate that the immunoprecipitable product is
largely L by tryptic peptide fingerprint analysis (14). The short-
comings of our translation assay have placed certain constraints

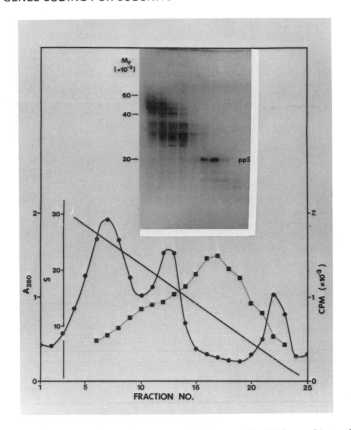

Figure 2.   Sedimentation characteristics of mRNA coding for ppS.
RNA was extracted from synchronous C. reinhardi at L4, and poly(A)+
RNA fractions were obtained as for Figure 1.   Ten μg of poly(A)+ RNA
was loaded onto a 15 to 30% sucrose density gradient (in 10 mM
Tris-HCl, pH 7.5, 100 mM NaCl, 1 mM EDTA, 0.1% sodium lauroyl
sarcosinate) and subjected to centrifugation at 38 krpm for 16 hr
at 2°C in a Beckman SW41 rotor.   The gradients were dripped from the
tube bottom, and RNA in each fraction was precipitated with ethanol.
RNA was resolubilized in water and translated in the wheat-germ system (27).   The entire reaction product labeled with ^{35}S-methionine
was denatured and subjected to electrophoresis on 11% SDS-polyacrylamide gels as described for Figure 1.   Shown is a fluorogram
of the gel in which the translated samples have been aligned over
the corresponding RNA fractions.   The sedimentation profiles are of
RNA samples sedimented in a parallel gradient.   The sedimentation
(A_{260}) of 25 S and 18 S ribosomal RNA and 4 S RNA from about 250 μg
of whole-cell RNA (●) obtained from L4 synchronous cells was used
to determine the sedimentation coefficient values (S) for poly(A)+
RNA (■) fractions obtained from cells labeled with ^{32}PO$_4$ from
L2 to L6.

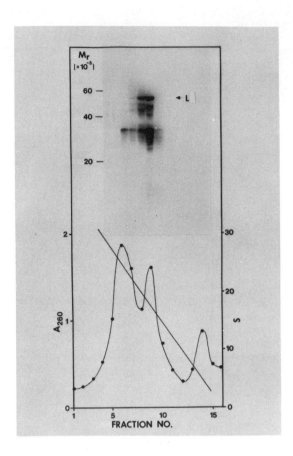

Figure 3. Sedimentation characteristics of mRNA coding for full-
sized L. RNA was extracted from synchronous C. reinhardi at L4,
and poly(A)- RNA fractions were obtained as for Figure 1. Poly(A)-
RNA (250 μg) was loaded onto a sucrose density gradient and subjected to centrifugation at 36 krpm for 14 hr at 2°C in a Beckman
SW 41 rotor. Half the RNA in each gradient fraction was used to
determine the sedimentation profile (●), and the other half was
translated in a wheat-germ system. The ^{35}S-methionine-labeled
translation products were subjected to gel electrophoresis as
described for Figure 2. Shown is a fluorogram in which the
translated samples have been aligned over the corresponding RNA
fractions.

on the interpretation of the results described below. Nonetheless, the assay has permitted us to learn about some of the properties of L mRNA and to purify it for use as a hybridization probe.

The separation of L mRNA from the very abundant cellular ribosomal RNAs is complicated by the fact that neither L mRNA nor ribosomal RNA binds efficiently to oligo(dT) cellulose (Figure 1). This suggests that L mRNA is not highly polyadenylated (14) and agrees with a similar observation made for Euglena L mRNA (12). The apparent state of polyadenylation of L mRNA is an operational observation, however, because there are uncertainties described below as to whether the messenger has been isolated in an intact form.

We have reported that when C. reinhardi whole-cell RNA is sedimented on sucrose density gradients under nondenaturing conditions, RNA coding for immunoprecipitable L sediments somewhat more slowly than ribosomal small subunit RNA (14). The estimated sedimentation coefficient for the messenger activity was about 14 S. Because the RNA in the 14 S fraction did not direct the synthesis of the complete L polypeptide in vitro, we could not be sure that this RNA represents the intact L message. When we have translated similar gradient fractions in a wheat-germ system, we have found that RNA coding for full-sized L (a protein that co-migrates with L on SDS-polyacrylamide gels) sediments at about 18 to 20 S (Figure 3). Nearly the same sedimentation pattern for Spirodela L mRNA translated in the wheat-germ system has been obtained by Rosner et al. (13). In C. reinhardi the possible difference in results between the two translation systems can probably be attributed to the L mRNA sedimenting close to or with ribosomal RNAs, and the presence of ribosomal RNA in translation assays may interfere with L synthesis. Therefore, translation assays alone leave some ambiguities about the size of L mRNA and whether it exists in a single homogeneous form.

Since both the apparent sedimentation and polyadenylation properties of L mRNA make it difficult to purify, we sought a different approach to isolate the messenger. We chose immunological techniques to isolate polyribosomes engaged in L synthesis. First we used radioactively labeled L specific antibody to identify the class of polyribosomes engages in L synthesis (15). It was surprising to find that labeled L antibody bound primarily to small (N = 2 to 3) polyribosomes and not to large ones (Figure 4). This was unexpected because L mRNA codes for a large polypeptide, and if L-synthesizing polyribosomes are packed with ribosomes to the same density as are other reported messages (15), then one might expect them to contain 10 to 15 ribosomes. We have reported many controls which demonstrate that this observation is probably not the result of RNA breakdown during polyribosome isolation or some other technical artifact (15). The observation suggests that translation initiation on L mRNA is slow and supports the idea that L mRNA may not be a highly active messenger in vivo.

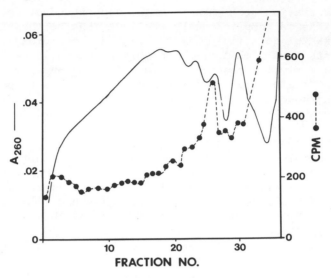

Figure 4

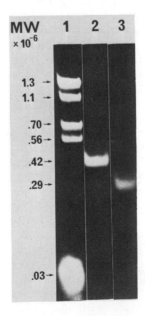

Figure 5

←Figure 4.   The binding of ^{125}I-labeled antibody specific for L
to C. reinhardi polyribosomes.   Polyribosomes (30 OD units) were
incubated with 3 x 10^5 cpm of labeled antibody for 2.5 hr at 2°C.
The incubated sample was layered onto a 15 to 30% sucrose gradient
(in 25 mM Tris-HCl pH 7.5, 10 mM Mg acetate, 200 mM KCl, 1 mM
2-mercaptoethanol, 10 μg/ml polyvinyl sulfate) and centrifuged at
40 krpm for 70 min at 4°C in a Beckman SW 41 rotor.   The polyribosome sedimentation profile (A$_{260}$) was obtained with a spectrophotometer having a flow cell, and distribution of labeled antibody was determined by counting fractions in a scintillation
counter.   Redrawn from Gelvin and Howell (15).

        To isolate immunoreactive polyribosomes bearing L mRNA, total-
cell polyribosomes were absorbed onto a Sepharose column to which L-
specific antibody had been linked, the bound polyribosomes were
eluted, and RNA was extracted and subjected to electrophoresis on
agarose gels (16).   Nonribosomal RNA species were identified, elec-
troeluted, and subjected again to electrophoresis.   Two discrete RNA
species were obtained (Figure 5), one with an apparent molecular
weight of 4.2 x 10^5 and a smaller species of 3 x 10^5 MW.   Both
species stimulated immunoprecipitable L synthesis in an E. coli system (16).   However, because neither species sponsored the synthesis
of full-sized L, it is possible that both RNAs are discrete L mRNA
fragments.   Nevertheless, we have taken advantage of this separation, for whatever reason it occurs, to obtain an L hybridization
probe in reasonably pure form.

            HYBRIDIZATION OF S AND L mRNA TO C. REINHARDI DNA

        The S and L mRNAs isolated as described above and obtained from
cells labeled with ^{32}PO$_4$ were used as hybridization probes to identify S and L genes.   S and L probes were hybridized to nuclear
(α-component) and chloroplast (β-component) DNA, respectively.   Because the nuclear genome of C. reinhardi is kinetically complex

←Figure 5.   Separation of RNAs from polyribosomes which bind to an
anti-L antibody linked Sepharose column.   C. reinhardi cells were
labeled with 20 μCi/ml ^{32}PO$_4$.   Polyribosomes were extracted, ad-
sorbed onto an antibody linked column, and specifically eluted
(16).   RNA was extracted from the eluted polyribosomes and subjected to electrophoresis on 2.5% polyacrylamide-0.5% agarose
gels (16).   Identifiable nonribosomal RNA species were extracted
from the gel and subjected again to electrophoresis in lanes 2
and 3.   In lane 1, rRNA and 4 S RNA from total-cell RNA are used
as molecular weight standards.   Shown is an autoradiogram of the
gel.   From Gelvin et al. (16).

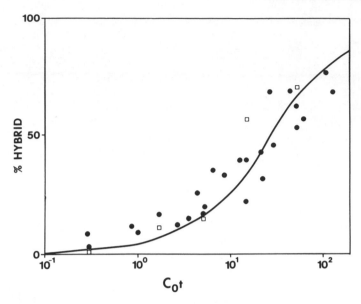

Figure 6. Hybridization of the S gene probe to C. reinhardi nuclear
DNA fragments. 11 S poly(A)+ RNA obtained from synchronous cells
labeled with $^{32}PO_4$ (50 μCi/ml) during the light phase was used as
an S gene probe (2000 cpm/reaction). Hybridization reactions
(0.2 ml) were carried out in solution (5 mM Tris-HCl, pH 7.5, 0.3 M
NaCl, 0.5 mM EDTA) at 65°C with an excess of sonicated nuclear DNA
fragments (17). (DNA was labeled by "nick translation" with E.
coli DNA polymerase I.) Percent hybrid formation was determined by
nuclease S_1 digestion (500 units, 37°C for 15 min). Line is a
theoretical single-component, second-order reassociation curve (k
= 0.33 liter mole^{-1} sec^{-1}) fit to the data for the reassociation
of ^{3}H-labeled nuclear DNA (●) and the hybridization of the ^{32}P-
labeled S probe RNA (□).

(1 x 10^5 kilo base pairs) (17), reassociation reactions were carried
out in solution with excess DNA. Under such conditions, nuclear DNA
reassociates nearly as a single component (Figure 6) with the kinetics expected for a genome composed primarily of single-copy DNA
(17). C. reinhardi nuclear DNA is unusual in that it is composed
mostly of nonrepeated DNA, except for ribosomal RNA cistrons, which
by and large have been relegated to the γ-component satellite (18).
The ^{32}P-labeled S hybridization probe (11 S RNA) hybridizes with
nuclear DNA (Figure 6), which indicates that the S genes, or at
least DNA sequences complementary to S mRNA, are found in the nuclear genome. The extent of probe reassociation is not complete
but is as great as expected when hybrid formation is detected by
nuclease S_1 digestion. The kinetics of the S gene-probe

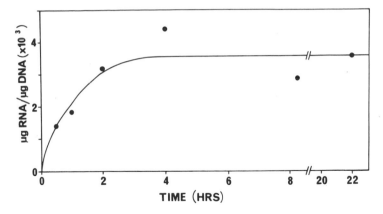

Figure 7.   Hybridization of the L gene probe to C. reinhardi
chloroplast DNA. ^{32}P-labeled L mRNA obtained from the gel electrophoresis separation of polyribosome RNA purified on an antibody
affinity column (described in Figure 5) was hybridized to [3]H-
labeled chloroplast DNA immobilized on nitrocellulose filters.   Hybridization was carried out as described by Gelvin et al. (16) in
the presence of competing RNA from both cytoplasmic and chloroplast ribosomes.   From Gelvin et al. (16).

reassociation are most interesting, because they are virtually the
same as that for the bulk nuclear DNA.   This suggests that the S
genes, like most C. reinhardi nuclear DNA sequences, are not highly
repeated.   Within the limits of accuracy for this experiment, we
can conclude that there are one to a few (~3) S gene copies per
haploid genome.
     The chloroplast genome in C. reinhardi, ~200 kilo base pairs
(17), is much less complex than the nuclear genome.   Therefore,
the hybridization of the L gene probe to chloroplast (β-component)
DNA can be done with filter-bound chloroplast DNA.   When this was
done (16), the ^{32}P-labeled L gene probe hybridized to chloroplast
DNA (Figure 7), and at saturation ~0.7% of the DNA was hybridized--
equivalent to one L gene per unit chloroplast genome.
     Therefore, both S and L gene appear to be present in one gene
(or a few genes for S) per unit nuclear or chloroplast genome.
This would seem to be an appropriate balance for genes whose gene
products are required in equimolar amounts.   However, in C. reinhardi, as in other plant cells, chloroplast unit genomes outnumber
nuclear genomes [by 50-70 to 1 in C. Reinhardi (17, 19, 20)].   L
genes must therefore outnumber S genes by about the same amount
(see Table 1).   If all L genes are functional, and there is no
reason to suspect they are not, then a large gene dosage imbalance
must exist between the two genes.

Table 1

S and L mRNA and Gene Properties

	S subunit	L subunit
mRNA hybridizes with:	nuclear DNA	chloroplast DNA
No. gene copies per unit genome	1 to 3 (?)	1
Total No. gene copies per cell	1 to 3 (?)	50 to 100
Apparent rate of transcription	high rel. to L gene; high rel. to other nuclear genes	low rel. to S gene; high rel. to most other chloroplast genes
mRNA size	11 S	14 to 20 S (?)
Polyadenylation [oligo(dT) cellulose binding]	polyadenylated	not highly polyadenylated
mRNA half-life (actinomycin D)	2½ hr (possibly linked)	2½ hr (possibly linked)
Apparent rate of translation per mRNA	normal (?)	low (initiation limited)
Translation product (in vitro)	20,000 MW polypeptide (precursor ?)	55,000 MW polypeptide

S AND L GENE DOSAGE IMBALANCE

The dosage imbalance between the S and L genes may not be a problem to the plant cell. In fact, it may be the solution to other problems. Mechanisms other than gene dosage balance must exist to ensure that equal numbers of S and L molecules are synthesized. The translation of S and L in plant cells appears to be tightly linked in that synthesis of one subunit depends on the concomitant synthesis of the other (21). In C. reinhardi there is no evidence that under normal conditions pools for free L and S accumulate (22). A linked synthesis mechanism might be able to correct for a great dosage imbalance between S and L genes, but if such an imbalance really existed then the synthesis of one subunit would severly limit synthesis of the other. That would seem to be a wasteful mechanism for regulating the synthesis of a protein such as RuBP carboxylase which is produced in great abundance.

We suspect that any mechanism which links the synthesis between the two subunits is probably performing a fine-tuning regulatory

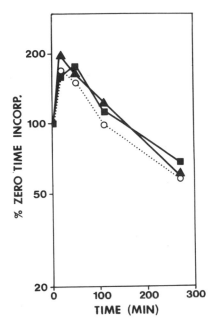

Figure 8. Half-life measurements of S and L mRNA. RNA synthesis
in C. reinhardi CW15 mt+ cells was blocked by the addition of
20 μg/ml actinomycin D at zero time. Cells were pulse-labeled (28)
for 15-min periods with $^{35}SO_4$ (40 μCi/ml) at various times, as
indicated, following addition of the inhibitor. Total acid-insoluble
^{35}S incorporation (o) was measured in an aliquot of the cell culture during each pulse-labeling period, and the remainder of the
sample was prepared for SDS-polyacrylamide gel electrophoresis.
Samples containing the same amounts of ^{35}S radioactivity were loaded
in separate gel channels. Polypeptide bands co-migrating with
purified S and L were cut from the gels, impregnated with 2,5-
diphenyloxazole, and counted in a scintillation counter. The amount
of radioactivity in each S or L band was multiplied by the total
incorporation in the total sample to obtain the total amount of
radioactivity incorporated (expressed as % of zero-time incorporation) into S (■) or L (▲) at the given time points.

function and is not correcting for a gross imbalance in available
S or L mRNAs. To make this assertion, we must assume that there
are features about the expression of S and L genes which compensate
for their different dosages. One feature, cited above, is the apparent slowness in the rate of translation initiation on L mRNA
as evidenced by the finding that polyribosomes engaged in L synthesis are sparsely packed with ribosomes (15) (as are almost all
chloroplast polyribosomes in C. reinhardi). The sparse ribosome

packing is not due to fast unloading of ribosomes or rapid polypeptide chain elongation rate, because chain elongation rates in
the cytoplasm are comparable with those in the chloroplast (23).
The slow rate of translation initiation on L mRNA would not completely compensate for the 25 to 75-fold imbalance in genes but
might reduce the imbalance to 5 to 25-fold (see Table 1). In any
case, this compensatory mechanism is not without costs. It means
that, in order to produce large quantities of L, a chloroplast
must synthesize large quantities of L mRNA (24).

To compensate for the remaining imbalance between genes and
to guarantee abundant RuBP carboxylase production, S mRNA must be
rapidly accumulated in the cell. The accumulation of an mRNA is a
function of its rates of both synthesis and breakdown. As yet we
do not know anything about the rate of S gene transcription. However, we have attempted to measure the rate of breakdown for both
S and L mRNA. We did this by clocking RNA synthesis with actinomycin D and by measuring the ability of cells to synthesize S and L
during pulse-labeling periods at various times after addition of
the inhibitor. The addition of actinomycin D to cells leads to
rapid super-induction of labeling followed by nearly exponential
decay. The labeling of both L and S decay at the same rate (half-
life = $2\frac{1}{2}$ hr), which is also the same as the rate of total protein
synthesis or ^{35}S incorporation (Figure 8). The similarity in decay rates between the two messages may not be real but may reflect
a possible translation coupling mechansim between S and L, as described above, in which case S and L mRNA decay curves may reflect
the decay characteristics only for the shortest lived message.
Nevertheless, in the presence of actinomycin D, S and L mRNA are
not uniquely spared but decay along with the other messages. Thus,
any greater rate of accumulation of S or L mRNA is probably due to
their greater rate of synthesis rather than their slower rate of
degradation.

In a previous report (24) we estimated that at any given time
the number of active L mRNA molecules per cell (or per chloroplast)
in C. reinhardi is about 6×10^3. The estimate was based on a
calculated rate of L synthesis during growth of 1×10^4 L or S
molecules/min. The same calculation can be made for the number of
S mRNA molecules. The time required for the synthesis of a single
ppS chain at 5 amino acids/sec (23) is about 35 sec. Therefore, the
number of S chains growing at any time must be about 5×10^3 chains
per cell. We assume that ppS is synthesized on polyribosomes with
4 to 6 ribosomes (bearing 4 to 6 nascent S chains). If so, about
1×10^3 S mRNA molecules are being translated in the cell at any
one time. Thus, at steady state, we might expect L mRNA molecules
to be 5 to 6 times as abundant as S mRNAs. If the turnover time or
half-life for the messengers is the same, then the total synthesis
rate of L mRNA must exceed that of S mRNA by the same factor (5
to 6). However, if all L genes are active and present in 20 to 60-
fold excess over S genes, the rate of mRNA synthesis per gene must
be 3 to 10 times higher on the S gene.

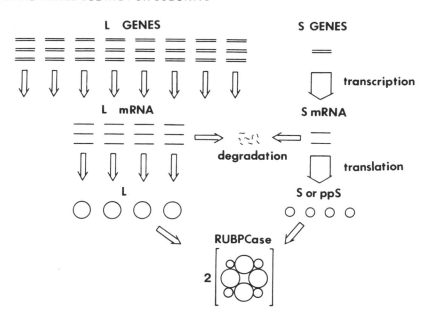

Figure 9.   Scheme for the synthesis of L from chloroplast genes
and S from nuclear gene(s).   Illustration indicates relative numbers
(not to scale) of genes, mRNAs, and gene products involved in the
synthesis of RuBP carboxylase.   Widths of arrows indicate relative
rates of gene transcription, mRNA degradation, and mRNA translation.   See text and Table 1 for explanation.

     The mechanism that emerges for RuBP carboxylase in C. reinhardi
is shown in Figure 9. The process of transcription of the L gene
and translation of L mRNA in the chloroplast is rather sluggish,
but this is compensated for by the presence of large numbers of L
genes in the chloroplast.   On the other hand, the translation of S
mRNA in the cytoplasm is an efficient process, as is the transcription of the few S gene(s) in the nucleus.   Thus, in the process of
RuBP carboxylase synthesis, the nuclear S genes appear to be far
more active in mRNA synthesis than the L genes, and further investigation of their regulatory functions may reveal how plant cells can
produce such large quantities of this enzyme.

                              REFERENCES

1.   Blair, G. E. and Ellis, R. J., Biochim. Biophys. Acta 319,
     223-34 (1973).
2.   Gooding, L. R., Roy, H., and Jagendorf, A. T., Arch. Biochem.
     Biophys. 159, 324-35 (1973).

3. Gray, J. C. and Kekwick, R. G. O., Eur. J. Biochem. 44, 491-500 (1974).

4. Roy, H., Patterson, R., and Jagendorf, A. T., Arch. Biochem. Biophys. 172, 64-73 (1976).

5. Chan, P-H. and Wildman, S. G., Biochim. Biophys. Acta 277, 677-80 (1972).

6. Kawashima, N. and Wildman, S. G., Biochim. Biophys. Acta 262, 42-9 (1972).

7. Kung, S-D., Science 191, 429-34 (1976).

8. Kolodner, R. D. and Tewari, K. K., Biochim. Biophys. Acta 402, 372-90 (1975).

9. Coen, D. M., Bedbrook, J. R., Bogorad, L., and Rich, A., Proc. Natl. Acad. Sci. USA 74, 5487-91 (1977).

10. Hartley, M. R., Wheeler, A., and Ellis, R. J., J. Mol. Biol. 91, 67-77 (1975).

11. Dobberstein, B., Blobel, G., and Chua, N-H., Proc. Natl. Acad. Sci. USA 74, 1082-5 (1977).

12. Sagher, D., Gosfeld, H., and Edelman, M., Proc. Natl. Acad. Sci. USA 73, 722-6 (1976).

13. Rosner, A., Reisfeld, A., Jakob, K. M., Gressel, J., and Edelman, M., Colloq. Int. CNRS 261, 561-7 (1977).

14. Howell, S. H., Heizmann, P., Gelvin, S., and Walker, L. L., Plant Physiol. 59, 464-70 (1977).

15. Gelvin, S. and Howell, S. H., Plant Physiol. 59, 471-7 (1977).

16. Gelvin, S., Heizmann, P., and Howell, S. H., Proc. Natl. Acad. Sci. USA 74, 3193-7, (1977).

17. Howell, S. H. and Walker, L. L., Biochim. Biophys. Acta 418, 249-56 (1976).

18. Howell, S. H., Nature London 240, 264-7 (1972).

19. Bastia, D., Chiang, K-S., Swift, H., and Siersma, P., Proc. Natl. Acad. Sci. USA 68, 1157-61 (1971).

20. Wells, R. and Sager, R., J. Mol. Biol. 58, 611-22 (1971).

21. Ellis, R. J., Phytochemistry 14, 89-93 (1975).

22. Iwanji, V., Chua, N-H., and Siekevitz, P., J. Cell Biol. 64, 572-85 (1975).

23. Baumgartel, D. M. and Howell, S. H., Biochemistry 16, 3182-9 (1977).

24. Howell, S., Heizmann, P., and Gelvin, S., in Biogenesis of Chloroplasts and Mitochondria, pp. 625-8, T. Bucher et al., Editors, North-Holland, Amsterdam, 1976.

25. Weeks, D. P. and Collis, P. S., Cell 9, 15-27 (1976).

26. Aviv, H. and Leder, P., Proc. Natl. Acad. Sci. USA 69, 1408-12 (1972).

27. Roman, R., Brooker, J. D., Seal, S., and Marcus, A., Nature London 260, 359-60 (1976).

28. Howell, S. H., Posakony, J. W., and Hill, K. R., J. Cell. Biol. 72, 223-41 (1977).

CATALYTIC MUTANTS OF RIBULOSE BISPHOSPHATE

CARBOXYLASE/OXYGENASE

K. Andersen, W. King, and R. C. Valentine

Plant Growth Laboratory and Department of Agronomy & Range Science
University of California, Davis, California   95616

Ribulose bisphosphate (RuBP) carboxylase/oxygenase, which may be the most abundant protein in nature, is recognized as the cardinal enzyme catalyzing carbon dioxide fixation yielding energy-rich photosynthate.

Biological nitrogen fixation is the primary research interest in our laboratory. Recently we have estimated that the minimum energy cost of N_2 fixation in vivo may be as high as 40 mole equivalents ATP consumed per mole N_2 reduced (1). During the past several years, evidence has accumulated that photosynthetic carbon dioxide fixation often represents a rate-limiting step for symbiotic nitrogen fixation in leguminous plants, agronomically the most significant form of nitrogen fixation (e.g., see Hardy et al., this volume). With this in mind, we have begun a biochemical/genetical study of RuBP carboxylase. We hope that this will give deeper insight into the mechanism of action and control of this enzyme and provide a basis for evaluation of the possibility of genetic engineering of carbon dioxide fixation.

CHOICE OF ORGANISM

Alcaligenes eutrophus is a gram-negative bacterium capable of growing autotrophically in an atmosphere of hydrogen, oxygen, and carbon dioxide, with hydrogen gas serving as energy source and carbon dioxide as the sole source of carbon. This organism was chosen because of its rapid growth and easy handling in the laboratory, and, more importantly, because it fixes CO_2 through the Calvin-Benson cycle, having a RuBP carboxylase very similar to that in plants. RuBP carboxylase from A. eutrophus has been purified to homogeneity (2, 3). The enzyme complex has about the same overall molecular weight (505,000) as the plant enzyme. It is composed of

large subunits (MW about 54,000) and small subunits (MW about
14,000) as is the plant enzyme. High-resolution electron microscopy
indicates that the quarternary arrangement of subunits may be similar
to that of the plant enzyme (2). The bacterial enzyme, like the
plant enzyme, catalyzes the oxygen-dependent cleavage of RuBP (oxygenase activity) (2, 3).

DEREPRESSION OF THE CARBOXYLASE GENES

The strategy for selecting CO_2 fixation (Cfx) mutants is based
on the ability of A. eutrophus to grow on a variety of fixed carbon
compounds such as organic acids, a heterotrophic lifestyle in which
RuBP carboxylase has a dispensable function. A. eutrophus has been
reported to produce some RuBP carboxylase also when grown with certain organic carbon sources (4, 5). We find the enzyme level to be
only 1 to 5% of the autotrophic level with "good" carbon sources
such as fructose, succinate, or glutamate, which all give rapid
growth with generation times of 2 to 3 hr. Isoleucine gave the
highest RuBP carboxylase level (about 20% of the autotrophic level)
and the slowest growth (generation time 8 hr). A. eutrophus synthesizes very large quantities of RuBP carboxylase under autotrophic
conditions, so that the subunits of this protein are readily visible
on denaturing acrylamide gels against a background of whole-cell
protein (see below). The use of gel electrophoresis (Figure 3) and

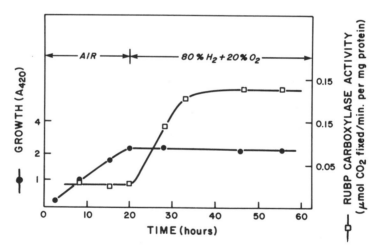

Figure 1.   Induction of RuBP carboxylase in A. eutrophus. The cells
were grown at 30°C in minimal medium (7) with 0.3% L-isoleucine as
carbon source. The gas phase was changed from air to 80% H_2 plus
20% O_2 at the time indicated. RuBP carboxylase activity in cells
made permeable by addition of 5% toluene was determined by measuring
RuBP-dependent incorporation of $^{14}CO_2$ into acid-stable products (2).

antibody against carboxylase (Figure 4) allowed us to rule out the
presence of an inactive state of the protein in cells having low
activity, and this supported the idea that the biosynthesis of RuBP
carboxylase is controlled by conventional enzyme induction/repres-
sion phenomena.

Hydrogen gas is known to inhibit the metabolism of a number of
carbon sources in A. eutrophus (6). When the gas phase is changed
from air to hydrogen plus oxygen (Figure 1), the cells stop growing.

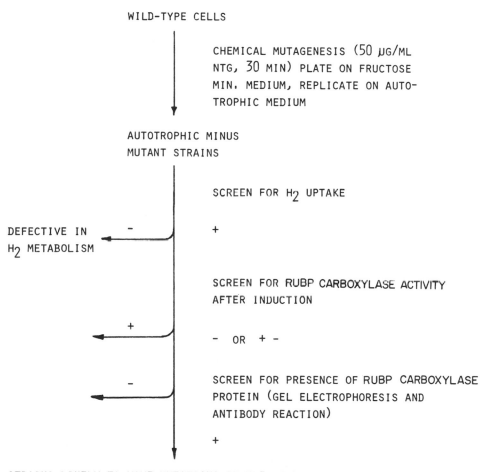

WILD-TYPE CELLS

CHEMICAL MUTAGENESIS (50 µG/ML
NTG, 30 MIN) PLATE ON FRUCTOSE
MIN. MEDIUM, REPLICATE ON AUTO-
TROPHIC MEDIUM

AUTOTROPHIC MINUS
MUTANT STRAINS

SCREEN FOR H₂ UPTAKE

DEFECTIVE IN   −   +
H₂ METABOLISM

SCREEN FOR RUBP CARBOXYLASE ACTIVITY
AFTER INDUCTION

                          +
                                  − OR + −

                          −   SCREEN FOR PRESENCE OF RUBP CARBOXYLASE
                                    PROTEIN (GEL ELECTROPHORESIS AND
                                    ANTIBODY REACTION)

                                    +

STRAINS LIKELY TO HAVE MUTATIONS IN THE STRUCTURAL GENES
FOR RUBP CARBOXYLASE

Figure 2.   Isolation and characterization of mutant strains
            of A. eutrophus blocked in CO_2 fixation.

Under these conditions they are presumably in a state of carbon
starvation while having an abundant supply of energy. The RuBP
carboxylase level reached under these conditions is as high as in
autotrophically growing cells.

This gives a convenient method for induction and study of RuBP
carboxylase in strains blocked in CO_2 fixation. The other enzymes
of the Calvin-Benson cycle are apparently also induced under these
conditions. Thus, when induced cells are supplied with CO_2, rapid
fixation starts immediately.

ISOLATION AND CHARACTERIZATION OF CO_2 FIXATION MUTANT STRAINS

The procedure used for isolation and characterization of mutant
strains is outlined in Figure 2. Conventional procedures developed
for construction of nutritional auxotrophs of E. coli and other
bacteria were easily adapted to A. eutrophus. A stationary-phase
culture was mutagenized chemically with N-methyl-N'-nitro-N-nitroso
guanidine (NTG). About 1% of the surviving cells after mutagenesis
were unable to grow autotrophically, although growing normally with
fructose as a carbon source. About 400 such mutant strains have
been isolated and characterized. One-third to one-half of the strains

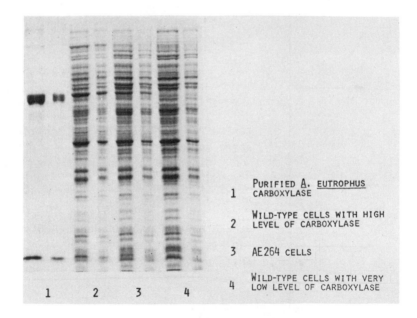

1	PURIFIED A. EUTROPHUS CARBOXYLASE
2	WILD-TYPE CELLS WITH HIGH LEVEL OF CARBOXYLASE
3	AE 264 CELLS
4	WILD-TYPE CELLS WITH VERY LOW LEVEL OF CARBOXYLASE

Figure 3. Electrophoresis of RuBP carboxylase in 12% acrylamide
gel in the presence of sodium dodecylsulfate. Samples were solubilized and subjected to electrophoresis as described by Ames (8).
Migration is toward the bottom.

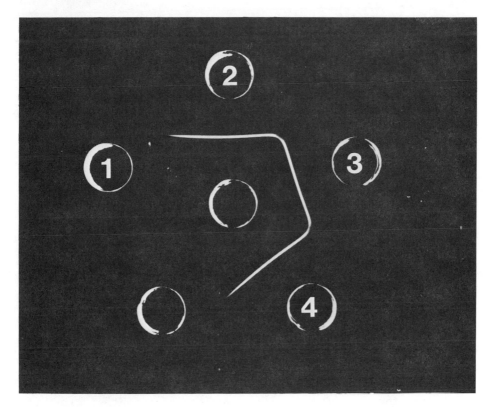

Figure 4. Immunodiffusion analysis for the presence of RuBP carboxylase protein. The center well contained antiserum against
purified A. eutrophus RuBP carboxylase. The other wells contained
the following: 1, extract of fructose-grown wild-type cells with a
very low level of RuBP carboxylase; 2, extract of wild-type cells
induced for RuBP carboxylase as described for Figure 1; 3, extract
of induced cells of strain AE264; 4, purified RuBP carboxylase
from wild-type cells. For experimental details, see ref. 9.

are autotrophic-minus because they can no longer metabolize hydrogen gas, the energy source for autotrophic growth. The remainder
of the mutant strains are induced for RuBP carboxylase as described
for Figure 1. They are then assayed for RuBP carboxylase activity.
This can conveniently be done by measuring RuBP-dependent incorporation of radioactive $^{14}CO_2$ in whole cells made permeable to the substrates with toluene.

     The cells are then screened for the content of RuBP carboxyl-
ase protein by polyacrylamide gel electrophoresis of whole cells
solubilized with sodium dodecylsulfate (8). Since RuBP carboxylase
is one of the major proteins of induced cells, it is possible to
distinguish both the large and small subunits of RuBP carboxylase

Table 1

Properties of Mutant Strains of A. eutrophus
Defective in CO_2 Fixation

Hydrogen uptake was measured in whole cells grown in fructose
minimal medium as described elsewhere (10). RuBP carboxylase
activity was measured in induced cells as described for Figure 1.
RuBP carboxylase protein level was estimated for immunodiffusion
analysis (Figure 4) or gel electrophoresis (Figure 3).

Mutant class	H_2 uptake activity, % of control	RuBP carboxylase activity, % of control	RuBP carboxylase protein level	Frequency, % of Cfx⁻ mutants
I	0		++	40
II	50-100	30-100	++	20
III	50-100	5- 30	+	25
IV	50-100	<5	-	15
V	50-100	<5	++	1

in samples of whole cells (Figure 3). The methods employed are
well suited for screening of large numbers of mutant strains, requiring only whole cells from a few ml of culture of each strain.
     Cell-free extracts of the various mutant strains were also
tested for the presence of protein cross-reacting with anti-RuBP
carboxylase serum (Figure 4). This serum was produced in the conventional way by injecting rabbits with highly purified RuBP carboxylase from wild-type A. eutrophus cells. The properties of Cfx
mutant strains isolated are summarized in Table 1. The strains are
conveniently divided into five classes.
     Class I is composed of uptake hydrogenase mutant strains and
is not further considered here. Mutant strains of this class appear
very frequently because of genetic instability of the genes govern-
ing H_2 catabolism; it has been suggested that the H_2 uptake genes
are harbored on a plasmid, since treatment of cells with plasmid-
curing agents such as mitomycin C yields many uptake hydrogenase
mutant strains (11, 12, and Andersen, unpublished results).
     Mutant strains of Class II are distinguished by (i) the pres-
ence of uptake hydrogenase, (ii) the absence of $^{14}CO_2$ incorporation
into whole cells, and (iii) a high activity of RuBP carboxylase in
broken cells. The simplest interpretation of the behavior of these
strains is that they are blocked in other enzymes of the Calvin-
Benson cycle but not in RuBP carboxylase itself. This class has not
yet been studied in detail.

Class III is something of a catch-all group since mutants with partial activities (leaky) are assigned to it. However, mutant strains from this group are expected to provide some interesting genetic and biochemical material. One of these strains (AE459) displays thermosensitivity of autotrophic growth. This strain produces a RuBP carboxylase of low specific activity (~10% that of the wild type). RuBP carboxylase of this strain is being investigated for its temperature denaturation profile.

Class IV mutant strains synthesize no detectable RuBP carboxylase catalytic activity or protein (both subunits missing). Although not yet characterized as to their specific defects, these strains represent potential deletion or regulatory mutants of RuBP carboxylase genes.

Table 2

Properties of RuBP Carboxylase Mutant Strains of A. Eutrophus

Oxygenase activity was measured as RuBP-dependent O_2 uptake catalyzed by purified, activated enzymes (13). $K_m(CO_2)$ was determined with activated enzyme (13). Antibody reaction was determined as described for Figure 4. The presence of large (L) and small (S) subunits was determined by gel electrophoresis as described for Figure 3. (P = present; ND = not determined.)

Strain	Parent strain	Frequency of spont. reversion	Auto-trophic growth	Anti-body[a] reac-tion	Sub-units L	Sub-units S	Activity, % of wild type Carbox-ylase	Activity, % of wild type Oxy-genase	$K_m(CO_2)$,[b] mM
Wild type			+++	+	P	P	100	100	0.06
AE264									
AE418	Wild type	8x10^{-9} to 5x10^{-9}	-	+	P	P	0	0[c]	
AE439									
AE463									
AE370	Wild type	8x10^{-9}	-	+	P	P	0	0	
AE370 R1	AE370		++	+	P	P	40	40	0.06
AE370 R2	AE370		+++	+	P	P	100	100	0.06
AE459	Wild type	5x10^{-9}	+ -	+	P	P	~10	ND	ND
AE459 R1	AE459		+	+	P	P	20	20	0.06

[a] Cross reaction with antibody prepared against purified A. eutrophus carboxylase.
[b] In the presence of 21% O_2.
[c] Oxygenase activity has been determined only for strain AE264.

Mutant strains of Class V, the rarest group, are of considerable interest because they represent extreme examples of catalytic alteration of RuBP carboxylase. Five strains belonging to this class have been isolated. RuBP carboxylase protein isolated from these strains displays no catalytic activity although it is synthesized in inactive form at levels similar to those in wild-type cells. Rare genetic revertants of mutant strains of Class V have also provided variants of RuBP carboxylase with partial activity.

The properties of some of the catalytically altered RuBP carboxylases are summarized in Table 2. Gel electrophoresis of whole solubilized cells indicates that both large and small subunits are present in all the mutant strains with altered RuBP carboxylase isolated so far. The RuBP carboxylase proteins of strains AE264, AE370, AE370 R1 and R2, and AE459 R1 have been purified as described by Bowien et al. (2) and partially characterized. All show strong cross-reaction with antibody prepared against wild-type RuBP carboxylase. All contain both large and small subunits which migrate like wild-type subunits in polyacrylamide gels with sodium dodecylsulfate. This indicates that the subunits are of the same size as in the wild-type enzyme. Work is in progress to determine which subunit is altered in these mutant enzymes. It seems likely that the mutations in these strains are in the structural genes for RuBP carboxylase, although we have no direct proof of this.

CARBOXYLASE AND OXYGENASE ACTIVITIES OF THE MUTANT ENZYMES

It is generally agreed that RuBP carboxylase, besides the carboxylation of RuBP, can catalyze also the oxygenation of RuBP yielding one molecule each of phosphoglycerate and phosphoglycolate (for review see Jensen and Bahr, 14). This seemingly wasteful reaction may be the main source of glycolate for photorespiration (for review, see Chollet, 15). It is of great interest to find out whether it is possible to separate the carboxylase and oxygenase functions of RuBP carboxylase by genetic means. The report of a mutant strain of Chlamydomonas reinhardi having a RuBP carboxylase with increased oxygenase activity indicates that this is possible (16). Our first approach to this problem, which might be called the "shotgun" approach, involves the construction and assay for oxygenase function of a very large number of RuBP carboxylase mutant strains carrying genetic alterations of the primary amino acid structure of RuBP carboxylase. The mutants described above are suitable for initiating this study since they have probably sustained amino acid changes altering catalysis.

Catalytic activities of purified RuBP carboxylase from mutant strains AE265, AE370, AE370 R1 and R2, and AE459 R1 have been determined (Table 2). Oxygenase and carboxylase activities were reduced to a similar extent in all these mutant enzymes. Strains AE264 and AE370 are simultaneously lacking both activities, and RuBP carboxylases with partial carboxylation activity display an equivalent loss of oxygenase activity. None of the mutant enzymes investigated had a K_m for CO_2 significantly different from that of wild-type enzyme (Table 2). The effect of these mutations is therefore on the maximum rate of catalysis. These results indicate that the carboxylase and oxygenase functions of RuBP carboxylase are closely linked. It has been reported by others that the carboxylase and oxygenase activities of RuBP carboxylase are affected in a similar manner by inhibitors, activators, and activation of the enzyme (14). It has been suggested that the same active site is involved in both reactions (17).

OXYGENASE ACTIVITY IN VIVO

Glycolate can serve as sole source of carbon and energy for growth of A. eutrophus. The magnitude of the oxygenase activity of RuBP carboxylase in vivo therefore cannot be assessed directly because the product is immediately metabolized. However, glycolate is excreted by autotrophically growing cells in the presence of 2-pyridylhydroxymethane sulfonic acid, which inhibits glycolate

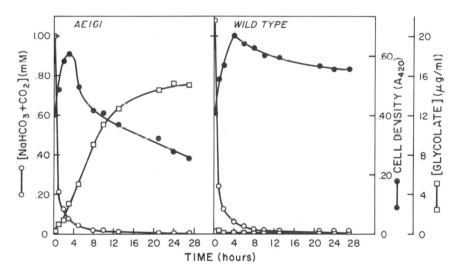

Figure 5. Glycolate excretion by A. eutrophus during autotrophic growth at high oxygen/low carbon dioxide concentration. Mutant strain AE161, which is blocked in glycolate metabolism, was isolated as described in the text. Cells grown autotrophically in the presence of 70% H_2, 20% O_2, plus 10% CO_2 were washed and transferred to autotrophic medium (pH 7) containing 1 mM $NaH^{14}CO_3$ under an atmosphere of 80% H_2 plus 20% O_2 at time zero. The concentration of $^{14}CO_2$ plus $H^{14}CO_3^-$ in the medium was determined as $^{14}CO_2$ in acidified samples. Glycolate was determined as described by Calkins (20).

metabolism (18). We have constructed a number of mutant strains of
A. eutrophus that are unable to grow on glycolate. The first enzyme in the metabolism of glycolate is glycolate oxidoreductase,
whose activity can be measured as glycolate-dependent glyoxylate
formation (19). One of the mutant strains, AE161, has no activity
of this enzyme when grown under conditions that induce high activities in the wild type. This strain, which is completely blocked
in glycolate metabolism, excretes large quantities of glycolate
under low CO_2 concentrations (Figure 5). No glycolate is excreted by
the wild type under the same conditions. The mutant strain grows
normally at high CO_2 concentration (10% CO_2), excreting very little
glycolate. This strain represents an interesting tool for determining the oxygenase activity of RuBP carboxylase under various
conditions in vivo. This type of mutant strain may also allow
scoring of the oxygenase activity of RuBP carboxylase at the colony
level by screening for glycolate excretion, permitting rapid visual
detection of catalytic mutants of oxygenase.

CONCLUSION

Use of a bacterial system has allowed the isolation of several
mutationally altered species of RuBP carboxylase having a wide
range of catalytic activities. In five mutant enzymes, the oxygenase and carboxylase activities were altered in a parallel manner.

REFERENCES

1. Shanmugam, K. T., O'Gara, F., Andersen, K., and Valentine,
    R. C., Annu. Rev. Plant Physiol., in press (1978).
2. Bowien, B., Mayer, F., Codd, G. A., and Schlegel, H. G.,
    Arch. Microbiol. 110, 157-66 (1976).
3. Purohit, K. and McFadden, B. A., J. Bacteriol. 129, 415-21
    (1977).
4. Kuehn, G. D. and McFadden, B. A., J. Bacteriol. 95, 937-46
    (1968).
5. Stukns, P. E. and DeCicco, B. T., J. Bacteriol. 101, 339-45
    (1970).
6. Bowien, B., Cook, A. M., and Schlegel, H. G., Arch. Microbiol.
    97, 273-80 (1974).
7. Repaske, R. and Repaske, A. C., Appl. Environ. Microbiol. 32,
    585-91 (1976).
8. Ames, G. F.-L., J. Biol. Chem. 249, 634-44 (1974).
9. Shanmugan, K. T., Chan, I., and Morandi, C., Biochim. Biophys.
    Acta 408, 101-11 (1975).
10. Andersen, K. and Shanmugam, K. T., J. Gen. Microbiol. 103,
    107-22 (1977).
11. Schlegel, H. G., Ant. van Leeuwenhoek 42, 181-201 (1976).
12. Pootjes, C. F., Biochem. Biophys. Res. Commun. 76, 1002-6
    (1977).

13.   Lorimer, G. H., Badger, M. R., and Andrews, T. J., Annal.
      Biochem. 78, 66-75 (1977).
14.   Jensen, R. G. and Bahr, J. T., Annu. Rev. Plant Physiol. 28,
      379-400 (1977).
15.   Chollet, R., Trends Biochem. Sci. 2, 155 (1977).
16.   Nelson, P. E. and Swizycki, S. J., Eur. J. Biochem. 61, 475-
      80 (1976).
17.   Lorimer, G. H. and Andrews, T. J., Nature London 243, 359-60
      (1973).
18.   Codd, G. A., Bowien, B., and Schlegel, H. G., Arch. Microbiol.
      110, 167-71 (1976).
19.   Lord, J. M., Biochim. Biophys. Acta 267, 227-37 (1972).
20.   Calkins, V. P., Ind. Eng. Chem. Anal. Ed. 15, 762-3 (1943).

DISCUSSION

HOWELL: What possibilities are there for genetic analysis of
these mutants as representatives of particular complementation
groups?   You talked about curing, and many of us are interested in
mobilization of the carboxylase gene.

VALENTINE: Our most important immediate goal is to develop
this gene transfer system.   The organism is gram negative.   Fortunately, it will mate with E. coli.   It is possible to use sex factors, and we have introduced them into these strains.   We think the
hydrogenase plasmid might be a large plasmid.   We have some problems
with compatibility, but I think the manipulation will be straightforward if we use the R factors and increase the rates of gene
transfer.   We have very active sexes in these organisms although
the rates of gene mobilization are below the desired level.   However, I am optimistic about the dual development of the genetic
system.

HOLDER: First, have you any other genetic markers that enable
you to do fine-structure mapping of the proposed regions; second,
have you detected any intrallelic complementation either in vivo
or in vitro?

VALENTINE: We have not reached that stage with this system
as Dr. Anderson has been working with it only for one year.   Our
goal is to tie these mutants, from his biochemical study, in with
the genetic system.   You asked key questions, which we would like
to answer when we get the gene mobilization system to work.   We
have not yet mapped the genes on the chromosome or the plasmid.

ZELITCH: I assume this organism is anaerobic.

VALENTINE: No.   It is strictly aerobic.

ZELITCH: In the wild type, how much of the carbon fixed
normally goes into glycolate?

K. ANDERSEN: What little information we have indicates that
at high CO_2 and 21% oxygen very little goes into glycolate, but
at low CO_2 the amount increases.   We plan to do quantitative
determinations of this increase as a function of CO_2 and oxygen
concentrations.

VALENTINE: That applies to the mutants.   The wild-type organism induces a glycolate oxidase pathway that is not rate limiting.   There is little or no glycolate accumulation.

SEPARATION OF RIBULOSE 1,5-BISPHOSPHATE CARBOXYLASE
AND OXYGENASE ACTIVITIES[*]

Rolf Bränden

Department of Biochemistry, Chalmers Institute of Technology
Fack S-402 20 Göteborg, Sweden

and

Carl-Ivan Bränden

Department of Chemistry, Swedish University of Agricultural Science
Uppsala, Sweden

INTRODUCTION

Recently it was shown that RuBP carboxylase and RuBP oxygenase activities were separated when the enzymes were prepared at a higher pH than normally used (1). Recent reports (2, 3), as well as some characteristics of the enzyme found earlier (4), e.g., the different behaviors of the two activities upon storage, suggest that two different enzymes are responsible for the two activities. We describe here three different methods for separating the two activities and two spectrophotometric methods for rapidly measuring the RuBP oxygenase activity.

RAPID ASSAY METHODS FOR THE RuBP OXYGENASE REACTION

1.   Oxygen Consumption Measured by The Clark Electrode

The rate of O_2 consumption can be used to measure RuBP oxygenase activity. However, O_2-consuming reactions other than the RuBP oxygenase activity also occur, especially in less purified systems, and these must be corrected for to obtain the actual rate of O_2 uptake of RuBP oxygenase.

---

[*]This study was supported by grants from the Swedish Natural Science Research Council.

RuBP carboxylase contains a large number of SH groups. In
the presence of reducing agents such as dithiothreitol, O_2 is
utilized by the oxidation of dithiothreitol and probably by the re-
oxidation of some of the SH groups of the enzyme. If the enzyme is
kept in an activating medium, the O_2 concentration is reduced after
a short time because the diffusion of O_2 into the solution is slow.
Addition of a small sample of this activated enzyme to the O_2-con-
taining assay medium, in which the O_2 concentration is much higher,
exposes the reduced SH groups of the enzyme to O_2. If some of these
SH groups are initially oxidized, there will be a concomitant initial rate of O_2 consumption. If RuBP is added to an assay mixture
containing enzyme that has already reached the steady state of O_2
consumption, RuBP interacting with the enzyme may expose buried SH
groups subject to a high rate of oxidation. Thus the initial O_2
consumption will be the sum of the O_2 consumed by the RuBP oxygenase and the O_2 used for oxidation of SH groups. We believe that
O_2 consumption as a measure of RuBP oxygenase activity is not entirely satisfactory. No evidence has yet been presented indicating
that the ratio between phosphoglycolate formed and O_2 consumed is
equal to 1.0 for the initial minute of the RuBP oxygenase reaction.

2.   Phosphoglycerate Formation in The Absence of CO_2

Since phosphoglycerate is formed in both the RuBP oxygenase
and carboxylase reactions, CO_2 must be eliminated before the oxygenase activity can be measured. "CO_2-free" buffers were prepared
with freshly boiled water, and a method similar to that used for
preparing "O_2-free" buffers (5). If activated enzyme is used, the
amount of enzyme added to the assay mixture must be so small that
the CO2 concentration in the assay is far below the K_m for the carboxylase reaction. When these precautions are taken, the phosphoglycerate formation can be followed spectrophotometrically at 340 nm
by the method of Racker (6). The limitation of this method is that
phosphoglycerate formation by the carboxylase reaction can probably
not be totally avoided.

3.   Phosphoglycolate Formation

The most reliable method for measuring the RuBP oxygenase reaction is to follow the formation of phosphoglycolate. A rapid
spectrophotometric method involves use of phosphoglycolate phosphatase and glycolate oxidase in the assay system for glycolate
oxidase of Duley et al. (7). It should be pointed out, however,
that this system does not function in the presence of reducing
agents such as dithiothreitol and ascorbate. The amount of H_2O_2
formed in the glycolate oxidase reaction is measured by reducing a
chromogen, o-dianisidine, with horseradish peroxidase. The reduction is followed spectrophotometrically at 515 nm.

RESULTS

1.   Separation of RuBP Carboxylase and Oxygenase Activities on a
Sepharose 6B Column

     The procedure for the separation is described elsewhere (1).
The fractions containing the RuBP oxygenase activity also contained
copper, and we have found that other transition metals (Fe and Mn)
accompany this activity. If the oxygenase-containing fractions
were again put on a Sepharose 6B column, the eluted fractions were
much lower both in activity and in metal content, which indicates
that transition metals are required for RuBP oxygenase activity.

2.   Separation of RuBP Carboxylase and Oxygenase Activities on a
Sephadex DEAE A25 Column

     Fresh parsley in 50 mM Tris-HCl buffer, pH 8.3, was macerated
in a Waring blender for 60 sec. The suspension was filtered

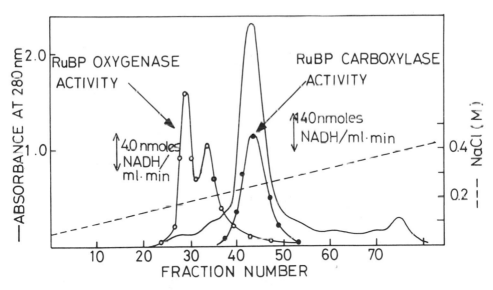

Figure 1. RuBP carboxylase and RuBP oxygenase activities from
parsley separated on a DEAE A25 Sephadex column at pH 8.3, as described in text. The buffer was adjusted at 25°C but the separation was performed at 5°C.

through 4 layers of cheesecloth and centrifuged for 10 min at
28,000 x g.   Solid $(NH_4)_2SO_4$ was added to the supernatant, and the
precipitate obtained at 30 to 50% saturation was dissolved in a
minimal amount of buffer.   The solution was applied to a Sephadex
G25 column equilibrated with 5 mM Tris-HCl buffer, pH 8.3 (adjusted
at 25°C).   The eluted protein fractions were put on a Sephadex
DEAE A25 column which was equilibrated with the same buffer and
eluted with a NaCl gradient from 50 to 400 mM (see Figure 1).   The
RuBP carboxylase and oxygenase activities were measured according
to the method of Racker with and without $NaHCO_3$, respectively.
Ascorbate (0.5 mM) was used as a reducing agent when RuBP oxygenase
activity was measured because it was found to be a better activating
agent than dithiothreitol.

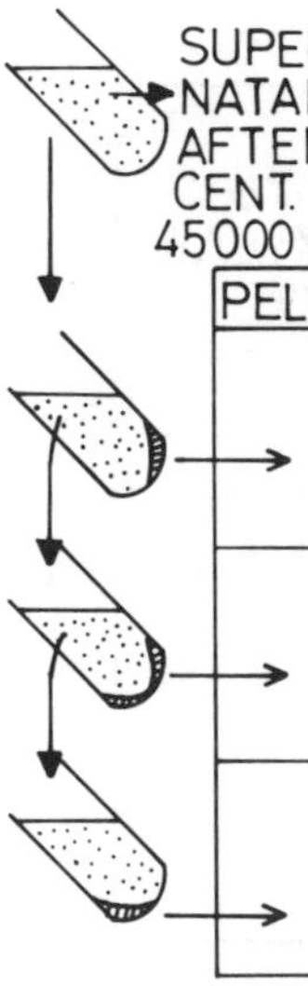

	TOTAL CAR-BOXYLASE ACTIVITY	TOTAL OXY-GENASE ACTIVITY		OXYGENASE ACT. CARBOXYLASE ACT.
		I	II	
SUPER-NATANT AFTER CENT. AT 45000 x g	1 0 0 0	2 0 0	2 0 0	0. 2
PELLET A	3 5 0	< 1 0	< 1 0	< 0.03
B	4 0 0	3 5	4 5	0.1 0
C	2 3 0	1 5 0	1 3 5	0.60

Figure 2.   The separation of RuBP carboxylase and RuBP oxygenase
activities by ultracentrifugation from parsley, as described in
test.   The total RuBP carboxylase activity was measured at pH 7.8
by the method of Racker (6) and normalized to 1000.   The RuBP oxygenase activity was normalized to 200 since it was found to be 20%
of the RuBP carboxylase activity by the method of Racker.   All
activity measurements were made at pH 6.8 and 25°C.   The O_2 concentration in the RuBP oxygenase assay was 0.2 mM.

The carboxylase peak showed very little oxygenase activity and also very small amounts of EPR detectable copper or iron. The fractions containing the oxygenase activity contained both copper and iron. However, the oxygenase activity and the metal content were low compared with those of the oxygenase-containing fractions obtained on a Sepharose 6B column. The carboxylase peak, on the contrary, contained high specific RuBP carboxylase activity, and very little phosphoglycolate was formed when these fractions were tested for RuBP oxygenase activity by the phosphoglycolate formation method.

3.   Separation of RuBP Carboxylase and Oxygenase Activities by Centrifugation

Fresh parsley was macerated in a minimal amount of deionized distilled water. The suspension was filtered through cheesecloth and adjusted to pH 7.4, and the filtrate was centrifuged at 45,000 x g for 30 min. RuBP carboxylase and oxygenase activities were measured in the supernatant. The ratio oxygenase activity/ carboxylase activity was 0.2 under the conditions used (see Figure 2). The supernatant was adjusted to pH 8.8 and centrifuged at 200,000 x g for 1 hr at 10°C. The pellet was resuspended in buffer, and the oxygenase activity/carboxylase activity was <0.03. The supernatant was adjusted to pH 6.8 and centrifuged at 200,000 x g for 1.5 hr at 10°C. A pellet was collected and resuspended; the oxygenase activity/carboxylase activity was 0.10. The supernatant was centrifuged at 200,000 x g for 7 hr at 10°C, and the pellet formed was resuspended. The oxygenase activity/carboxylase activity had increased to 0.6, which is 20 times as high as the ratio found for the pellet formed at pH 8.8. The RuBP oxygenase activity/ RuBP carboxylase activity for the original supernatant, before ultracentrifugation, showed no change upon storage for 10 hr.

DISCUSSION

We have presented three methods for differentiating RuBP carboxylase and oxygenase activities. Oxygen consumption was not used as a measure of RuBP oxygenase activity; instead, the products formed in the reaction were followed spectrophotometrically.

The fractions containing the carboxylase activity, eluted from a Sephadex DEAE A25 column (Figure 1), showed a high specific RuBP carboxylase activity but only a minor amount of RuBP oxygenase activity. The RuBP oxygenase activity was nicely separated from the RuBP carboxylase activity. However, the total amount of RuBP oxygenase activity was very low compared with that of the starting material, which means that most of the RuBP oxygenase activity is lost during this procedure. Transition metals were also lost, both when this method was used and when the fractions containing RuBP oxygenase activity, eluted from a Sepharose 6B column, were gel-filtered again on the same column. They are apparently necessary for full RuBP oxygenase activity.

The simplest way to separate the two activities is to ultra-centrifuge the original supernatant at high pH (see Figure 2). The redissolved pellet shows high RuBP carboxylase activity but only minor RuBP oxygenase activity. If pH is dropped to 6.8, the ratio oxygenase activity/carboxylase activity increases with increasing centrifugation time. A reasonable explanation for this separation is that RuBP oxygenase, in contrast to RuBP carboxylase, separates into smaller units at pH 8.8 but not at pH 6.8. This can also explain why it is possible to separate the two activities on a gel filtration column at high pH.

The ratio RuBP oxygenase activity/RuBP carboxylase activity is about the same regardless of the spectrophotometric method used to measure the oxygenase activity (Figure 2). However, the ratio phosphoglycerate formed/phosphoglycolate formed is uncertain since the extinction coefficient for the absorbance change in the phosphoglycolate assay was not precisely determined. Furthermore, dithiothreitol was used only in Racker's method. For these reasons the absorbance change in the assays containing the original supernatant was normalized to the same value.

Since the oxygen consumption was not followed, the ratio between O_2 consumed and phosphoglycolate formed for the first 60 sec is not known. The importance of this ratio is obvious. A ratio of 1.0 is an absolute requirement for correct interpretation of data when O_2 consumption is used as a measure of RuBP oxygenase activity.

REFERENCES

1. Brändén, R., Biochem. Biophys. Res. Commun. 81, 539-46 (1978).
2. Bravdo, B.-A., Palgi, A., and Lurie, S., Plant Physiol. 60, 309-12 (1977).
3. Harris, G. C. and Stern, A. I., Plant Physiol. 60, 697-702 (1977).
4. Andrews, T. J., Lorimer, G. H., and Tolbert, N. E., Biochemistry 12, 11-18 (1973).
5. Andréasson, L.-E., Malmström, B. G., and Vänngård, T., Eur. J. Biochem. 34, 434-9 (1973).
6. Racker, E., in Methods of Enzymatic Analysis, p. 188, H. Bergmeyer, Editor, Academic Press, New York, 1963.
7. Duley, J. and Holmes, R. S., Anal. Biochem. 69, 164-9 (1975).

DISCUSSION

TOLBERT: One must always be alert to new data and new ideas. This is not a new idea, but the interpretation of data could be very important to plant breeders if it were possible that there is an oxygenase. When this information came out, a lot of us in this audience, and particularly in our lab, tried to reproduce the data exactly as published in ref. 1, and we could not do so. After talking to Dr. Brändén, we realized that certain things were not in the publication, and we had not done the experiment identically the way he did. I would like to mention some of the differences. One is that he was using a pH probably around 9 rather than around 8.3 because pH was adjusted at one temperature and buffer used at another temperature. Another is that he was using a very low ionic strength buffer in the grinding medium, which could be very important because it could be dissociating the protein and he could be looking at a subfraction. One of the greatest criticisms of his work, which he is now well aware of, is that his oxygenase activity assay was invalid, as all the rest of us run it, as was pointed out today by several of us, including Bill Ogren and myself. When the enzyme is assayed after activation according to the procedure of Lorimer, it becomes inactive within 2 min and therefore the assay must be done within the first minute or so. He sees high endogenous rates, which none of the rest of us sees. He sees a blue color in his protein which none of us has seen (from Sam Wildman in the early '40's on down through the years) in spite of the immense amount of work on ultracentrifugation and purity of this enzyme; this is indeed a challenge, although I prefer to think the blue color represents some sort of association or contaminating copper protein. I do not question that he is measuring oxygen uptake and that there is a copper protein present. The question is what the copper protein is and whether or not it is the RuBP carboxylase. Perhaps it is a fragment, and that would be a very exciting and important contribution. Frankly, and I believe I speak for most of us, Dr. Bränden, we find it very difficult to expect there to be another large protein fraction to account for all of this photorespiration. We have prepared a paper for publication: Ribulose 1,5-Bisphosphate Carboxylase/Oxygenase from Paisley, S. D. McCurry, N. Hall, J. Pierce, C. Paech, and N. E. Tolbert, submitted to Biochem. Biophys. Res. Commun.

ROUND TABLE DISCUSSION:

LONG-RANGE RESEARCH PLANNING IN CARBON ASSIMILATION

H. W. Siegelman, Chairman, Brookhaven National Laboratory

J. T. Bahr, Mobil Chemical Company
J. A. Bassham, University of California
C. C. Black, University of Georgia
M. Gibbs, Brandeis University
M. Lamborg, C. F. Kettering Research Laboratory
W. L. Ogren, USDA, University of Illinois
H. C. Reeves, National Science Foundation
N. E. Tolbert, Michigan State University
I. Zelich, Connecticut Agricultural Experiment Station

LAMBORG: Exciting things have been happening in nitrogen-
fixation research within the last three years and are happening
now.   Regarding nitrogen fixation by free-living prokaryotes,
Winston Brill and his colleagues have reported isolation and stabilization of the molybdenum-iron cofactor extracted with dimethyl-
formamide. Since this cofactor can now be obtained in a form that
is stable at room temperature for long periods, I think the likelihood is high that it will soon be crystallized and x-ray diffraction patterns will be taken. This has enormous implications for
the understanding of the active center of nitrogenase and for the
ability to synthesize similar compounds.
    In the field of legume research, many areas are being studied.
Mapping of the genomes of a number of species of Rhizobium is being
done by the Berringer group and by Ethan Signer at MIT, and cloning
of nif genes and related genes is being studied by F. Ausubel and
Fred Cannon. Continued lectin research by Frank Dazzo and W. D.
Baur at our lab, concerned with the presumed recognition between
the exo-polysaccharide of Rhizobium and the lectin compounds, which
presumably are on the root hair cells of legumes, needs to be confirmed. The recognition reactions that seem to be taking place
show high biological specificity and beyond them lies the potential
for a whole new slant with regard to plant pathology based on
lectin proteins.

An observation made simultaneously in several laboratories within six months was that nitrogen fixation by Rhizobium, mostly the cowpea strain but also Rhizobium japonicum species, is affected by microaerophilic conditions. This may seem simplistic, but it has resulted in an enormous burst of research activity because it has allowed separation of phenomenology associated with nodule ontogeny from that associated with nitrogen fixation itself; and from that, other things have developed. For example, in some strains of Rhizobium that are incapable of being turned on to fix nitrogen as free-living organisms, simply creating microaerophilic conditions apparently makes it possible for humoral agents from soybean cells grown in culture to elicit nitrogen fixation.

What I find most exciting is the work by Shanmugam and O'Gara, described at the joint U.S. and International Solar Energy Society Conference at Davis just last month, on a mutation of Rhizobium trifoliae for glutamine synthetase. The organism was converted from a nonfixing free-liver to a fixing free-liver, and from fast growing to slow growing, apparently with an altered host range specificity for infection; and upon reversion the organism regained its original characteristics. The changes are not simple; they are apparently multiple. Ray Valentine knows more about this than I, but it has enormous implications not only for nitrogen fixation and infection but also for the recognition reaction associated with legume nitrogen fixation and for strain specificity in general.

Bob Darrow in our lab has made an observation accepted by a number of researchers including Ray Valentine, that glutamine synthetase is involved in repression of the nif gene. He showed that in Rhizobium there are two glutamine synthetases: one having the regulatory characteristics of the E. coli enzyme (that is, it is adenylated and deadenylated) and another having no regulatory characteristics at all. One of these enzymes appears to be under regulation by the oxygen tension, that is, at P_{O_2} below 0.4, that glutamine synthetase is lost, and it is regained at an oxygen tension of 0.45%; this is a very narrow range. What the functional need is for two glutamine synthetases remains to be elucidated; it is certainly unusual.

Regarding nitrogen fixation in legumes, the whole question of efficiency remains, having to do with the evolution of hydrogen and the question of whether or not uptake hydrogenases are present in legumes, and the degree to which energy is lost as a consequence of the concomitant interrelationship between nitrogen fixation on one hand and hydrogen evolution on the other. There has been an enormous increase in activity in the last two or three years, and there will be much more, because this has become an area of intense interest, at least partly because we can now study nitrogen fixation in free-living organisms.

Nonlegume nitrogen fixation is also of interest. In the Azolla-Anabaena azollae relationship, Azolla is a water fern and Anabaena azollae is a blue-green alga; both are photosynthetic and one is a nitrogen fixer. Bill Rains, a colleague of Ray Valentine, has reinvented the nitrogen fixing wheel, in that he has investigated dual culture of this photosynthetic nitrogen fixer and rice in paddies. He finds that the requirement for fixed nitrogen fertilizer can be ameliorated by this organism used both as a green manure and as a combination of weed growth repressor and perhaps nitrogen excretor-that remains to be clearly dissociated. Within the last six months or so, John Torrey at Harvard has identified the organism that infects nonlegume nitrogen fixers such as Comptonia and showed it to be an actinomycete-like microorganism; he fulfilled Koch's postulate in that he infected the plant and showed that nitrogen fixation takes place. Maurice Lalond in our lab has shown that the same actinomycete infects not only Comptonia but Ulmus, which is not very closely related; this suggests that biological specificity in nonlegume nitrogen fixers may be different from that in legume systems. Johanna Döbereiner has pioneered work with Spirillum lipoferum which is important because she showed that a number of soil microorganisms have potential major importance as nitrogen fixers, and she developed the methodology for both identification and assessment.

Last spring at a Gordon Research Conference on plant cell culture research, one whole day was devoted to Agrobacterium tumefaciens and Ti plasmid activity. An enormous amount of research is being done, especially regarding the potential of Ti light plasmids as natural vectors for gene cloning. Schilperoort reported that the Ti plasmid, in concert with RP41, could infect Rhizobium and could produce either tumors on the shoot, or nodules on the root that are capable of nitrogen fixation.

What does all this mean? The Kettering Foundation is concerned with the world food problem and with increasing productivity of food. The problem relates not so much to this country as to the rest of the world, and it is becoming obvious that developing countries will have to become much more self-sufficient in food production than they are now. Nitrogen-fixing plants will be of major importance to their becoming self-sufficient both in the near term and in the far future. Now let me proselytise. Knowledge about nitrogen fixation is increasing rapidly, especially regarding plant, legume, and nonlegume associations. What is desperately needed is input from photosynthesis people. The crisis concerns increased CO_2 fixation to maximize agronomic yield, as discussed by Ralph Hardy. Young researchers who have not yet made a commitment to a specialty should think seriously about the interface between nitrogen fixation and photosynthesis. That is where a major research endeavor is needed, especially on blue-green algae.

TOLBERT: The shoe was on the other foot at the meeting here in March 1977 (Genetic Engineering for Nitrogen Fixation, Plenum, 1977), where a few people, including myself, came to represent photosynthesis to the nitrogen fixers; this year you are representing nitrogen fixation, and your presentation of the exciting developments

is very appropriate because nitrogen fixation and photosynthesis
are the two major areas in research on crop productivity and food.
The nitrogen fixation people have a better press than we have
and are doing a good job of promoting nitrogen fixation as being
more important than photosynthesis although we all admit that both
are equally important. Ralph Hardy, when he said he thought photosynthesis was perhaps a bigger challenge and certainly a more
immediate challenge was saying the same thing Dr. Lamborg just said.
But we are not organized and are not putting out the cry for funds
the way the nitrogen-fixation people have done; they have, at the
same time, helped us a great deal with their excellent publications
in the past few years. When Ralph Hardy (who, like Dr. Lamborg,
stands between the two fields) said that a 6% per year crop increase in corn and soybeans is needed from now on just to stay even,
and asked how we thought we could do that through photosynthesis research, he was emphasizing the same thing as Dr. Lamborg. And the
fact that we have to have more photosynthesis to provide the energy
for more nitrogen fixation is the point, Dr. Reeves, Dr. Rabson,
and Dr. Senich, that we really want to get across to you, even
though the people here have not, in my judgment, been able to put
the message across to the granting agencies as the nitrogen-fixation people have.

     BAHR: Among the participants here, except Ray Chollet, I am
probably in a unique situation, as I have been involved for the
last two years in a group within an industrial organization whose
ultimate goal is to define agricultural chemicals that could increase carbon assimilation. With that target one rapidly dis-
covers what information about carbon assimilation is lacking that
is needed in order to reach the goal. In assessing the current
state of the literature, we find major gaps, things we must know,
at least from a chemical point of view as opposed to a genetic
point of view, if we are to proceed to a practical solution to
the problem of increasing carbon assimilation.
     We need a much clearer resolution of the rate-limiting steps
in CO_2 fixation, with particular focus on physiological conditions
rather than the more convenient high-CO_2 conditions. Work should
perhaps start at the chloroplast level but should extend in both
directions and include enzymic, chemical and physiological studies.
Hardy showed that for nonlegume grains there wasn't one limiting
step to higher yield but in fact two: both carbon and nitrogen
had to be increased at the same time; increasing just one wasn't
enough. The situation may be similar in chloroplasts, where more
than one step or process within the organelle must be speeded up
to improve the CO_2 fixation rate. If we knew which ones, we'd
know what chemistry to investigate.
     Very little is known about the chemistry of even the carboxyl-
ase, let alone any of the other enzymes of the system. In working

with chemists it becomes immediately obvious that we don't have
enough chemical information about the processes involving carboxylase and oxygenase even to know where to begin looking for a chemical
approach to the problem of increasing carbon assimilation. Regarding the catalytic site, the greatest lack is any idea of the mechanism by which either oxygen or CO_2 interacts with the enzyme. Regarding the activator sites, which should be equally important, we
lack by and large, any knowledge of the chemistry or the kinetics
of the activators that have been described for carbon assimilation:
NADPH, PGA, 6-phosphogluconate, dihydroxyacetone and at this meet-
ing--inorganic phosphate. These may not necessarily be physiologically important, but they take part in a chemical process involving
the enzyme that we need to know more about.

Our group is constantly trying to assess the physiological
significance and importance of various sorts of phenomena reported
in the literature. We can distinguish between good and bad physiology and between good and bad biochemistry, but even good biochemistry
isn't necessarily physiological. How are we to assess the importance
of carboxylase activation, changes in RuBP levels, phosphate levels,
regulation of the chloroplast by ADP, as discussed by Dr. Walker,
not in the chloroplast but in the whole plant under physiological
and perhaps even field conditions? Which of the various factors
that can influence the rate of photosynthesis in the intact organelle in fact influence it in the whole plant? One becomes aware
that whole plants regulate the level of photosynthesis and perhaps
of the carboxylase. We know, for example, that net photosynthesis
is regulated by translocation and by source-sink relationships,
by a mechanism which is totally undescribed. We know that regulation varies with the stage of growth. Ogren found that the photosynthesis rate in soybeans is more or less constant until flowering
and then takes a significant jump. What has the plant done chemically to change the photosynthesis rate? Has it made more chloroplasts, has it made more carboxylase, has it changed the cytoplasmic
phosphate level--what precisely has happened? Huffaker said that
the carboxylase levels in the intact leaf are stable throughout the
life cycle until senescence and that turnover of the enzyme is very
slight or zero. We know that the level is determined by the light
intensity at which the plant is grown and by its nitrogen nutrition, but we know nothing about how it is regulated. Is the level
reached under maximum light and best possible nitrogen nutrition
an optimal level for maximizing the photosynthesis rate, or would
it be useful actually to have more carboxylase?

More active interchange is needed between researchers ranging
from chemists, who know something about the mechanism of the carboxylase, to people who work with chloroplasts, with protoplasts,
with leaves; by getting together, not once a year at meetings like

this, but more frequently on a regular basis, they can assess the
significance at a physiological level of events and phenomena
occurring at a chemical level and at all levels in between.

OGREN: I normally don't talk at this kind of meeting because
I prefer to stay at my bench, but I have always been interested in
how the field of photosynthesis is developing and where the research money is being spent. Maybe I've been a bit paranoid because of my own situation, but it seems to me the emphasis has not
generally been in the right place. I have worked in a laboratory
where electron transport was being studied. Everybody else there
worked with light reactions and wouldn't consider using plants;
all the work was done on Anabaena. I am still not sure why I ended
up working on the kidney bean, looking at the very end of the light
reaction, that is, reduction of pyridine nucleotides. From Wayne
State I came to the University of Illinois into a similar situation,
if not worse, with a high-powered group of electron transport people,
led at the time by Rabinowitz and Govindjee and continuing today
with Charlie Arntzen, Tony Crofts, Colin Wraight, and other people
in other departments.
Primary emphasis has been on electron transport for a long
time, as seen at international meetings. For example, I went to
the Photosynthesis Congress in 1974 in Israel, thinking I would find
out what's going on in photosynthesis, but it was just more of the
same; nine out of ten sessions were concerned with electron transport. I think it is very clear that dark reactions and CO_2-fixing
reactions are very important in productivity, and they are the
things that should be studied. About that time the Pound report
had come out, and it said the USDA didn't have anyone working on
photosynthesis and something should be done about this. Dr. Gibbs
made an exceedingly appropriate comment, "When we have photosynthesis congresses, we ought to have a little less photo and a little
more synthesis." Apparently his advice was heeded because at the
Reading Congress, in September 1977, other aspects of photosynthesis
were discussed. It seems this area of photosynthesis is being re-
vitalized by the discovery of C_4 photosynthesis.
Although I recognize that many people disagree with some of
the things I've said, I think we have pretty well established what
we have to do if we want to increase CO_2 fixation by green plants.
We have to get the carboxylase to turn over more rapidly and we
have to reduce photorespiration. These are recognized objectives,
and administrators are begining to see that progress can be made
and are starting to provide support. The Photosynthesis Congress
at Madison, organized by Clanton Black and Bob Burris in 1976, and
this Symposium, devoted to CO_2 fixation, emphasize that this particular aspect of photosynthesis is now recognized, and I am confident
that the money will come and some of the problems will be solved.
We don't know whether it is possible to get rid of photorespiration,
but within the next five years we expect at least a clue.

The 1960's was the era of electron transport and the 1970's
the era of carbon; looking ahead, I think the next big breakthrough
will be in whole-plant photosynthesis. We know almost nothing
about how the plant develops photosynthetic capacity or what it does
to change photosynthetic capacity through the growing season. Some
of the still unanswered questions appear simple enough to provide
the potential for making big breakthroughs. The simplest is, what
is limiting photosynthesis? Is it source limitation or sink limitation? As Jim Bahr mentioned, a soybean plant, regardless of variety, adapted to its area, has a certain photosynthesis rate up
until the time of flowering, but at that time differences in photosynthesis appear. Depending on maturity group, the increase may
be faster or slower, and we haven't the slightest clue why the soybean does not show any genetic variability up to this time and
then suddenly does. The source-sink problem has been discussed;
someone like myself, working on carbon metabolism, thinks there is
a source limitation; a person interested in translocation normally
considers that there is a sink limitation. But this problem should
be simple to answer if people interested in mechanisms will observe
what is actually happening at the time of the changes in photosynthesis.

Another question is whether we can get the plant to partition
its energy more efficiently. This has been discussed at this
Symposium at the chloroplast level, but we are considering whether
the entire plant is doing an efficient job and establishing its
canopy as soon as it can. With many crops you have to establish
the canopy at the start of the season; first you build the factory
and then you produce what you want--the bean. You can alter the
partitioning of photosynthates to get the factory built quicker,
e.g., by using narrower rows, which is in fact a common practice,
and then you have a longer production time. These are things on
which data are available, and data mean potential application. Now
that people are interested in carbon, the money will come in because people are impressed by something they can see and will support it. I think the next area that should receive emphasis is the
whole plant, and observations made in the field should be integrated
with work in the laboratory to find out what the plant's mechanism
is and how it can be improved. At Urbana we are trying to establish
a multidisciplinary group of people working in the field, on leaf
photosynthesis, on carbon, on electron transport; reproducing field
observations in the growth chamber; and integrating the various
results.

GIBBS: My subject is long-range planning of research on carbon assimilation. The title of my short speech is "The Old and
the Young." The theme came up during dinner tonight in a statement
by Andre Jagendorf. I take the theme from two slides I have recently seen. One was prepared by Bill Hillman for Anton Lang's
retirement party at Michigan State, and it showed the growth in
the number of papers per year in the field of photosynthetic carbon

metabolism compared with that in other fields in plant physiology; the former had a positive slope in contrast to the latter. This shows that work on the whole of plant physiology is growing, but work on photosynthetic carbon metabolism is growing faster. This is obviously an exciting field and is attracting people; it also must be attracting money. The other slide was shown to me by Akazawa, and it was a plot of the number of papers per year in the field of RuBP carboxylase. The plot shows a lag period, being rather flat for a few years, and then shows a linear increase. Another indication of rapid growth in the field of RuBP carboxylase is this Symposium. These are signs that the field is exciting, that people must be coming into it, and, since it takes dollars to do science, that, in spite of what might have been said here about competition for funds, it is being supported.

A few years ago, Sylvan Wittwer asked me to do a survey on the amount of money going to photosynthesis research from federal agencies including the U.S. Department of Agriculture (through the Agricultural Research Service and through Experiment Stations), the National Science Foundation, the Department of Energy (then ERDA), and all others. I obtained a rough figure of about $10 million/year for photosynthesis, besides about $5 to 7 million/year for nitrogen fixation. These were rather handsome sums 2 or 3 years ago. I have not done a calculation since, but I assume the amount has gone up with inflation.

My field is teaching, and in the 10 or 12 years that I have been teaching undergraduates, I think only one has gone on to get a Ph.D. degree. The others have gone on to medical school. This hurts a little. It is obvious that they have become medical doctors not because the field of medicine is exciting but because they see a future in it for themselves.

I thought it was good that Phillip Handler, in his yearly report to the National Academy, made the novel suggestion that dollars be used to protect these young people, to keep them from leaving the field of science. He said there should be an infusion of dollars, perhaps massive (although he didn't say where they would come from), to support young people in assistant professorships or apprenticeships (or whatever you want to call them) in the universities while waiting up to a decade until the oldsters reached retirement and positions became available. Funding is needed to support young people, to make sure they can sustain themselves, and I think this is more important at the moment than doubling or tripling the funding for research on plant science and carbon assimilation.

BLACK: My main concern is teaching graduate and undergraduate students. I try to stress to them in a positive way that in plant biology quantitative work is needed, and I illustrate this in several ways. I ask them, for example, how much of a reactant is present, how fast a reaction is going, where in the plant a reaction takes place, or when in the lifetime of a plant it takes place. These are some of the major questions that we have left unanswered in dealing with plants; and I am greatly concerned that we must study plants in a quantitative way. For example, the topic of this Symposium is the regulation of RuBP carboxylase, and yet, if we ask what the RuBP concentration is inside a chloroplast or inside a cell, we don't get an answer. The goal of quantitative biology is not only to ask those sorts of questions (how much, how fast, where, and when) but also to integrate that information in order to understand the whole organism.

The integrative approach needs to be stressed even in the field of carbon assimilation. I was interested in having two nitrogen-fixation people probe us a little at this Symposium, but in fact many other aspects of plant biology need to be integrated with carbon metabolism, and to think in isolation about carbon metabolism is a big mistake. Plants assimilate at least four major oxides, those of phosphorus, nitrogen, carbon, and sulfur. They reduce three of them, but not that of P. Therefore, if we consider carbon alone, without the others, we are mistaken in thinking we can regulate plants or increase their production.

Before I mention some areas of carbon metabolism where we need some input and will make some progress in the next two years, I'll give a personal illustration of what I'm talking about. I think one of the reasons we have been successful at the University of Georgia is that I work with a colleague in the Agronomy Department, Harold Brown, who thinks quite differently from the way I do, and we have on campus other people concerned about plants and trying to understand them from their own particular disciplines. Yet we are able to put our individual thoughts together and come out with some level of understanding about plants. Of course our understanding is not total in any area, but I believe this integration of various disciplines and thought patterns, along with the available information about plants at various levels of organization, is the way to improve it. We need clarification at the level of enzymes, of organelles - chloroplasts, mitochondria and peroxisomes, and of vacuoles (we have been concerned with that), and also of different phases in a plant's lifetime.

A more specific problem is whether some understanding of photosynthesis can be applied to genetics. It turns out that we are almost totally ignorant of the genetics of photosynthesis. We have no idea how photosynthetic characteristics can be transferred from one plant to another. A few years ago, when C_4 plants were discovered, some people wanted to do magic experiments like making C_4 plants out of all the C_3 plants (which was a pretty childish idea), and yet information is still totally lacking about the genetics of the transfer of a characteristic such as PEP carboxylase from one plant to another. Knowledge of the genetics of photosynthesis is very deficient, and I think research is needed in this area.

Let me finish by repeating that I believe the only way we can make substantial progress in plant biology is by an integrated approach to understanding the totality of a plant, from every

level of organization. This will be done not by individuals but
by groups working together, and the granting agencies will have to
recognize this.

BASSHAM: As a plant biochemist, I have noted that several
questions related to RuBP carboxylase have not yet been resolved.
The reports, especially the one from Ed Tolbert, about the impurity
in RuBP carboxylase raised the question (which had already bothered
me) of what the dissociation constant for RuBP really is. According to some reports the K_m for RuBP is ≤ 1 μM, but according
to several more recent ones it is 35 to 100 μM. It occurred to me
that in the old experiments, which in one case were done with radioactive RuBP, maybe people were measuring binding of the impurity.
I don't mean to do an injustice to the people who did the studies,
but I do think it's important to establish just what the biochemical
parameters of this enzyme are; and, although we've talked a lot here
about K_m for CO_2, we haven't talked much about K_m for RuBP.
Another problem is the levels of metabolites. Different reports have been presented about the concentrations of metabolites
in chloroplasts, and it is important to establish what the physiological levels are, particularly for things that might be effectors
fine-tuning the activity of the enzyme. Agreement is general that
magnesium, pH, and carbon dioxide are the principal regulators
of the enzyme, but considerable doubt remains about the roles of
components such as sugar phosphates, 6-phosphogluconate, and NADPH.
We need to establish first, the levels of those compounds under
various physiological conditions and, second, their effects on
properly activated and assayed enzyme.
A major question (which some people think is resolved, but not
everybody agrees) is the origin of glycolate for photorespiration.
Controversy has continued over the years as to whether all the
glycolate comes via the oxygenase activity or whether RuBP leads
to the formation of phosphoglycolate, which is converted to glycolate, which then serves as the substrate for photorespiration. Few
would argue that the latter at least is one major pathway, but some
think there may be other sources of glycolate, and some even consider that photorespiration may to some extent involve (as yet un-
identified) substances besides glycolate. This certainly is a
question that should be resolved.
Now I'd like to describe my own flight of fancy about nitrogen
fixation and carbon dioxide fixation, which relates to the papers
by Sam Wildman and Ralph Hardy. It has to do with the production
of protein in green leaves as a source of food for the growing
population of the world. In the case of tobacco I think the protein
would be particularly useful as a special food supplement, as a
high quality protein perhaps for those who can't tolerate milk
protein, and so forth. But I have a broader view, as do others,
that protein in green leaves represents the primary protein production in the world. I do not advocate abolishing animal husbandry

or abandoning the use of soy protein, but I think that protein extracted from leaves is very important for the future, and we should
begin to think now about how to get the most protein economically,
what type of plants to use, and how to integrate the entire growing and production process.

How does this relate to RuBP carboxylase? Ralph Hardy described
the effect of carbon dioxide enrichment, particularly on legumes,
and noted that the increase in annual or seasonal yield was greater
for legumes than for anything else because, when CO_2 fixation is
increased by CO_2 enrichment, more photosynthate is left over to
go down to the root nodule bacteria and provide a substrate
for nitrogen fixation. I think this will be of increasing importance in the future. In fact, what I have been advocating for a
year or so is that we consider discarding our prejudices about its
not being economic, and so forth, and give some serious thought
to the possibility of carbon dioxide enrichment for agronomic crops.
Hardy says it will not be an economic proposition in any foreseeable
future, and he may be right, but I don't think we have considered
all the factors.

Large areas of this country are severly limited in future
potential because of their limited water supply, and the only method
I can see for significantly increasing the efficiency of water utilization is covered agriculture on a grand, vast scale covering
thousands of square miles. I suspect that we could increase the
efficiency of water use by a factor of maybe 10. A related consideration is the prediction by some people that the climate of
the world will change because the carbon dioxide produced by burn-
ing fossil fuels will raise the world's temperature by several
degrees. This would cause the wet and dry zones to expand away from
from the equator. The wet equatorial zomes would move to the dry
zones now in Mexico and the Sahara, the dry zones would move to
North America and Southern Europe, and the temperate agricultural
zones would go further north. This might be great for Canada, but
it would not be good for the United States, and perhaps we should
do some planning now about how we would cope with this situation
if it were to arise. That may be fanciful, but we actually have
large areas that could benefit from water conservation.

If we cover agriculture, then we can consider seriously the
possiblity of using carbon dioxide enrichment. This brings us back
to research on RuBP carboxylase. We are trying to eliminate photorespiration, and, although we cannot make all plants C_4 plants, this
would be approached by simply raising the carbon dioxide level.
Hardy questions the economics of covered agriculture, but all his
crops were cyclical crops that go through their season and become
senescent; these have a limit programmed into them as to how much
their yield can be increased by added CO_2. Legumes like alfalfa,
however, can be harvested repeatedly throughout the year with
enough leaves left to form a new canopy each time, and there is no
reason why the sink can't be the production of new leaves rather

than the formation of a seed or a tuber. In this case I suspect, although we lack the data to prove it, that the improvement in CO_2 fixation and biomass production could be as much as 300% or more. Other obvious advantages would include year-round growing seasons. Although plants, through evolution, have had to adapt to air levels of CO_2 and O_2 and therefore have had to make this unfavorable (to us) compromise between RuBP oxygenase and RuBP carboxylase so that (as far as we know, although there are some counterindications) oxygen binds competitively with CO_2, still, we might have great scope for increasing yields by breeding plants that could make use of carbon dioxide enrichment.

Norman Terry at the University of California, Berkeley, has told me privately that some of his plants can benefit from CO_2 enrichment only up to about 0.11%, (about threefold) and at higher $[CO_2]$ actually show a drop in the rate of photosynthesis. I think this is probably because the plants are adapted by evolution to air levels. CO_2 enrichment presents many problems: how to make a suitable plastic; how to generate the CO_2; what the economics are; the temperature; the greenhouse effect and how to overcome it. I have discussed some of these in some of my publications. I think a basic program in plant physiology addressed to developing plants that can make use of CO_2 enrichment would be worth while, with emphasis on legumes, which show the greatest promise for increased nitrogen fixation.

PALLAS: Quite a commercial venture in CO_2 fertilization is under way in our West, through irrigation water. It is apparently based on some Russian research indicating that a fair amount of CO_2 for carbon fixation actually comes from the soil. I have not found any publications in English, originating in the United States. It appears that the system works very well, as some people are making pretty good money with it, but we need some absolute facts as to whether it is based on fiction or reality.

ZELITCH: There are about 52 USDA Agricultural Experiment Stations, and I am at one of the smallest, having only about 45 scientists who work essentially in plant sciences in seven departments; perhaps this will bias my remarks. As I look at the world of photosynthesis from this small institution, certain things seem evident. One is that the science in this country and the agriculture are both very vigorous. It is obvious that both must become even more vigorous or we will be in deep trouble within the next 25 years. The strength of this country in science and agriculture, the scientific basis of its agriculture, is due to our having a pluralistic system. The field of photosynthesis, which is our major concern, is also pluralistic in terms of biological organization. We have people working on enzymes, organelles, leaf tissues, and crops; we have people working on molecular biology, biochemistry, and plant physiology; and we have people working on plant breeding and agronomy. It is obviously necessary to

have people working at all these levels of organization and in all
these disciplines.

One problem is that since about 1966, in real dollars, the
total amount of funding for research relating to agriculture has
actually diminished. By about 1970 it had become obvious even to
the man in the street that there was a world food shortage and it
was likely to get much worse before it got better. People scurried
around deciding what should be done about this, while in fact for
some years the funding for agricultural research had actually been
diminishing.

One of the advantages of Agricultural Experiment Station organization is that it provides an opportunity for people in all the
disciplines I mentioned to serve either under the same roof or in
close proximity. Organization by itself is no guarantee of excellence, but scientists do not work in a vacuum, and an organization
is needed that facilitates cooperation between scientists at
different levels and from different disciplines, as has already
been pointed out here. The Experiment Station system is not the
only way to accomplish this, but it has been in existence for more
than 100 years, and it seems to work. Obviously this kind of system should be encouraged and in fact used as a model for other
groups who want to cooperate in both practical and laboratory science. Unless theory and practice are united, the food supply will
be the loser, as will all of us. The kind of work I see offering
the greatest opportunity is the union of plant physiology and biochemistry with genetics. The participants in this Symposium have
been impressed and perhaps frustrated by the apparent discrepancies
in our knowledge about carboxylation, oxygenation, the mechanisms of
glycolate synthesis, and the production of CO_2 in a photorespiration pathway. Great uncertainties are always encountered at the
frontiers of knowledge and are a cause not for worry but for further
work, and meetings like this one stimulate people to be more
critical of their work and to push on.

Genetics has a particularly important role in providing the
tools for approaching some of these vexing problems: not only
do mutants provide opportunities for testing some of the hypotheses,
but also some mutants may themselves comprise valuable material
from which plant breeders may obtain crops with higher yields. I
emphasize the importance of geneticists and plant physiologists and
biochemists working together although I recognize the need for
individual excellence. Most granting agencies are realizing that
they must create vehicles for compatible groups to work together
for a common purpose.

There are many bottlenecks to plant productivity. Photorespiration is an obvious one. Dark respiration may be an important
one that is hardly being worked on at all, especially if the alternative pathway producing less ATP turns out to be important in many
forms of green plants. Transport also has an important regulatory
role in net CO_2 fixation. These uncertainties are a challenge to

all of us, and I hope many of us will take available opportunities
to work together with people in other disciplines, so that our
ideas can come to realization in the real world out in the field.

REEVES: I'd like to make a few general comments about the
National Science Foundation budget and some specific comments about
the Division I am responsible for. The latest word on the NSF
1979 budget is that the House Appropriations Committee cut our re-
quest by $40 million. It now goes to the Senate Appropriations
Committee, and they usually put some money back. If the $40 million
cut were taken proportionately away from all the programs, then
the Physiology, Cellular, and Molecular Biology (PCM) division
would get roughly 4% more than it got this year. The original re-
quest was for about a 7% increase after all the zero-base budgeting
exercises we went through last year. I expect we will end up with
a 4 to 7% increase, which is not a very bright prospect. Our
Division has had problems, perhaps more than other Divisions,
not only with insufficient dollars but also with an overload of
proposals coming in. We now have about 42% more proposals than we
had last year at this same time. In the first half of fiscal year
78 we reviewed 78% as many proposals as the total number reviewed
in fiscal 77. The money is about holding level and the number of
proposals is increasing fantastically.

An inconsistency here bothers me a little concerning the funding for plant sciences. In our Division several programs are actively involved in support of plant sciences: the metabolic biology
program, the genetics program, and the developmental biology program.
Last year Mary Clutter, the Program Director for developmental biology, organized an interagency group on plant sciences that meets
every month to keep everyone aware of what is happening. The inconsistency is that last week it became known that some one-half
million dollars had to be spent on nitrogen fixation, as mandated
in the NSF budget, and it was sent to our Division. Only two
divisions could have competed for that money, the PCM Division and
the Division of Environmental Biology (DEB), which deals more with
whole-plant systems. DEB could not spend that money and I was con-
vinced that we could, so I managed to get the whole $500,000 that
was earmarked for nitrogen fixation, but I had trouble spending it
because we didn't have the right proposals. So at the same time
I'm being told that we've got to pump more money into plant sciences,
I'm having trouble spending the $500,000 I got two weeks ago. Since
this is money I have to spend in fiscal 78, I looked back over the
year to see how much we spend on nitrogen fixation, and it wasn't
much more than $500,000 because we hadn't received suitable
proposals.

Our fiscal year is from October through September, but, since
the Foundation's is from October through June, we have to spend all
our money by the first of July, unlike NIH. We have only four more
weeks to finish spending, so it is too late to put in proposals for
this year, but the matter of proposals is of great concern to me.

HILLMAN: M. Gibbs referred to a slide of mine that showed carbon assimilation research going up as a percentage of work in plant physiology and flowering research going drastically down. This is relevant to some of the points raised here. Many studies now are directed toward the whole plant, which means essentially towards development. In fact, as Bill Ogren and Clanton Black pointed out, development controls, to some degree, many of the other events in carbon assimilation. My interpretation of the reason for the big drop in flowering research (and many other fields of developmental work in plants) is not that ideas or interest are lacking, but (and I simply make this assertion without being able to prove it) that the time constant for getting, say, one reliable datum point in developmental study may be one to two orders of magnitude greater than in a more biochemical type of study. This means that the gamble on the part of the investigator, particulary the young investigator, going into a developmental field is much greater, and the funding has to be given over a considerably longer time before you can ask for payment in the form of results. It is important for the granting agencies to realize this. Perhaps too much money is now mandated for the fast fields (you can't even spend your $500,000 this year for nitrogen fixation), and there may be work that needs doing in plant development that nobody even dares make proposals for because results can't be expected within the two- or three-year term of a grant. We must realize that certain fields that should become more important again have a totally different time scale for the research support required, and that they are relevant to the direction that all plant physiology research will take as it integrates work on the whole organism.

REEVES: Certain amounts of money are mandated but they are relatively limited. Besides the nitrogen-fixation mandate this year, we had a Congressional mandate concerning equipment and instrumentation; occasionally line items are put in, but generally they are not a very large percentage of the budget. At least in our Division, and I think in other areas at NSF, we're trying to get away from being a two-year funding agency and are pushing very hard now to make three-, four-, and even five-year grants. Your points are well taken, and the program officers are aware of some of them.

HOWELL: I don't want to disparage the whole-plant biologists, but we molecular biologists have been pretty well underrepresented here, and we are entering an exciting period, particularly with regard to RuBP carboxylase. This Symposium included a report of an attempt to unravel some of the primary amino acid sequence of RuBP carboxylase. Within the next year or two, through the cloning of the RuBP carboxylase large subunit gene, which has already been done, the DNA sequence certainly will be determined by one group

or another.   That will make it possible to cleave out, and to re-
synthesize, certain portions of that DNA sequence, and thus to
substitute amino acids into the primary sequence of the gene at
will.

My question to the people studying the enzymology is what kind
of changes they are looking for.   Do they want us to build changes
into this enzyme to increase the turnover number, to improve the
binding efficiency for substrates, to improve the efficiency of
a particular gene or a particular enzymic activity?   We may be able
to make such modifications in the test tube, but introducing them
back into the plant could be another matter entirely.   Nonetheless,
we are at an exciting point, and we will see great strides within
the next couple of years.   It is time for the people studying the
enzymology and the kinetic features of the enzyme to address themselves to these issues and communicate with the people working on
the molecular biology aspects.

O'LEARY: I am one of the few people here who never worked on
RuBP carboxylase, although I may in the future.   I came here to
be educated, and that has certainly happened.   I am an organic
chemist and enzymologist, and from my point of view, i.e., regarding RuBP carboxylase as simply an enzyme rather than a fundamental
part of the system, we know a number of things about this enzyme
now that we didn't know before.   I would like to share some prejudices because I think that people who do research, who referee
papers, and who edit journals can no longer afford to ignore them.
First, we must be careful about the activation state of the enzyme in
work like that of J. Bahr, G. Lorimer, R. Jensen, and others.   Anyone who purports to study purified enzyme without very carefully
defining the activation state is doing us a disservice.   Second,
we have to be certain about the purity of the substrates.   It was
reported here how RuBP decomposes; we don't know what that implies
except that a lot of old work has to be carefully reexamined.
Third, my own work on phosphoenolpyruvate carboxylase has taught
me that people should be more careful than they have been about
the background CO_2 levels in their solutions.   Simply bubbling
nitrogen through for an hour or so may or may not remove all the
CO_2, which has to be accounted for very carefully in any study involving CO_2 activation.   Fourth, we must consider the concentration
of the enzyme in our experiments vs. the concentration in a cell;
I can only worry about this, I don't know what to do about it.
Fifth, with people doing assays for 10 sec, 30 sec, 5 min, or
30 min, is the assay really linear?   Bahr and Jensen, in particular,
have shown that we have to be careful about that.   And finally, we
have seen how many other things might be present in the solution
that can affect the activity of the RuBP carboxylase.   From the
point of view of an enzymologist who has not yet started working
on this enzyme, putting this all together is scary--there is so much
to worry about.   But we have to do it; otherwise nobody will (a)
believe or (b) reproduce our results.

ABSTRACTS OF POSTER DISCUSSIONS

1. Differences in Catalytic Activities of Ribulose 1,5-Bis-
phosphate Carboxylase in Various Genetic Lines of Barley. W. R.
Andersen, S. V. Tingey, and C. A. Rinehart, Dept. of Botany & Range
Sci., Brigham Young U., Provo, UT 84602.
     We have screened several barley lines for genetic variation in
RuBP carboxylase activities. The enzyme is rapidly extracted from
7-day-old barley seedling leaves into an activation mixture. The
reactions for a given extract are measured under two atmospheric
systems consisting of compressed air or nitrogen. The reactions
under compressed air are inhibited to 30% by oxygen. Aliquots of
the reactions are terminated after 30 sec with 1/4 volumes of acetic
acid. The ratio of the 30-sec reaction rates under nitrogen and under air provides a selective index value for the response of a given
plant extract to inhibition by atmospheric oxygen. The continuing
reaction is finally terminated after 2 min. The activities at the
2-min level are compared with the expected activities projected linearly from the 30 sec levels. This ratio provides an index value
which measures the degree of inactivation of each enzyme extract in
the presence of RuBP substrate during the reaction. We thus measured
two selective index values for a given plant extract: (i) the relative degree of oxygen inhibition of activated enzyme, and (ii) the
relative degree of inactivation of each enzyme during the reaction
at chloroplast levels of CO_2 and RuBP. Our data indicate that different catalytic forms of the enzyme exist in the genetic lines of
the barley breeding pool.

2. Activation of RuBP Carboxylase by Cyanate, $MnCl_2$, or $CaCl_2$.
James T. Bahr, Dept. of Nutrition & Food Sci., U. of Arizona,
Tucson, AZ 85721. (Present address: Mobil Chemical Co., Edison,
NJ 08817.)
     RuBP carboxylase is normally activated by formation of a re-
versible enzyme-CO_2-Mg^{2+} complex. It has been found that cyanate,
an analog of CO_2, can substitute for CO_2 during activation of RuBP

carboxylase. At pH 8.1 and in the presence of 10 mM $MgCl_2$, the carboxylase was fully activated by 40 mM KNCO. Half-maximal activation was obtained at about 5 to 10 mM KNCO, compared with 2 to 3 mM $NaHCO_3$ under these conditions. The specific activity of the fully activated enzyme was the same whether saturating HCO_3^- or KNCO was used during activation. Activation by KNCO was reversible. Cyanate also inhibited the carboxylase when present during the assay. The inhibition was competitive with respect to HCO_3^-, with a K_i of about 3 to 4 mM. With incubations of enzyme with KNCO for up to 40 min, no evidence for irreversible inhibition of the enzyme was obtained. $MnCl_2$ and $CaCl_2$ could replace $MgCl_2$ during activation of RuBP carboxylase in the presence of CO_2. The apparent activity of enzyme activated by Mg^{2+} and CO_2 was independent of Mg^{2+} concentration in the assay medium. The enzyme activated by Mn^{2+} and CO_2 was about 25% as active, regardless of the Mg^{2+} concentration in the assay medium. The enzyme activated by Ca^{2+} and CO_2 was 70% as active when assayed with Mg^{2+} present, but only 1% when assayed without Mg^{2+}. Enzyme incubated with CO_2 only was 12% as active with Mg^{2+} in the assay, and 2% without Mg^{2+} in the assay.

3.   Breakdown of RuBP Carboxylase During Dark-Induced Senescence.   Vernon A. Wittenbach, Central Research & Development Dept., Experimental Station, E. I. duPont de Nemours & Co., Wilmington, DE 19898.

When 8-day-old wheat seedlings are placed in the dark, the fully expanded primary leaves undergo the normal changes associated with senescence, e.g., loss of chlorophyll, soluble protein, and photosynthetic capacity (Wittenbach, Plant Physiol. 59, 1039, 1977). This senescence is completely reversible if plants are transferred to the light during the first 2 days but thereafter becomes irreversible. During the reversible stage, the loss of RuBP carboxylase, quantitated immunochemically, accounts for 80% of the total loss of soluble protein. No significant change in activity per mg of antibody-recognized carboxylase occurs during this stage despite an apparent decline in specific activity on the basis of mg soluble protein. With the onset of the irreversible stage of senescence there is a rapid decline in activity per mg carboxylase, possibly indicating a loss of active sites, and a correlated increase in total proteinase activity. The main class of wheat proteinases responsible for the increase in activity are the thiol proteinases, which have a high affinity for RuBP carboxylase, exhibiting an apparent K_m at $38°C$ of 1.8×10^{-7} M. The K_m for casein was 1.1×10^{-6} M. Thus, if casein is representative of noncarboxylase protein, then the higher affinity for carboxylase may help provide an explanation for its apparent preferential loss during the reversible stage of senescence.

4.   Crystallographic and X-ray Diffraction Studies on Glycolate Oxidase From Spinach.   Carl-Ivar Brändén and Ylva Lindqvist, Dept. of Chemistry, Swedish U. of Agriculture Sci., Upsala, Sweden.

Glycolate oxidase was prepared from spinach by the method of Kerr and Groves for the enzyme from pea leaves (Phytochemistry 14, 359, 1975) with some modification.   Crystals were obtained by concentrating a solution of the enzyme in 0.05 M Tris-HCl buffer at pH 8.3.   The crystal is a tetragonal bipyramid with cell dimensions of 104 x 104 x 133 Å.   The space group is tetragonal p422, with one dimer having a molecular weight of 100,000 per asymmetric unit.

5.   Identification of Two Different Chromosomes Which Code for Specific Small Subunit Polypepties of Fraction I Protein. Kevin Chen, Dept. of Biology, U. of California, Los Angeles, CA 90024.

The material analyzed was developed by Sand and Christoff (J. Heredity 64, 24, 1973) from crossing N. debneyi ♀ x N. tabacum ♂, doubling the chromosome number, and backcrossing the amphiploid through nine successive generations with pollen of N. tabacum. The progeny produced flowers that varied greatly in morphology, and four male sterile types were selected.   Plants with flower types 1, 2, and 3 had all 48 N. debneyi chromosomes replaced by 48 N. tabacum chromosomes.   Their two Fraction I protein small subunit polypeptides were identical to those of normal N. tabacum.   Plants with flower type 4 have three polypeptides, one being identical to a small subunit polypeptide found in N. debneyi.   Burns, Gerstel, and Sand (in press) have shown that type 4 plants have 49 chromosomes, the extra chromosome being derived from N. debneyi.   Therefore, the presence of a third polypeptide depends on the presence of chromosome "d." Two nullisomic cultivars of N. tabacum, provided by D. U. Gerstel, each had one of the 24 pairs of chromosomes missing.   Absence of the E chromosome pair reduces the amount of one of the two small subunit polypeptides, but absence of the S chromosome pair does not.

6.   Chemical Modification of Tobacco RuBP Carboxylase by 2,3-Butanedione.   Raymond Chollet, Lab. of Agricultural Biochemistry, U. of Nebraska, Lincoln, NE 68583.

With the recent development and use of α-dicarbonyl compounds, including 2,3-butanedione and phenylglyoxal, as selective reagents for the chemical modification of arginyl groups under mild conditions, there is an increasing awareness of the importance of arginine residues in binding phosphorylated substrates, coenzymes, and effectors in a wide variety of enzymes.   (J. F. Riordan et al., Science 195, 884, 1977).   Since RuBP carboxylase not only acts on a phosphorylated substrate but also is modulated by anionic effectors including NADPH and 6-PGlcA, we are currently investigating the possible importance of arginyl residues in the catalytic and regulatory properties of this bifunctional enzyme by modification with 2,3-

butanedione. Treatment of crystalline tobacco RuBP carboxylase
with butanedione results in a time- and concentration-dependent loss
of enzymic activity. Inactivation is markedly enhanced by borate
buffer and alkaline pH and is partially reversed upon removal of
excess reagent and borate by gel filtration, which strongly suggests
that the inactivation is due to modification of essential arginyl
residues. When the modification reaction is performed in the presence of various phosphorylated ligands, including RuBP, carbamyl-P,
NADPH, 6-PGlcA, 3-PGA or P-glycolate, only RuBP significantly decreases the rate of inactivation. CO_2-Mg^{2+} activation of the enzyme
(in the absence or presence of 6-PGlcA or NADPH) prior to modification has little influence on inactivation. Butanedione modification
of RuBP carboxylase (15 to 80% loss of activity) does <u>not</u> alter modulation by phophorylated effectors, including NADPH, 6-PGlcA, Fru-P_2,
3-PGA or Rib-5-P.

        7.   Chemical Modification of Tobacco RuBP Carboxylase by Cyanate.
Raymond Chollet, Lab. of Agricultural Biochemistry, U. of Nebraska,
Lincoln, NE 68583.
        Cyanate has been shown to inhibit a variety of enzymes irrevers-
ibly under neutral or weakly alkaline conditions by carbamylating
the ϵ-amino group of lysyl residues. In addition, it has been suggested that isocyanic acid (O=C=NH), the reactive tautomer of cyanate, is both a structural and an electronic analogue of O=C=O.
We reasoned that cyanate might be a useful chemical modifier for
further investigating the role of lysyl residues in RuBP carboxylase,
an enzyme for which CO_2 is both an activator and a substrate. Crystalline tobacco RuBP carboxylase is irreversibly inactivated by incubation with potassium cyanate at pH 7.4. The rate of inactivation
is pseudo-first-order and linearly dependent on reagent concentration.
In the presence of RuBP or high levels of CO_2 and Mg^{2+} the rate constant for inactivation is reduced, which suggests that chemical modification occurs in the active-site region of the enzyme. In contrast,
neither the effector NADPH nor the activator Mg^{2+} alone significantly
affects the rate of inactivation by cyanate; however, NADPH markedly
enhances the protective effect of CO_2 and Mg^{2+}. Incubation of the
carboxylase with potassium ^{14}C-cyanate in the absence or presence
of RuBP revealed that the substrate specifically reduces cyanate incorporation into the large catalytic subunits of the enzyme. Analysis of acid hydrolysates of the radioactive carboxylase indicated
that the reagent carbamylates both NH_2-terminal groups and lysyl
residues in the large and small subunits. Comparison of substrate-
protected enzyme with inactivated carboxylase revealed that RuBP
preferentially reduces lysyl modification within the large subunit.
The data here presented indicate that inactivation of RuBP carboxylase by cyanate or its reactive tautomer, isocyanic acid, results

from the modification of lysyl residues within the catalytic subunit, presumably at the activator and substrate CO_2 binding sites
on the enzyme (Biochim. Biophys. Acta, in press),

RuBP Carboxylase Activation by CO_2: $E-NH_2 + O{=}C{=}O \rightleftharpoons E-N-C-OH$ (with H, O, and double-bond O on the C)

Carbamylation by Cyanate: $E-NH_2 + O{=}C{=}NH \xrightarrow{pH>6} E-N-C-NH_2$ (with H, O, and double-bond O on the C)

(Research performed in the Central Research and Development Department of E. I. duPont de Nemours & Co.)

8. The Mechanical Instability of Ribulose 1,5-Bisphosphate
Carboxylase. Gregory C. Gibbons, Dept. of Biotechnology, Carlsberg
Lab., DK-2500 Copenhagen, Valby, Denmark.
    The intrinsic mechanical instability of ribulose 1,5-bisphos-
phate carboxylase has been analyzed with highly purified enzyme
from Nicotiana sylvestris and N. tabacum. Purified RuBP carboxylase
in solution, when examined with strong backlighting, often exhibits
inclusions which are normally removed by a centrifugation step and
are frequently referred to as undissolved protein. The nature of
these inclusions and their formation were studied. The yield of
RuBP carboxylase during isolation of the protein on a large scale
can be greatly affected by the amount of this precipitate formed.
At enzyme concentrations >0.125 mg/ml, mechanical agitation of the
native protein in solution leads to the formation of insoluble
fibrillar precipitates of high molecular weight. At high protein
concentrations, precipitate formation is reduced by the protective
effect of a stable protein foam. Both the large and the small subunits of the enzyme are found in the fibrillar precipitate and in
the supernatant in the same ratio as in the native protein. This
strongly indicates that during mechanical disturbance aggregation
occurs between native RuBP carboxylase molecules and does not result from dissociation of subunits and subsequent separate aggregation of either large or small subunits. The molecular nature of
the mechanically induced fibrillar aggregates of RuBP carboxylase
may involve conformational changes of the molecule. It is of interest that aggregation of RuBP carboxylase occurs in situ inside the
chloroplast under conditions of stress. Further electron microscope studies will aid in determining whether the molecular organization of the two types of aggregates is the same.

9. Cross-linking of Ribulose 1,5-Bisphosphate Carboxylase
From Pisum sativum. A. Grebanier, D. Champagne and H. Roy, Dept.
of Biology, Rensselaer Polytechnic Inst., Troy, NY 12181.

RuBP carboxylase from peas was treated with tetranitromethane (TNM) or with dimethylsuberimidate (DMS). Cross-linked subunits in the treated enzyme were detected by SDS polyacrylamide gel electrophoresis. When the enzyme was incubated with 10 mM $NaHCO_3$ and 20 mM $MgCl_2$ before treatment with TNM or DMS, slightly greater amounts of large subunit dimers and small subunit dimers were formed than when the enzyme was not treated before cross-linking. A more striking increase was seen in material that migrated slightly more rapidly than the large subunit during electrophoresis. This material was formed only by treatment with TNM and was derived from the large subunit. $NaHCO_3$ alone had no effect on cross-linking with TNM or DMS. RuBP diminished the extent of cross-linking. Treatment with $NaHCO_3$ and $MgCl_2$ activates pea carboxylase as it does carboxylases of other species. Possibly the cross-linking reactions are monitoring conformational changes associated with such activation.

10. Peptide Mapping of the Large Subunit of Ribulose Bisphosphate Carboxylase: Application to the Enzyme From the Genus Oenothera. Anthony A. Holder, Dept. of Physiology, Carlsberg Lab., DK-2500 Copenhagen, Valby, Denmark.

The chloroplast genome has been shown to play a role in evolution within the genus Oenothera. On the basis of incompatibility between the chloroplast genome and certain nuclear genomes, which leads to defective chloroplasts, the chloroplast genomes can be assigned to a number of classes (W. Stubbe, Genetica 35, 28, 1964). This report describes the fingerprint analysis of peptides derived by chymotrypsin digestion of the RuBP carboxylase large subunit polypeptides from a number of Oenothera species, done in order to clarify the relationship between the chloroplast genomes. RuBP carboxylase was prepared from Oenothera by a procedure that prevents modification of the enzyme by polyphenols (A. A. Holder, Carlsberg Res. Commun. 41, 321, 1976). The S-carboxymethylated large subunit was isolated by gel filtration and then digested with chymotrypsin. The resultant soluble peptides were characterized by a peptide mapping procedure involving a combination of ion-exchange chromatography on Dowex AG50W-X4 resin and thin-layer chromatography on silica gel G coated plates, and the use of ninhydrin and amino acid-specific reagents to detect the peptides. This procedure resolved ~80 peptides from each large subunit. Comparison of the resultant peptide maps from different species showed the presence of most of the peptides in each large subunit. A few peptide differences were detected between some of the species, confirming that evolutionary divergence in the chloroplast DNA has occurred within the genus Oenothera.

11.   The Effect of Low Temperature Adaptation of Puma Rye on the Structure and Function of RuBP Carboxylase/Oxygenase.   N. P. A. Huner, Chemistry & Biology Research Inst., Agriculture Canada, Ottawa K1A OC6, Canada.

The in vitro effects of low temperature on the structure and function of RuBP carboxylase have been studied in detail in N. tabacum.   Recently Chollet and Anderson showed that the reversible in vitro cold-inactivation of RuBP carboxylase from tobacco was associated with a loosening of the native structure (Biochim. Biophys. Acta 482, 228, 1977).   In contrast to these in vitro results, changes in RuBP carboxylase from Puma rye have been observed during in vivo temperature adaptation.   It was found that during the cold-hardening of rye plants, in which growth at $4°C$ for several weeks increased their freeze resistance from $-4°C$ to $-30°C$, their RuBP carboxylase changed in charge but not molecular weight or amino acid composition.   The enzyme from cold-hardened plants was more stable to low temperature in vitro and had a more compact native structure, determined by SH-group titration with 5,5-dithiobis(2-nitrobenzoic acid), than the enzyme from plants grown at $25°C$ (unhardened).   That the tertiary structures were different was supported by the differential susceptibility of the enzymes from cold-hardened and unhardened rye to denaturation by SDS.   Concomitant changes in catalytic function were also evident.   In general, as the temperature increased, the apparent affinity of both enzymes for HCO_3^- decreased.   However, below $10°C$, the apparent affinity of RuBP carboxylase from cold-hardened plants for HCO_3 was consistently higher than that of the enzyme from unhardened plants.   Furthermore, above $10°C$, the apparent affinity of RuBP carboxylase from cold-hardened rye for HCO_3^- was consistently lower than that of the enzyme from unhardened rye.   Differential effects of pH on subunit structure and $K_m(HCO_3^-)$ were observed.   It is concluded that growth of Puma rye at cold-hardening temperatures resulted in a change in tertiary structure and catalytic properties due to in vivo adaptation to low temperature.

12.   Enzymatically Active Crystalline RuBP Carboxylase From Spinach.   Sarjit Johal and Don P. Bourque, Dept. of Nutrition & Food Sci. and Dept. of Biochemistry, U. of Arizona, Tucson, AZ 85721.

A unique feature of tobacco Fraction I protein (ribulose 1,5-bisphosphate carboxylase/oxygenase) is the ease with which it can be isolated in high yield as a pure crystalline protein.   Though F-I-p has been isolated and characterized from a wide variety of plants and certain photoautotrophic microorganisms, only species of the genus Nicotiana have yielded crystalline F-I-p.   Recent improvements in technology for crystallization of macromolecules have provided protein crystals useful for structural determinations.

The application of a microcrystallization technique, vapor diffusion, has resulted in crystallization of spinach F-I-p. Identity
of the crystalline material with F-I-p was shown by SDS gel electrophoresis, immunological properties, and RuBP carboxylase activity
assay. (This research was supported by NSF-RANN grant AER76-15618
A01 to DPB.)

        13. RuBP Carboxylase/Oxygenase Activity and Gas Exchange
Rates During Leaf and Fruit Development. Susan Lurie, Nahman Paz,
Naomi Struck, and Ben-Ami Bravdo, Faculty of Agriculture, Hebrew
U. of Jerusalem, Rehovot, Israel. (BAB is on sabbatical at USDA-
SEA-FR, Southern Piedmont Conservation Research Center, Wakinsville,
GA 30677.)
        Four-month-old tobacco plants (<u>Nicotiana</u> <u>rustica</u>) containing
seven leaves were examined leaf by leaf for net photosynthesis
and CO_2 compensation point. The stomatal and mesophyll resistances
were calculated. Each leaf was then ground separately, and the
homogenates were assayed for RuBP carboxylase and oxygenase activity. The highest net photosynthesis activity was found in the
middle leaves although the youngest leaves had the highest RuBP
carboxylase activity. The latter, however, also had higher stomatal
resistance, which would lower their net photosynthesis. The RuBP
oxygenase activity did not parallel the carboxylase activity, being
lowest in the youngest and oldest leaves and maximal in the middle
leaves. The ratio carboxylase activity/oxygenase activity increased
steadily from the oldest to the youngest leaves and showed a good
inverse correlation with the compensation point measured in whole
leaves. In developing tomato fruits, the oxygenase activity again
did not parallel the carboxylase activity. The former was highest
in the climacteric stage, paralleling CO_2 and ethylene evolution,
whereas the latter decreased steadily. This pattern was not found
in Rin and Nor, which are known to be nonripening tomato mutants.

        14. Inhibition of RuBP Carboxylase/Oxygenase by RuBP Epimeriza-
tion and Degradation Products. Christian Paech, Stephen D. McCurry,
John Pierce, and N. E. Tolbert, Dept. of Biochemistry, Michigan
State U., East Lansing, MI 48824.
        Xylulose 1,5-bisphosphate, arising in preparations of ribulose
1,5-bisphosphate (RuBP) via nonenzymic epimerization, inhibits the
enzyme. Another inhibitor, a diketo degradation product of RuBP,
is also present. Both compounds simulate the substrate inhibition
of RuBP carboxylase/oxygenase previously reported for RuBP. Freshly
prepared RuBP has little inhibitor activity but is nearly impossible
to store without inhibitory impurities arising. This chemical instability may be one reason for a high level of RuBP carboxylase in
chloroplasts, where the concentration of active sites exceeds the

total concentration of RuBP. Because the dissociation constant of
the enzyme-substrate complex is <1 μM, it is concluded that nearly
all RuBP generated in situ would be complexed by the carboxylase
and prevented from deteriorating to inhibitors.

     15. Isolation and Characterization of Cyanogen Bromide
Fragments From the Large Subunit of Ribulose Bisphosphate Carboxylase From Barley. Carsten Poulsen, Dept. of Physiology, Carlsberg
Lab., DK-2500 Copenhagen, Valby, Denmark.
     Polypeptide fragments obtained by CNBr cleavage of S-pyridyl-
ethylated large subunit of RuBP carboxylase from barley were partially purified by ion-exchange chromatography and gel filtration.
On the basis of amino acid compositions, the eight isolated fragments were found to account for ~425 of the 490 residues in the
polypeptide chain. One fragment (CBP-CM6) was blocked to N-terminal sequencing, which suggests that it is the N-terminal region
of the polypeptide. A second fragment (CBP-CM8) contained more
than one equivalent of homoserine(-lactone), which indicates the
presence of an uncleaved Met-Ser or Met-Thr peptide bond. A third
fragment occurred in two forms with identical N-terminal sequences
and very similar amino acid compositions (CBP-CM1 and CBP-CM5X).
The two forms eluted differently during gel filtration on Bio-Gel
P-30, possibly because of secondary modifications. The absence of
homoserine(-lactone) indicates that this fragment is the C-terminal
region. The N-terminal sequence of CBP-CM2-II was identical to a
part of one of the proposed active-site peptides isolated from the
spinach enzyme by Stringer and Hartman (Biochem. Biophys. Res.
Commun. 80, 1043, 1978). Two fragments (CB-VIA and CB-VIB) could
not be purified by ion-exchange chromatography. From the Bio-Gel
P-30 column they eluted as a mixture together with fragment CBP-CM3.
The latter could be isolated by ion-exchange chromatography, and
its sequence was determined. Of the two smallest fragments obtained
from the large subunit of spinach, one was homologous to CB-VIA and
the other (CB-VID) was partially homologous to CBP-CM8. The different fragmentation patterns in the two species indicate that some
of the methionines are not in equivalent positions in the primary
structure.

CBP-CM1	:	PALTEIFGDDSVLQFGGGTLGHPGWNAPGA	(30 of 125)
CBP-CM2-I	:	AGVCDYCLTYYTPEYETCDTDFLAAFRVSPQPGV	(34 of 90)
CBP-CM2-II	:	SGGDHIHSGTVVGKLEGERE	(20 of 55)
CBP-CM3	:	PGVIPVASGGIHVWWHM	(17)
CBP-CM8	:	IKGAVFARQLGVPEKDGA?SITF	(22 of 45)
CB-VIA	:	KAVIDHRQ	(8 of 14)

(We thank Drs. Brian Martin and Ib Svendsen and Ms. Lone Sørensen
of the Dept. of Chemistry, Carlsberg Lab., for the automatic
sequence determinations.)

16. PEP Carboxylase Reduces Photorespiration in <u>Panicum</u>
<u>miliodes</u>. C. K. M. Rathnam and Raymond Chollet, Lab. of Agriculture,
U. of Nebraska, Lincoln, NE 68583.

<u>Panicum</u> <u>milioides</u> represents the first well documented example
of a higher plant species with reduced photorespiration and O_2 inhibition of net photosynthesis. We have investigated the biochemical mechanism(s) involved in reduced O_2 sensitivity of photosynthesis in this species by parallel inhibitor experiments with thin
leaf slices of <u>P. milioides</u>, <u>P. bisulcatum</u> (C_3), and <u>P. miliaceum</u>,
(C_4). We reasoned that a leaf slice system would facilitate the
use of specific inhibitors of C_3 and C_4 photosynthesis as metabolic
probes, while preserving the <u>in vivo</u> arrangement of different leaf
cell types and any associated inter- and intracellular compart-
mentation of photosynthetic-photorespiratory carbon metabolism.
The inhibitory effect of 21% O_2 on net photosynthesis in <u>P. milioides</u>
at 30°C and 10 μM CO_2 (17% inhibition) was gradually increased with
increasing concentrations of the specific PEP carboxylase inhibitors,
malonate and maleate. At saturating levels of inhibitor, photosynthesis in 2% O_2 was decreased by about 18% (compared with 0% in
<u>P. bisulcatum</u> and 85% in <u>P. miliaceum</u>), and the inhibitory effect
of 21% O_2 was identical to that observed with <u>P. bisulcatum</u> in
the absence or presence of inhibitor (29% inhibition). A similar
reversion of the reduced O_2 sensitivity of photosynthesis in
<u>P. milioides</u> to a C_3 value by specific PEP carboxylase inhibition
was observed at elevated pO_2. A significant potential for C_4
photosynthesis in <u>P. milioides</u>, compared with its complete absence
in <u>P. bisulcatum</u>, was demonstrated on the basis of (a) a tight
coupling of leaf slice CO_2 fixation by PEP carboxylase with the
C_3 cycle; (b) NAD-malic enzyme-dependent C_4 acid (aspartate and
malate) decarboxylation in leaf slices; (c) substantial activity
of a full complement of C_4 cycle enzymes (including PEP carboxylase,
aspartate and alanine aminotransferases, NAD-malic enzyme, and
pyruvate, P_i dikinase) in crude leaf extracts; and (d) Kranz-like
leaf anatomy with numerous plasmodesmata traversing the mesophyll-
bundle sheath interfacial cell wall. Our data indicate that the
reduced photorespiration and O_2 inhibition of net photosynthesis
in <u>P. milioides</u> is due to PEP carboxylase participation, possibly
by creation of a limited C_4-like CO_2 pump, rather than to an
altered RuBP carboxylase/oxygenase as previously suggested.

17. Increased Levels of Ribulose Bisphosphate Carboxylase in
<u>Rhodospirillum</u> <u>rubrum</u>. John V. Schloss, E. F. Phares, Mary V. Long,
Claude D. Stringer, and Fred C. Hartman, U. of Tennessee - Oak
Ridge Graduate School of Biomedical Sci., and Biology Div., Oak
Ridge Nat. Lab., Oak Ridge, TN 37830.

Ribulose bisphosphate carboxylase from <u>R. rubrum</u> (a photosynthetic purple nonsulfur bacterium) is a dimer of identical subunits

and may thus serve as a prototype for the more complex analogous enzymes found in higher organisms. Serial culture of R. rubrum under autotrophic conditions results in greatly increased activity of ribulose bisphosphate carboxylase (about 6-fold greater than the highest previously reported). Purification and characterization of the enzyme from these cells reveal that the increased level of activity is due to an increase in quantity of the enzyme rather than to production of a species with altered structural and kinetic properties. The carboxylase from R. rubrum grown serially on CO_2 in H_2 can represent ~40% of the soluble protein. Thus, levels of the enzyme obtained in the current study are comparable with those of Fraction I protein found in higher plants. With two 50-liter fermentors, sufficient bacteria can be grown every two weeks to provide about 1.5 g of pure enzyme (28% recovery). The availability of large quantities of this enzyme provides an impetus for sequence and X-ray crystallographic studies. Edman degradation of the enzyme reveals that the NH_2-terminus is free (in contrast to the catalytic subunit of other carboxylases examined) and that the NH_2-terminal sequence is Met-Asx-Glx-. The purified enzyme can be crystallized from distilled water. (Research sponsored jointly by the National Institute of General Medical Sciences, NIH, Grant GM-1974, and by the Division of Biological and Environmental Research, DOE, under Contract W-7405-eng-26 with Union Carbide Corp.)

18. Light Regulation of mRNA Coding for the Small Subunit of RuBP Carboxylase. Elaine Tobin, Biology Dept., U. of California, Los Angeles, CA 90024.

In Lemna gibba G-3, the synthesis of RuBP carboxylase is a light-regulated process. I have examined some effects of white light on the synthesis of this enzyme and on the level of mRNA coding for its small subunit. Synthesis of the small subunit declines drastically when the plants are kept in darkness and resumes when they are returned to light. When mRNA was isolated from plants that had been in the dark for 4 days, no translation product corresponding to the small subunit could be detected whether the RNA was isolated from polysomes or from total cellular RNA. When the plants were returned to light for 18 hr, the amount of at least two translation products coded for by the poly(A) RNA increased dramatically. One major band that increased in response to light had a molecular weight of 32,000. The second of these bands was shown by immunoprecipitation to be the 20,000-dalton precursor of the RuBP carboxylase small subunit. This increase could be seen as early as 3 hr after light treatment. The 32,000-dalton product might be a precursor of the protein moiety of the light-harvesting chlorophyll a/b protein. This protein is the major thylakoid membrane protein and its amount is known to be modulated by light. The light treatment does not simply affect

mobilization of mRNA onto polysomes because no differences in
translation products were found whether the RNA was isolated from
polysomes or from total cellular RNA of dark-grown tissue. The
lack of RuBP carboxylase synthesis in dark-treated tissue is consistent with the apparent lack of translatable poly(A) RNA coding
for the small subunit of the enzyme.

19. Differential Reactivation of RuBP Oxygenase With Low
Carboxylase Activity. G. F. Wildner and J. Henkel, Dept. of Plant
Biochemistry, Ruhr U., D-4630 Bochum 1, Germany.

RuBP carboxylase/oxygenase is deactivated by dialysis against
EDTA. Incubation of the dialyzed enzyme with $MgCl_2$ alone enhances
the activity of carboxylase (67% of initial activity) and oxygenase
(56%), which can be further stimulated by the addition of $NaHCO_3$
(93% and 73% respectively). Treatment of the dialyzed enzyme with
$MnCl_2$ restored only the oxygenase activity (111%). The carboxylase
activity was low (10%) and remained unchanged even after addition
of $NaHCO_3$. These different degrees of reactivation of oxygenase
and carboxylase by $MnCl_2$ suggest different roles for the divalent
cations. $MgCl_2$ (plus CO_2) restores the catalytic sites of the
enzyme molecules, whereas $MnCl_2$ (or other divalent transition-
metal ions) can directly participate in the mechanisms of the
RuBP oxygenase, but not the carboxylase, reaction. The role of
$MnCl_2$ in the enzyme mechanism may be to serve as the site for
oxygen activation. RuBP alone cannot be oxidized by $MnCl_2$ under
similar conditions, which indicates that the enzyme-RuBP-Mn^{2+}
complex is the reactive species. The "artificial" Mn^{2+}-dependent
RuBP oxygenase reaction can serve as a tool for studying the
enzyme's properties (interaction of subunits, binding of RuBP)
without preincubation in the presence of $MgCl_2$ and $NaHCO_3$. Inhibition and reactivation studies may also help in formulating a
plausible model of the reaction mechanism of RuBP oxygenase.

20. The Effect of Glycidate on the Translocation of Photo-
assimilates. G. F. Wildner, Ruhr U., D-4630 Bochum 1, Germany,
and C. Larsson, U. of Lund, Sweden.

Glycidate (2,3-epoxypropionate) stimulated CO_2 fixation in
isolated spinach (Spinacia oleracea) chloroplasts up to 100%.
In the presence of glycidate the chloroplasts excreted mainly
3-phosphoglycerate instead of dihydroxyacetone phosphate and
therefore required less energy per carbon fixed. Glycolate
formation was not inhibited by glycidate, which is in agreement
with the observation that $NaHCO_3$-preactivated ribulose 1,5-bis-
phosphate oxygenase is not inhibited by this compound. Furthermore, glycidate had no effect on electron transport (NADP reduction)
and photophosphorylation, and no change in the coupling ratio was
observed. The change in the end product of CO_2 fixation could be
explained by a correlation between the translocation process and

the potential of the inner envelope membrane. The countertransport, 3-phosphoglycerate^{3-} (out) against phosphate^{2-} (in), is an electrogenic step creating a membrane potential; therefore, dihydroxyacetone phosphate^{2-} is the preferred export compound. However, depolarization of this membrane potential by the influx of glycidate^{1-} into the chloroplasts would make possible the excretion of 3-phosphoglycerate^{3-} instead of dihydroxyacetone phosphate^{2-}.

PARTICIPANTS IN THE SYMPOSIUM

Akazawa, T.
  School of Agriculture
  Nagoya U.
  Chikusa, Nagoya 464, Japan
Andersen, K.
  Plant Growth Lab.
  Dept. of Agronomy & Range Sci.
  U. of California
  Davis, CA 95616
Andersen, W. R.
  Dept. of Botany & Range Sci.
  Brigham Young U.
  Provo, UT 84602
Anderson, C. W.
  Biology Dept., BNL
Anderson, L.
  Dept. of Biological Sci.
  U. of Illinois
  Chicago, IL 60680

Bahr, J. T.
  Mobil Chemical Co.
  Edison, NJ 08817
Ballantyne, G. P.
  Biology Dept., BNL
Bassham, J. A.
  Lawrence Berkeley Laboratory
  U. of California
  Berkeley, CA 94720
Berlyn, M.
  Dept. of Biochemistry
  Connecticut Agr. Exp. Sta.
  New Haven, CT 06504

Beyer, E., Jr.
  Central R&D Dept., Exp. Sta.
  E. I. duPont de Nemours & Co.
  Wilmington, DE 19898
Black, C. C., Jr.
  Dept. of Biochemistry
  U. of Georgia
  Athens, GA 30602
Bond, V. P.
  Associate Director, BNL
Bourque, D. P.
  Dept. of Biochemistry
  U. of Arizona
  Tucson, AZ 85721
Bränden, R.
  Dept. of Biochemistry
  Chalmers Inst. of Technology
  Fack S-402 20 Göteborg, Sweden
Brasil, O.
  Biochemistry Dept.
  U. of Georgia
  Atlanta, GA 30602
Bravdo, B-A.
  Faculty of Agriculture
  Hebrew U. of Jerusalem
  Rehovot, Israel
Brown, R. H.
  Dept. of Agronomy
  U. of Georgia
  Athens, GA 30602
Burr, B.
  Biology Dept., BNL
Burr, F. A.
  Biology Dept., BNL

Buzash-Pollert, E. A.
  Biology Dept., BNL

Canvin, D. T.
  Dept. of Biology
  Queen's U.
  Kingston, Ont. KL7 3N6, Canada

Chen, K.
  Dept. of Biology
  U. of California
  Los Angeles, CA 90024

Chia, C.
  Biology Dept., BNL

Chollet, R.
  Lab. of Agr. Biochemistry
  U. of Nebraska
  Lincoln, NE 68583

Chowdhry, V.
  Central R&D Dept., Exp. Sta.
  E. I. duPont de Nemours & Co.
  Wilmington, DE 19898

Chua, N-H.
  Rockefeller U.
  New York, NY 10021

Combatti, N.
  Biology Dept., BNL

Crockett, W.
  Biology Dept., BNL

Crowther, D.
  Biology Dept., BNL

Dunn, J.
  Biology Dept., BNL

Eisenberg, D. S.
  Molecular Biology Inst.
  U. of California
  Los Angeles, CA 90024

Elzina, M.
  Biology Dept., BNL

Ferguson, H.
  Biochemistry Dept.
  U. of Georgia
  Atlanta, GA 30602

Ganu, V.
  Biology Dept., BNL

Garrett, M. K.
  Agr. & Food Sci. Center
  Dept. of Agriculture
  Belfast, N. Ireland

Giaquinta, R.
  Central R&D Dept., Exp. Sta.
  E. I. duPont de Nemours & Co.
  Wilmington, DE 19898

Gibbons, G. C.
  Dept. of Biotechnology
  Carlsberg Lab.
  DK-2500 Copenhagen, Valby,
  Denmark

Gibbs, M.
  Inst. for Photobiology
    of Cells & Organelles
  Brandeis U.
  Waltham, MA 02154

Glover, M.
  Bigelow Lab.
  West Boothbay Harbor, ME 04575

Green, G.
  Biology Dept., BNL

Gross, G.
  Biology Dept., BNL

Gugliemelli, L. A.
  Biology Dept., BNL

Hall, N. P.
  Biochemistry Dept.
  Michigan State U.
  East Lansing, MI 48824

Hardy, R. W. F.
  Central R&D Dept., Exp. Sta.
  E. I. duPont de Nemours & Co.
  Wilmington, DE 19898

Harrison, P.
  Biochemistry Dept.
  U. of Georgia
  Atlanta, GA 30602

Hartley, R. W.
  Nat. Inst. Arthritis & Metab. Dis.
  NIH
  Bethesda, MD 20014

Hartman, F. C.
  Biology Div.
  Oak Ridge Nat. Lab.
  Oak Ridge, TN 37830

Hendricks, S. B.
   1118 Dale Drive
   Silver Spring, MD 20910
Hill, J.
   Biochemistry Dept.
   U. of Georgia
   Atlanta, GA 30602
Hill, R.
   Biology Dept.
   Fisk U.
   Nashville, TN 37203
Hillman, W.
   Biology Dept., BNL
Hind, G.
   Biology Dept., BNL
Hoch, G. E.
   Dept. of Biology
   U. of Rochester
   Rochester, NY 14627
Holder, A. A.
   Dept. of Physiology
   Carlsberg Lab.
   DK-2500 Copenhagen, Valby,
   Denmark
Hollaender, A.
   Associated Universities, Inc.
   Washington, DC 20036
Houston, W.
   Biology Dept.
   Atlanta U.
   Atlanta, GA 30314
Howell, S.
   Dept. of Biology
   U. of California at San Diego
   La Jolla, CA 92093
Hudis, J.
   Biology Dept.
   Suffolk Co. Community College
   Selden, NY 11784
Huffaker, R. C.
   Plant Growth Lab.
   Dept. of Agronomy & Range Sci.
   U. of California
   Davis, CA 95616
Huner, N. P. A.
   Chemistry & Biology Res. Inst.
   Agriculture Canada
   Ottawa, Ont. K1A 0C6, Canada

Jagendorf, A. T.
   Section of Genet., Devel. &
      Physiology
   Cornell U.
   Ithaca, NY 14853
Jensen, R. G.
   Dept. of Chemistry
   U. of Arizona
   Tucson, AZ 85721

Kenyon, W.
   Biochemistry Dept.
   U. of Georgia
   Atlanta, GA 30602
Kettner, C.
   Biology Dept., BNL
Klessig, D. F.
   Cold Spring Harbor Lab.
   Cold Spring Harbor, NY 11724
Koivuniemi, P. J.
   Biochemistry Dept.
   Michigan State U.
   East Lansing, MI 48824
Kolson, D. L.
   Dept. of Biology
   Pennsylvania State U.
   University Park, PA 16801
Kossiakoff, A.
   Biology Dept., BNL
Kringstad, R.
   Biochemistry Dept.
   U. of Georgia
   Atlanta, GA 30602
Ku, M. S. B.
   Horticulture Dept.
   U. of Wisconsin
   Madison, WI 53706
Kung, S. D.
   Dept. of Biological Sci.
   U. of Maryland Baltimore Co.
   Catonsville, MD 21228
Kudirka, D.
   Biology Dept., BNL
Kycia, J.
   Biology Dept., BNL

Lacks, S.
   Biology Dept., BNL

Lade, B.
  Biology Dept., BNL
Lamborg, M. R.
  Enhancement of Plant Productivity
  C. F. Kettering Lab.
  Yellow Springs, OH 45387
Lamm, S.
  Biology Dept., BNL
Lane, M. D.
  Dept. of Physiological Chemistry
  Johns Hopkins U.
  School of Medicine
  Baltimore, MD 21205
Lawyer, A. L.
  Dept. of Biochemistry
  Connecticut Agr. Exp. Sta.
  New Haven, CT 06504
Ledbetter, M.
  Biology Dept., BNL
Link, G. L.
  Biological Labs.
  Harvard U.
  Cambridge, MA 02138
Lorimer, G.
  Inst. für Toxikologie & Biochemie
  Gesellschaft für Strahlen-
    & Umweltforschung mbH
  8000 München 2, West Germany

Matlin, K. S.
  Rockefeller U.
  New York, NY 10021
Mawson, B. T.
  Dept. of Biology
  Queen's U.
  Kingston, Ont. K7L 3N6, Canada
McCashin, B. G.
  Dept. of Biology
  Queen's U.
  Kingston, Ont. K7L 3N6, Canada
McCurry, S. D.
  Dept. of Biochemistry
  Michigan State U.
  East Lansing, MI 48824
McFadden, B. A.
  Dept. of Chemistry
  Washington State U.
  Pullman, WA 99163

Mills, J.
  Biology Dept., BNL
Miziorko, H. M.
  Biochemistry Dept.
  Medical College of Wisconsin
  Milwaukee, WI 53233
Moradshahi, A.
  Biochemistry Dept.
  U. of Georgia
  Atlanta, GA 30602
Morris, I.
  Bigelow Lab.
  West Boothbay Harbor, ME 04575
Muckerman, C.
  Biology Dept.
  Suffolk Co. Community College
  Selden, NY 11784
Mulready, P. D.
  Biology Dept., BNL

Nauman, A.
  Biology Dept., BNL
Neuberger, M.
  Biology Dept., BNL
Noggle, G. R.
  9650 Rockville Pike
  Bethesda, MD 20014

Ogren, W. L.
  Dept. of Agronomy
  U. of Illinois
  Urbana, IL 61801
O'Leary, M. H.
  Dept. of Chemistry
  U. of Wisconsin
  Madison, WI 53706
Oliver, D. J.
  Dept. of Biochemistry
  Connecticut Agr. Exp. Sta.
  New Haven, CT 06504
Olson, J.
  Biology Dept., BNL
Owens, O.
  Competitive Grants Office
  U. S. Dept. of Agriculture
  Arlington, VA 22209

Paech, C.
  Dept. of Biochemistry
  Michigan State U.
  East Lansing, MI 48824
Pallas, J. E., Jr.
  Sci. & Ed. Administration, USDA
  Watkinsville, GA 30677
Peavey, D. G.
  Dept. of Biology
  Woods Hole Oceanographic Inst.
  Woods Hole, MA 02543
Pierce, J.
  Dept. of Biochemistry
  Michigan State U.
  East Lansing, MI 48824
Porter, C. A.
  Monsanto Co.
  St. Louis, MO 63166
Poulsen, C.
  Dept. of Physiology
  Carlsberg Lab.
  DK-2500 Copenhagen, Valby, Denmark
Price, C. A.
  Waksman Inst. of Microbiology
  Rutgers U.
  Piscataway, NJ 08854
Punnett, T.
  Biology Dept.
  Temple U.
  Philadelphia, PA 19122

Quebedeaux, B.
  Central R&D Dept. Exp. Sta.
  E. I. duPont de Nemours & Co.
  Wilmington, DE 19898

Rabson, R.
  Div. of Biomedical &
    Environmental Research
  Dept. of Energy
  Washington, DC 20545
Randall, D. D.
  Dept. of Biochemistry
  U. of Missouri
  Columbia, MO 65201
Reardon, E. M.
  Waksman Inst. of Microbiology
  Rutgers U.
  Piscataway, NJ 08854

Reeves, H. C.
  National Science Foundation
  Washington, DC 20550
Reger, B.
  Photosynthesis Research Unit
  U.S. Dept. of Agriculture
  Athens, GA 30605
Rejda, J. M.
  Lab. of Agr. Biochemistry
  U. of Nebraska
  Lincoln, NE 68583
Rhodes, P. R.
  Dept. of Biological Sci.
  U. of Maryland Baltimore Co.
  Catonsville, MD 21228
Rinehart, C. A.
  Dept. of Botany & Range Sci.
  Brigham Young U.
  Provo, UT 84602
Robinson, J. M.
  Inst. for Photobiology
  Brandeis U.
  Waltham, MA 02154
Rosenberg, A.
  Biology Dept., BNL
Rosenthal, A.
  Biology Dept., BNL
Roy, H.
  Dept. of Biology
  Rensselaer Polytechnic Inst.
  Troy, NY 12181
Rushizky, G.
  Nat. Inst. Arthritis & Metab. Dis.
  NIH
  Bethesda, MD 20014

Schloss, J. V.
  Biology Div.
  Oak Ridge Nat. Lab.
  Oak Ridge, TN 37830
Schmidt, G. W.
  Rockefeller U.
  New York, NY 10021
Senich, D.
  Div. Director for Integrated
    Basic Research
  National Science Foundation
  Washington, DC 20550

Shelp, B. J.
  Dept. of Biology
  Queen's U.
  Kingston, Ont. K7L 3N6, Canada
Siegelman, H. W.
  Biology Dept., BNL
Simon, M.
  Biology Dept., BNL
Simpson, E.
  Biology Dept.
  Washington U.
  St. Louis, MO 63130
Slovacek, R.
  Biology Dept., BNL
Smith, E. W.
  Medical Products Div.
  Union Carbide Corp.
  Tarrytown, NY 10591
Smith, H. H.
  Biology Dept., BNL
Smith, J. H.
  Corporate Research Dept.
  Imperial Chem. Industries Amer. Inc
  Wilmington, DE 19897
Smith, W. W.
  Molecular Biology Inst.
  U. of California
  Los Angeles, CA 90024
Spellman, M.
  Biochemistry Dept.
  Michigan State U.
  East Lansing, MI 48824
Stevenson, E.
  Biology Dept.
  Atlanta U.
  Atlanta, GA 30314
Stokes, B.
  Biology Dept.
  Atlanta U.
  Atlanta, GA 30314
Stoner, S.
  Biology Dept., BNL
Studier, F. W.
  Biology Dept., BNL
Suh, S. W.
  Dept. of Chemistry
  U. of California
  Los Angeles, CA 90024

Tingey, S.
  Dept. of Botany & Range Sci.
  Brigham Young U.
  Provo, UT 84602
Tobin, E.
  Biology Dept.
  U. of California
  Los Angeles, CA 90024
Tolbert, N. E.
  Dept. of Biochemistry
  Michigan State U.
  East Lansing, MI 48824

Valentine, R. C.
  Plant Growth Lab.
  Dept. of Agronomy & Range Sci.
  U. of California
  Davis, CA 95616
Vineyard, G. H.
  Director, BNL

Wagner, G.
  Biology Dept., BNL
Walbot, V.
  Dept. of Biology
  Washington U.
  St. Louis, MO 63130
Walker, D. A.
  Dept. of Botany
  U. of Sheffield
  Sheffield S10 2TN, England
Wildman, S. G.
  Dept. of Biology
  U. of California
  Los Angeles, CA 90024
Wildner, G. F.
  Dept. of Plant Biochemistry
  RUHR U.
  D-4630 Bochum 1, Germany
Wittenbach, V. A.
  Central R&D Dept., Exp. Sta.
  E. I. duPont de Nemours & Co.
  Wilmington, DE 19898
Woo, K. C.
  Dept. of Biology
  Queen's U.
  Kingston, Ont. K7L 3N6, Canada

Yadov, N.
  Plant Research Lab.
  Michigan State U.
  East Lansing, MI 48824
Yates, I.
  U. S. Dept. of Agriculture
  Athens, GA 30605

Zalik, S.
  Dept. of Plant Sci.
  U. of Alberta
  Edmonton, Alb. T6G 2E3, Canada

Zelitch, I.
  Dept. of Biochemistry
  Connecticut Agr. Exp. Sta.
  New Haven, CT 06504
Zimmer, E. A.
  2620 Nichols Street
  Spencerport, NY 14559

436

INDEX OF SPEAKERS

---

*Symposium paper.
†Abstract.

INDEX OF GENERA

SUBJECT INDEX

Actinomycin D, 376
ADP glucose pyrophosphorylase,
   54
Affinity labeling, 245-6
Agricultural Experiment Sta-
   tions, 411
Aldolase, 108
Alfalfa, see genus Medicago
Allosteric activator, 25
Allosteric site, 232
Amino acid analog, 161
Aminopeptidase, 142
Amphidiploidy, 13
Antibody, 281,285,326
Antibody-binding sites, 20
Arginine, 234
ATP/ADP ratio, 45,49-51,53-5

Bacterial plasmids, 349
Barley, see genus Hordeum
Beet, see genus Beta
Bicarbonate, 23,231
Bromoacetyldiethanolamine
   bisphosphate, 247
Bromacetylethanolamine
   phosphate, 246,259
Bromacetylphosphoserine, 247,
   265
3-Bromo-1,4-dihydroxy-2-butanone
   1,4-bisphosphate, 116,233,246,
   251
Bundle sheath cells, 354
2,3-Butanedione, 234,246,417

Carbamylation, 418
Carbon assimilation, 43,95

Carbonic anhydrase, 24
2-Carboxy-D-arabinitol 1,5-
   bisphosphate, 235
2-Carboxy-3-ketoribitol 1,5-
   bisphosphate (3-keto-CRBP),
   33,36-7
2-Carboxy-D-lyxitol 1,5-
   bisphosphate, 235
2-Carboxy-D-xylitol 1,5-
   bisphosphate, 235
Carboxylation, 22,37-9,228
Carboxypeptidase, 142
2-Carboxyribitol 1,5-
   bisphosphate (CRBP), 27,30-6,
   38,41,292
Castor oil plant, see genus
   Ricinus
Catalase, 220
p-Chloromercuribenzoic acid,
   232,245
Chloroplast, 83-4,86-7,90,95
   chromosome, 349
   DNA maps, 350
   genome, 307,364,373
   reconstituted, 52-3,84
   stroma, 48,88
Chymotrypsin, 420
Cloned DNA, 351
CO_2
   activation, 98,284
   assimilation, 43,95
   availability, 95
   compensation point, 135,
      422
   concentration, 86
   enrichment, 178,409

HOW WE KNOW ABOUT

THE ROMANS

JOHN AND LOUISE JAMES

PETER BEDRICK BOOKS
NEW YORK

Cop. 1

17.95

ACKNOWLEDGEMENTS

The publishers would like to thank John Orna-Ornstein for his advice and assistance in
the preparation of this book, and the organizations that have given their permission to
reproduce the following pictures:

AKG Photo, London: 8 (fresco).
Ancient Art & Architecture Collection: 10 (dogs), 24 (lamp), 25 (mosaic), 26 (boots).
Bath Archaeological Trust: 14 (mask).
Bayerisches Landesamt Für Denkmalpflege/K. Leidorf: 6 (fort).
Courtesy of the Trustees of the British Museum: 6 (diploma), 8 (figurine), 11 (slave tag),
12 (gaming pieces), 13 (strigils), 14 (bronze boar), 16 (plowman), 18 (mould), 19 (pendant), 23 (wheel), 25 (figure), 28 (medallion).
Colchester Museums: 10 (figures).
C.M.Dixon: 21 (relief).
Deutsches Bergbau-Museum, Bochum/Astrid Opel: 22 (miners).
English Heritage, Chesters Museum: 16 (measure).
Gloucester City Museum: 26 (gravestone).
Robert Harding Picture Library: 28-29 (Trajan's column).
Michael Holford: 9 (inkpot).
Landesmuseum, Trier: 16-17 (relief).
Musée de Bourges: 19 (gravestone).
Musée Fenaille-Rodez, collection Société des Lettres, Sciences et Arts de l'Aveyron) © Gilles Torjeman: 19 (tally).
Museum of London: 13 (strigils and jars), 20 (coin), 20 (vase).
Rheinisches Landesmuseum, Mainz: 26 (dagger in scabbard).
Scala Fotografico: 8 (bracelet), 8 (glassware), 9 (doll), 15 (altar), 24-25 (relief), 26 (sword), 29 (standard).
Service Photographique de la Reunion des Musées Nationaux: 14-15 (relief).
St Albans Museum: 15 (figurine).
Yale University Art Gallery/courtesy Oxford University Press: 6 (letter).

Written by: Louise James
Illustrated by: John James
Editor: Andrew Farrow
Design: John James and Alec Slatter
Art Director: Cathy Tincknell
Production Controller: Lorraine Stebbing
Consultant: John Orna-Ornstein

Published by
PETER BEDRICK BOOKS
2112 Broadway
New York, N.Y. 10023.

Library of Congress Cataloging-in-Publication Data
James, John. 1959-
How we know about the Romans / John and Louise James.
p. cm
Includes index.
Summary: An Illustrated survey of the history, culture, politics, warfare, and religion of the ancient Romans.
ISBN 0-87226-534-X
1. Rose--Civilization--Juvenile literature [1. Rose--Civilazation.] I. James, Louise. II Title.
DG77.J26 1997
937 .6--dc21
96-37885 CIP AC

Printed and bound in Italy
First Edition, 1997

CONTENTS

This scene of a busy market shows Romans and Celts going about their daily business.

SETTING THE SCENE

FRANCE AD 150

According to legend, the city of Rome was founded in 753 BC. Through the efforts of its people and the conquests of its army, it became a great empire that lasted for hundreds of years. The empire was vast, with frontiers as far north as Britain, west into Portugal, south into Africa and east to parts of Asia. There the Romans took their way of life, language and their culture. They also adapted and adopted the cultures and religions of the peoples they now ruled, and even made some of them Roman citizens.

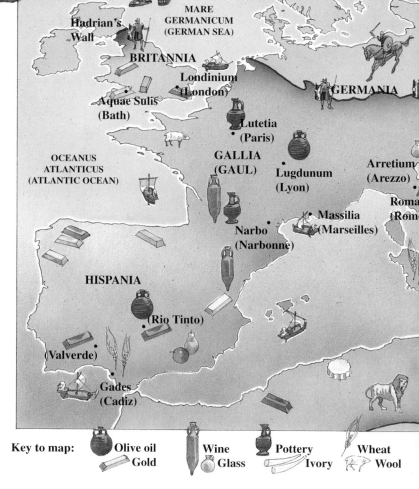

MARE GERMANICUM (GERMAN SEA)

Hadrian's Wall

BRITANNIA

Londinium (London)

Aquae Sulis (Bath)

GERMANIA

Lutetia (Paris)

OCEANUS ATLANTICUS (ATLANTIC OCEAN)

GALLIA (GAUL)

Lugdunum (Lyon)

Arretium (Arezzo)

Roma (Rome)

Massilia (Marseilles)

Narbo (Narbonne)

HISPANIA

(Rio Tinto)

(Valverde)

Gades (Cadiz)

Key to map: Olive oil | Wine | Pottery | Wheat
Gold | Glass | Ivory | Wool

About 850 BC Greece emerges from Dark Ages   About 753 BC Foundation of Rome   332 BC Egypt falls to Alexander the Great   140 BC Greece falls to Roman armies

EGYPT

GREECE

ROMANS

1000 BC   500 BC   0

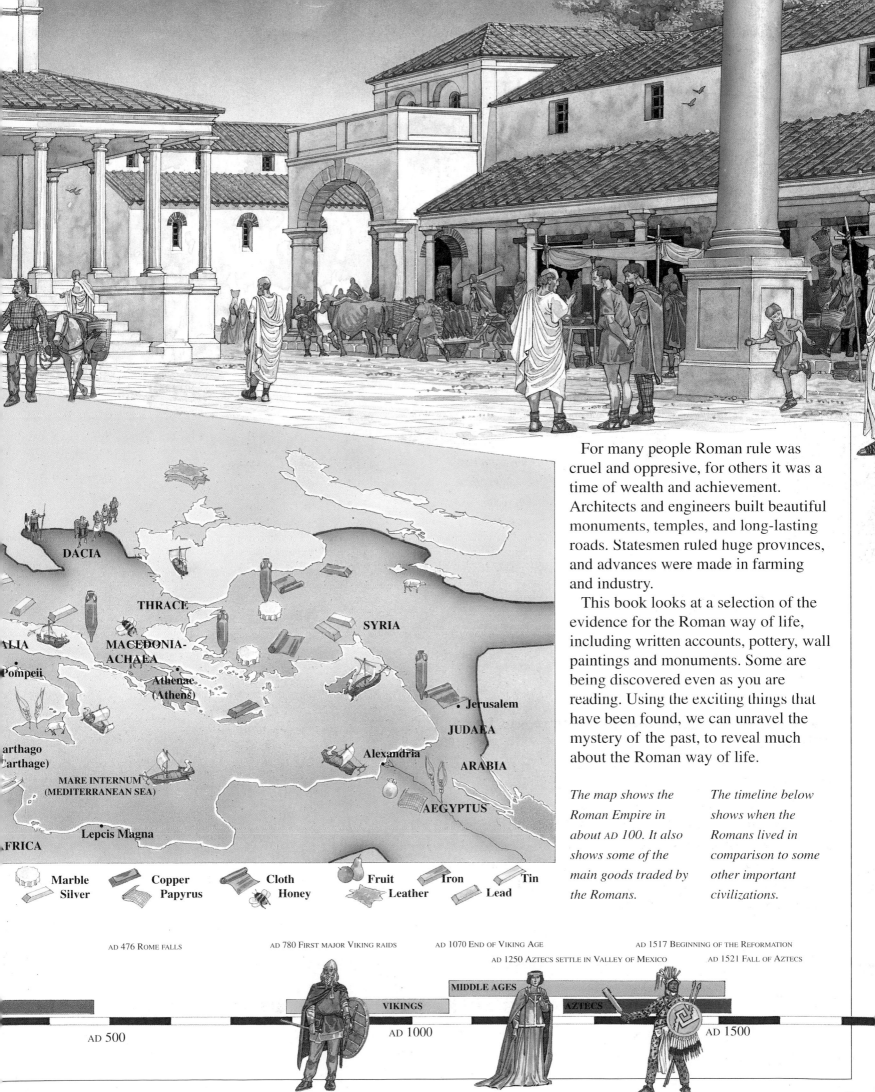

For many people Roman rule was cruel and oppresive, for others it was a time of wealth and achievement. Architects and engineers built beautiful monuments, temples, and long-lasting roads. Statesmen ruled huge provinces, and advances were made in farming and industry.

This book looks at a selection of the evidence for the Roman way of life, including written accounts, pottery, wall paintings and monuments. Some are being discovered even as you are reading. Using the exciting things that have been found, we can unravel the mystery of the past, to reveal much about the Roman way of life.

The map shows the Roman Empire in about AD 100. It also shows some of the main goods traded by the Romans.

The timeline below shows when the Romans lived in comparison to some other important civilizations.

DACIA

THRACE

SYRIA

MACEDONIA-ACHAEA

...ALIA

Pompeii

Athenae (Athens)

Jerusalem

JUDAEA

...arthago (Carthage)

Alexandria

ARABIA

MARE INTERNUM (MEDITERRANEAN SEA)

AEGYPTUS

Lepcis Magna

...FRICA

Marble
Silver

Copper
Papyrus

Cloth
Honey

Fruit
Leather

Iron

Tin

Lead

AD 476 ROME FALLS

AD 780 FIRST MAJOR VIKING RAIDS

AD 1070 END OF VIKING AGE

AD 1517 BEGINNING OF THE REFORMATION

AD 1250 AZTECS SETTLE IN VALLEY OF MEXICO

AD 1521 FALL OF AZTECS

MIDDLE AGES

VIKINGS

AZTECS

AD 500

AD 1000

AD 1500

HISTORY IN EVIDENCE

How do we know about the Romans? We have learned much about their way of life because archaeologists and historians study the things that have survived from past times. They are a lot like detectives, looking for evidence and working out how and why things happened.

Some information comes from books written by Roman citizens. The politician and general Julius Caesar wrote about the wars he waged. Others described the common events of daily life. For example, the writer Juvenal described a street in the city of Rome in the second century AD:

'However fast we hurry there's a huge crowd ahead and a mob behind pushing and shoving... You get dug in the ribs by someone's elbow... The street are filthy – our legs are plastered with mud, someone tramples your feet, or a soldier's hob-nailed boot lands right on your toe.'

This photograph shows the remains of a Roman fort. Archaeologists use aerial photographs of the area they will be working on. From the air it is often possible to see the outlines of buildings demolished long ago.

Archaeologists use a variety of tools in their work, from digging tools and trowels, like those shown on the left, to pencils, paper and lap-top computers for recording finds. There are hundreds of Roman sites that have not yet been studied.

Written evidence such as diaries and military records help us to understand the lives of the Romans. Above is part of a letter found in North Africa, telling of a reception for an important visitor. Left is a bronze diploma giving a soldier an honorable discharge from the Roman army.

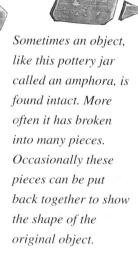

Sometimes an object, like this pottery jar called an amphora, is found intact. More often it has broken into many pieces. Occasionally these pieces can be put back together to show the shape of the original object.

Archaeologists usually study the places where people lived and worked. Before 'digging' begins, accurate details of a site are taken, including photographs at ground level and from the air. Sometimes, special electronic surveys are also carried out, as they can detect things hidden underground, such as a ditch or stone wall. As the site is 'dug', everything that is found is recorded and labeled. Each item can give clues to the uses of rooms and buildings, and about who lived there.

The soil taken from an excavation is checked in case any small items have been missed. Microscopic examination of the soil can also show what type of plants grew in it, what animals there were and even the food that people ate.

The style of buildings has changed as new civilizations have built on top of earlier sites, especially towns and cities. Here we can see a range of buildings from an ancient Bronze Age hut to a modern block of offices.

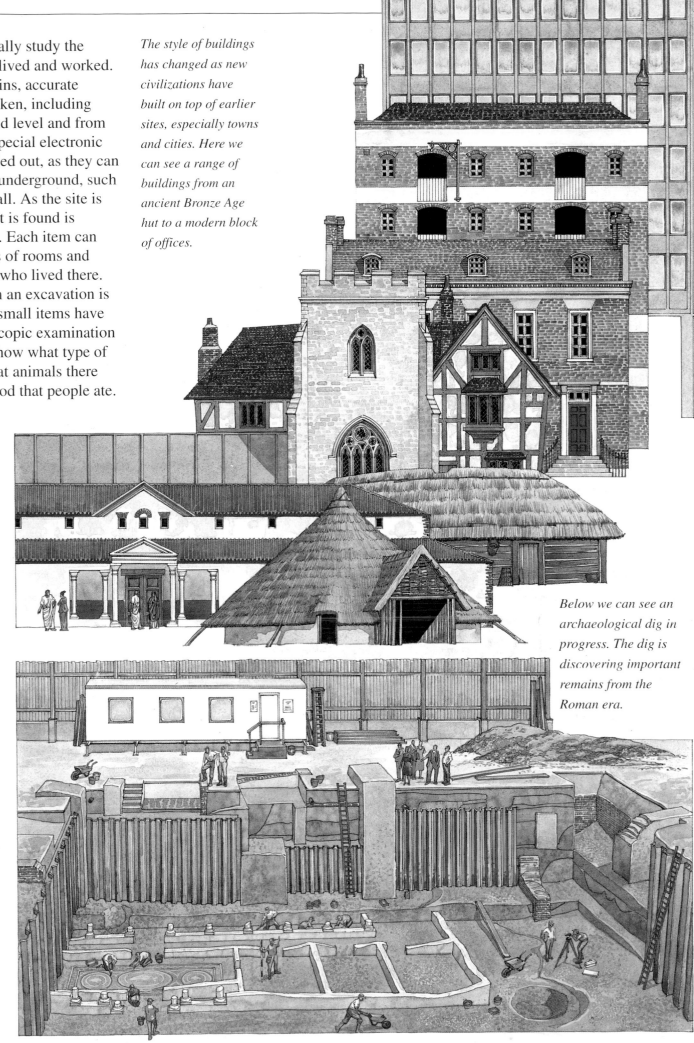

Type I, until AD 90

Type II, AD 40–100

Type III, AD 70–200

Type IV, AD 150–400

Typology is the study of how the design of an object changes over time. This builds up a sequence that can be used to date other artifacts. Above is a sequence of oil lamps and their main periods of use.

Below we can see an archaeological dig in progress. The dig is discovering important remains from the Roman era.

ROMANS AT HOME

In the Roman Empire, the sort of home people lived in depended on whether they were rich or poor. In a Roman city or town, poorer people lived in apartment buildings called *insulae* (islands), which were four or five stories high. The apartments were always crowded and often dirty. Two whole families, maybe more, might be jammed into a single room.

Rich people and the patricians, the Roman nobles, had magnificent homes. They had their own bathrooms, piped water and many luxuries. Some people lived in town houses. Others, who lived out in the country, had villas, like the one shown on the right. The main entrance led into a reception room called an *atrium*. The *atrium* often had beautifully carved columns and a pool called an *impluvium* in the middle.

Family life centered around the *atrium*. Doors off it might lead into a dining room, the kitchen, bedrooms and maybe a library. Families kept a small shrine with a statuette of their *lar familiaris*, their guardian spirit (below right). Every day, the family said prayers at the shrine.

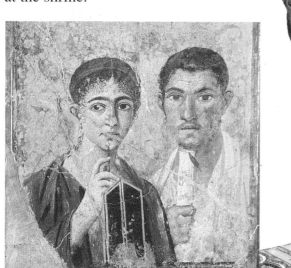

The writer Juvenal said that the ideal Roman mother (shown in this fresco from Pompeii) had to be hardworking, loyal to her husband and family, and make sure the home was a comfortable and happy place to be.

The houses of wealthy Romans were full of treasures such as the lustrous glassware (right), brightly-colored pottery, or jewelry made from gold or silver, like the serpent bracelet on the left. Jewelry was often decorated with portraits of people. Ordinary people, the plebeians, could not afford things like these.

The marble floors of villas were covered in beautiful mosaics (above) made from tessarae, tiny pieces of stone or colored glass (see page 21).

The scene on the left shows some of the people in a wealthy family. The father, seated, was the *paterfamilias*, Latin for 'father of the family'. Children had to obey their father without question. The mother was also important: she was in charge of the smooth running of the household. Just as the father trained the boys, the mother trained the girls. Each delegated, or gave, work to servants.

Some of the people working in Roman families did not want to be there. These were the men, women and children who had been sold as slaves. In the middle of the first century AD, there were millions of slaves in Italy. Slaves did all kinds of work – as servants in private homes, laborers on farms, or oarsmen in Roman ships.

In colder areas, the walls of some villas were made hollow so that heat from the hypocaust (below), the heating system beneath the house, could warm the rooms. A fire was lit in the basement, and the heat traveled up through the walls.

Children played with toys that are still popular generations later. This wooden doll had movable arms and legs, with metal pins to attach them to the body. Female dolls had their own sets of clothes.

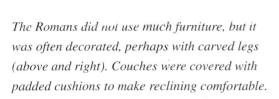

The Romans did not use much furniture, but it was often decorated, perhaps with carved legs (above and right). Couches were covered with padded cushions to make reclining comfortable.

One of the family's most trusted servants kept the household records, using an inkpot and pen like these.

IN THE GARDEN

Near the center of each villa was a garden. The garden was an important place where the family could relax. Sometimes it was also used as a summer dining room. In the picture on the right we can see an informal meal being enjoyed by the family and their guests. This is probably the main meal of the day, which might start in the afternoon and last late into the evening.

Roman diners did not sit on chairs, but reclined on couches. We also know, from written sources and archaeological evidence, what types of food and drink were popular. The empire was vast, so there was a wide choice of food and fine wines for the few who could afford them. However, most people ate simple dishes with strong tasting sauces like *garum*, a fish sauce. A Roman meal was not hurried — people took time to enjoy many courses. Meal time was meant to be a social event, and politics, business and art were often discussed.

Below is a large bronze jug. The base of the handle is in the shape of a man's head, with flowing hair and long beard.

These funny terracotta (clay) people were found in a child's grave. One of them is telling a story to the other diners.

Wealthy diners used things made by the finest craftsmen, like the terracotta dish and plate (top), the bronze spoon (above) and the silver dish and plate (left).

The Romans kept dogs as pets and as guard dogs. This marble statue (above) comes from Italy.

Below is a slave tag, which was lost until discovered by archaeologists. It reads 'Hold me lest I flee, and return me to my master Viventius on the estate of Callistus'.

These keys, found in a Roman garden, were for locking away the family's valuables.

The dig (left) shows the discovery of an outdoor dining area in France. Only in warm climates did people dine outside.

Diners ate and drank while laying on couches arranged around three sides of a table. A good host might entertain his guests with dancing and music played on instruments like these bronze cymbals, flute and bell (left).

The garden was a beautiful place, with flowers, trees, a fountain and ornate walkways. Sometimes walls were decorated with paintings called frescos, which were painted directly on to the plaster. Many show scenes of the gods or animals, or beautiful patterns. The Romans also painted their statues, although the paint has now worn away.

The children of the household are not present at the meal. Perhaps they are at a small private school or are being taught at home, by a slave who had been well educated himself. Not all children had the chance to be educated. Many children who were poor, or even slaves, had to work long hours.

AT THE BATHS

The Romans were very clean. Many of them bathed daily at the public baths, the *thermae*, which were not very expensive to use. But the baths were not just for getting clean, they were like a club where people could get together to relax and enjoy themselves. They swapped news and jokes, discussed business matters, placed bets on wrestling matches or chariot races, played games, or exercised hard in a gymnasium. The writer Lucius Seneca described some of the people at one bath-house:

'First there are the "strongmen" doing their exercises... with grunts and groans. Next there are lazy ones having a massage... And what about the ones who leap into the pool making a huge splash as they hit the water!'

The baths were a place for socializing. People met there to relax, talk and enjoy themselves. These counters and game pieces (above, left) would have been used in a variety of games. Gambling and betting, although not officially legal, were common.

Busy towns had two or three public baths, some small, some large. Most baths had a system to heat the water, floors and air. Hot air from a furnace was channeled under the floor and up through spaces in the walls.

Men and boys usually bathed at different times than women and girls. Soap like we use today had not been invented. People cleaned their bodies by pouring perfumed oil or a kind of soap made from tallow (animal fats) onto their skin. They would perhaps exercise then relax in a warm pool called a *tepidarium*. Later they would move to the hot and steamy *caldarium*, where the heat opened the pores of the skin, and then scrape off the oil and dirt using a curved tool called a *strigil*. After a swim in a hot pool, the bather would plunge in a cold pool (the *frigidarium*) to close the skin's pores.

Below right are an oil flask and strigiles. *Baths had many slaves to scrape the customers clean and serve food and drink.*

The remains of Roman baths can be found in places like Bath in southern England. Bath was known as Aquae Sulis in Roman times. The ruined baths were discovered and restored in the 19th century (left).

Left, this green glass flask probably held a perfumed oil which was rubbed into the body after bathing. Below is a bronze toilet set, with a nail cleaner and tweezers. Most bathers had their own set.

Below are an ear scoop, tweezers and a nail cleaner. In the scene a bather is using a scoop to clean out his ear.

The bronze dish (left) was used for pouring water over a bather. Sometimes the floor beneath bathers' feet was so hot they had to wear wooden clogs, called pattens (above).

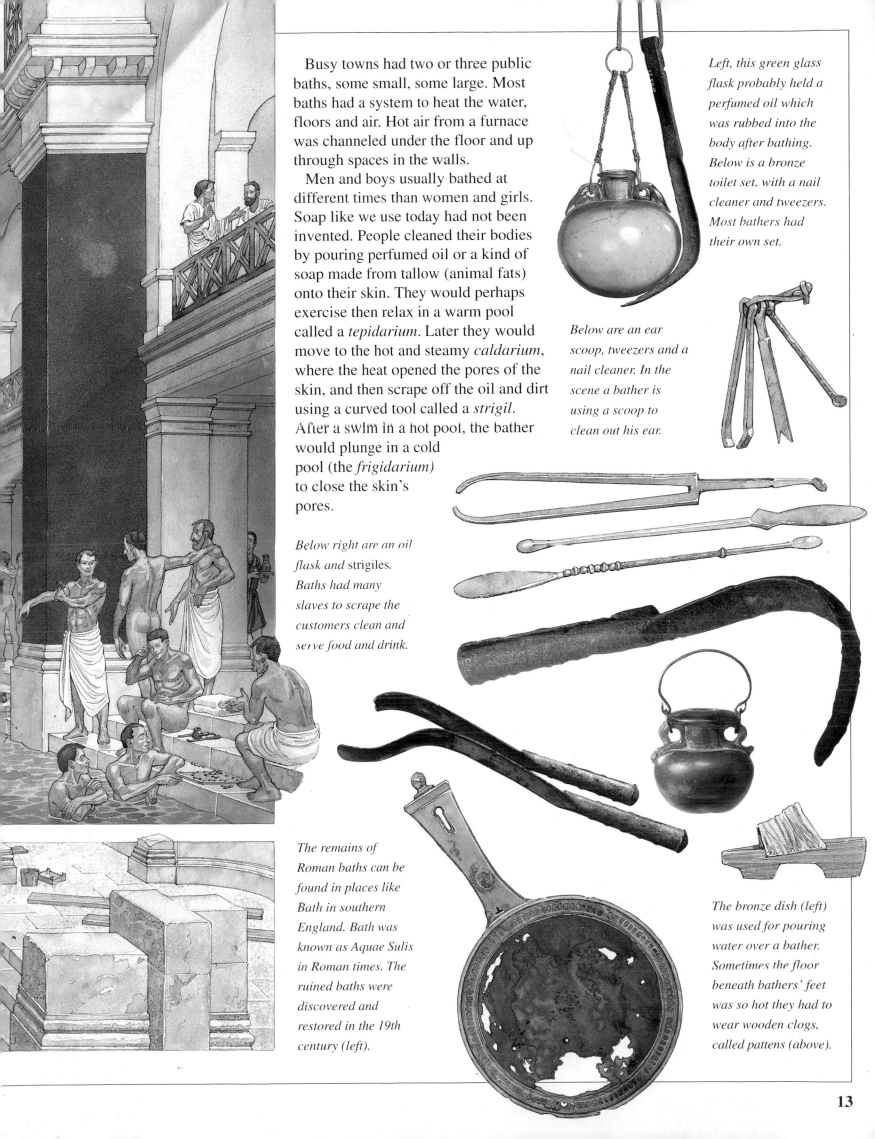

13

SACRIFICE AT THE TEMPLE

ITALY AD 125

The Romans worshiped many gods. They thought that each god influenced certain aspects of the world, and each would help them in specific areas of their lives. For example, people who were sick or infirm prayed to Asclepios, the god of healing.

Everyone had to worship the official Roman gods. Most official gods were the same as Greek gods, but known by different names. The Greek god of war, Ares, was known as Mars, Aphrodite became Venus, and the father of the gods, Zeus, became Jupiter. However, as their empire expanded, the Romans often adopted the beliefs and honored the gods of the people they conquered, such as the Egyptian god Isis and the Persian god Mithras.

The Romans built magnificent temples for their gods. Many were dedicated to just one god, though some large temples were for many, such as the Pantheon in Rome. Sometimes a special day was dedicated to a god. This day was treated as a holiday and there would be a festival.

The larger-than-life tin mask below was found in a spring. It was probably used in a sacrificial ritual.

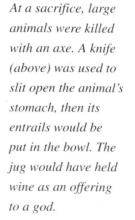

At a sacrifice, large animals were killed with an axe. A knife (above) was used to slit open the animal's stomach, then its entrails would be put in the bowl. The jug would have held wine as an offering to a god.

This bronze model shows a man leading a boar to be sacrificed. The animal will be cut open so its entrails can be burned and examined for omens.

*The Persian god
Mithras (top) became
popular among
slaves and soldiers.
Priests of Mithras
sacrificed bulls in
secret ceremonies.
The goddess Venus,
above, was associated
with gardens, beauty,
love and fertility.*

*On the left is part of a
marble relief from
Rome showing a
sacrifice to the god
Mars. Altars were
often decorated, such
as the one in Rome's
main forum, above.*

In the scene above, priests are
sacrificing live animals on an altar
outside a temple. People believed that
sacrificing animals pleased the gods.

In the first century AD, Christianity
became popular. Christians worship
only one god and follow the teachings
of Jesus Christ. This brought them into
conflict with the emperors, who thought
this would anger Rome's official gods
and bring ill-fortune to the empire. The
Christians were forced to worship in
secret and many were persecuted. For
example, the emperor Nero outlawed
them and treated them as criminals.
Some were paraded in the amphitheatre
and forced to battle with wild animals.

FIRE AT THE BAKERY

BRITAIN AD 155

One of the most important foods was bread. Freshly baked bread was part of most Romans' daily diet, so in every town there was at least one bakery. Wheat was ground there by huge stones called querns. Then the bread was made and baked in ovens, before being sold at the shop counter.

Wheat for bread making was grown in many parts of the empire. Every year, the emperor had to import shiploads of wheat from Egypt to feed the poorer people of Rome. If the grain ships did not arrive in time, the people might riot. The army, too, needed wheat, to feed its soldiers, and so it controlled many farms. Other large farms were owned by wealthy nobles who had hundreds of slaves to prepare the soil, plant seeds and harvest the crops.

A shortage of water could be a big problem for farmers. Therefore the Roman army built waterways, called aqueducts, to take water to the places it was needed, and the farmers were able to irrigate (water) the land.

As the bronze model shows, the design of plows used by the Romans remained virtually unchanged until replaced by modern machinery. The sickle shown on the right would have been used to cut down the fully grown wheat.

On the right is the top part of a quern, which was used to grind wheat into flour. Every farm and estate would have had one.

Above, this bronze model shows oxen pulling a plow. The farm worker, who is wearing a hooded cloak, is guiding the cattle and plow.

Sometimes farmers had to give part of their harvest to the army, especially when the legions were fighting a long campaign. On the left is a first century bronze corn measure used to collect this corn levy (tax), called the annona.

The German relief above shows bakers preparing and mixing their ingredients and baking bread. Most bakeries had a mill room. The grain was poured into the top of the quern, then the upper stone was turned to grind the grain into flour.

The scene above shows a bakery that has caught fire. A fire in a large timber-framed building could spread quickly and burn down the whole town. Here the townsmen are filling their buckets from a water trough. Fires were such a problem in Rome that in AD 6 the emperor Augustus created a group of fire-fighters called the *vigiles*. The *vigiles* were freed slaves who won the right to citizenship after six years' service.

This loaf of bread was found at Pompeii in Italy, buried by ash when the volcano Vesuvius erupted.

Tradesmen used a tool called a steelyard (above) to weigh corn and wheat. The item was hung on one of the hooks, and the weight on the left was moved along the bar until it balanced the load. The tray on the left was used for baking small buns.

Romans paid for goods with coins. Each had the head of the emperor on one side. It could take many years for new coins to circulate to the distant provinces.

THE POTTER'S WORKSHOP

FRANCE AD 175

Many things that we use today are made from glass, metal or plastic. In Roman times they were made from pottery. For storing food, large jars with narrow necks kept food or liquid cool and fresh. People ate from pottery plates, and lit their homes using oil lamps made from clay.

A master potter could produce a range of items, from luxury goods to every-day cooking and drinking vessels. He would have learned about potting by serving an apprenticeship with an experienced craftsman.

Sometimes a pottery was a small family business. Other potteries were large workshops that could produce thousands of items in a week. In a large workshop, the master potter's assistants were slaves who worked hard, in dirty conditions and for long hours.

The mold above was used for making terracotta statuettes of a female figure. It is about 12 in. high.

Right is a model of a mime artist holding a bag of money. Actors are usually shown wearing masks.

For about 2,000 years this flask in the shape of a rabbit has survived unbroken. It once contained soothing ointment.

The excavation above shows the remains of a kiln found at Narbonne in southern France. Roman potteries were some of the world's first factories.

This terracotta ink pot (above left) has three small holes in the top, so it can be hung up. Below it are some marbles for children to play with.

18

This small pendant, made from a black stone called jet, shows two cupids making a pot.

Above is a gravestone from France. The hanging pots show that the man was either a potter in his shop or a merchant selling pottery on a market stall.

There were many forms of pottery in the Empire. Here are a third century Casterware beaker (left) and a Samian bowl and jug (below).

The clay used for potting varied from region to region, and this meant that many types of pottery, or ware, were produced. At Arretium (Arezzo) in Italy and Lugdunum (Lyon) in France, a shiny red pottery commonly known as Samian ware was made. The pots, which often had raised patterns on them, were usually formed in a mold. The mold was stamped with designs carved from wooden blocks. Samian ware was very popular and was exported throughout the empire.

The pots were fired (heated) in an oven called a kiln. Most pots were not decorated, but some were covered with colored glazes made with powdered glass and fired again.

Each potter put their trademark on their work (above left). Most marks were made by pressing a carved wooden block into the wet clay.

Left is a potter's tally (list) from France. The plate lists the names of potters, type of pottery and number of items that were fired – an incredible 27,945!

19

THE SINKING SHIP

BRITAIN AD 210

In AD **43 the Romans invaded England. Soon the town of Londinium (London) became an important and busy port, visited by ships and merchants from many parts of the empire.**

The town's river banks were lined with quays and small warehouses for storing goods. Ships took the goods around the coast to other ports, and also from country to country. Their cargoes included building materials, tin, wheat, pottery, wine and olive oil. One quay-side site was used for bottling fish, perhaps for making the popular fish sauce called *garum* (see page 10).

In the River Thames at London, the wreck of the Roman trading vessel shown in the scene has been found by archaeologists (above). The timbers had been burrowed into by shipworms, so we know that the boat often sailed at sea.

Above, this coin was found in the socket of the ship's mast. It features the goddess Fortuna, and would have been put there as a good luck token.

The remains of several ships and their cargo have been found in the River Thames, including this Samian vase.

These iron nails were bent to hold the ship's overlapping planks together. The cone heads were hollow, which is very unusual.

As trade increased, the River Thames became crowded with vessels, all trying to get into the safe channels. But sailing was never easy, even in the calm conditions of the river. In the scene on the left we can see how, in a mist, a trading ship has been rammed by another boat. The ship is loaded with a heavy cargo of building stone, which has moved in the collision, and it is capsizing and sinking rapidly. The sailors, some of whom are slaves, are desperately trying to swim away before they are sucked under.

Archaeologists have found the remains of this ship in waterlogged silt, which preserved the wood, European oak, from rotting. The cargo included an unfinished millstone from Yorkshire, over 300 miles to the north.

Using a method of studying wood called dendrochronology, (see page 30) the ship has been dated to AD 140. Microscopic study shows that a variety of both fresh and sea water bacteria had lived in the timbers, so we know that the ship traveled widely, perhaps as far as the Mediterranean Sea.

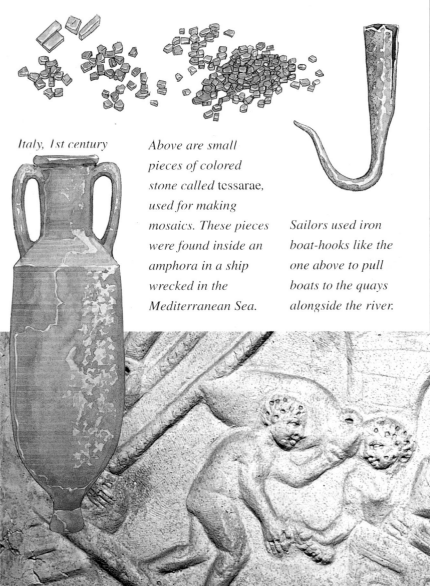

Italy, 1st century

Above are small pieces of colored stone called tessarae, *used for making mosaics. These pieces were found inside an amphora in a ship wrecked in the Mediterranean Sea.*

Sailors used iron boat-hooks like the one above to pull boats to the quays alongside the river.

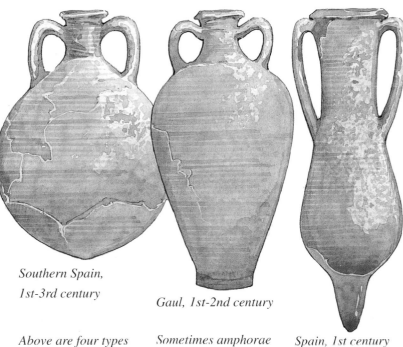

Southern Spain, 1st-3rd century

Gaul, 1st-2nd century

Spain, 1st century

Above are four types of amphorae. They are of a similar style, but with variations in size and type of handle. The jars were used to store or carry goods such as wine or olive oil.

Sometimes amphorae are found with traces of their original contents. Scientists can work out where the goods came from and therefore how widespread and common a trade was.

This third century relief (right) shows clerks recording the number of amphorae being unloaded from a ship in the Italian port of Ostia.

Down the Mine

SPAIN AD 250

The emperors paid for the Roman army, which conquered new lands and defended the frontiers. They ordered lavish games to keep the people entertained. The government built thousands of magnificent buildings, temples and baths. So where did they find the money to do all of this?

Some money came from conquests and taxes. The rest came from businesses owned by the government, such as the trade in grain and metals. Mining was important throughout the empire. In Greece, lead and silver were often found in the same mines – the lead was used for making water pipes, the silver for coins and jewelry. Other important metals and minerals were iron, copper, tin, gold and marble.

Below we can see archaeologists discovering the remains of a wooden waterwheel in a Spanish copper mine. Some waterwheels were made like model kits, ready to be assembled at the mines. In Spain, the skeletons of 18 miners killed and buried by a rock fall have also been found.

This stone carving shows Spanish miners going to work. They are wearing leather aprons and carrying hammers, picks and oil lamps.

22

Iron ankle restraints were used to stop slaves escaping. This one (right) still holds the remains of a slave's foot!

Slaves had to power wooden waterwheels like this one found in a copper mine at Rio Tinto in Spain. The wheels scooped up water and lifted it out of the galleries.

Above are an iron hammer and pick, and two buckets, one made from wood and woven grass, the other from bronze.

This force pump was found in a mine at Valverde in Spain. Like all machinery in the mines, it was powered by slaves. It was probably used to pump fresh water to the workers.

This small bronze figurine of a young slave was found in England. She has been tied by a rope at the neck, wrists and ankles, making it impossible for her to escape.

The mines were planned carefully by engineers who surveyed the landscape and designed ways of getting the valuable materials out of the ground. They also used complex machinery to remove water, to prevent flooding. In the scene on the left an engineer is surveying a new tunnel using an instrument called a *groma*. Below him, slaves are driving waterwheels. The wheels are lifting water into troughs and up to the surface, so the mine doesn't flood.

Most of the hard work was done by slaves, who had to toil in cold, dirty and dangerous conditions. In the 1st century BC, the historian Diodorus Siculus described slaves in a mine:

'*... as a result of their underground excavations day and night they become physical wrecks, and because of their extremely bad conditions the mortality [death] rate is high.... Some...prefer dying to surviving.*'

AT THE RACES

LIBYA AD 118

All Romans enjoyed their public holidays. They flocked to the theatre to cheer or jeer at the latest comedy. At the huge amphitheatres they watched bloodthirsty gladiators fighting each other to the death. But perhaps the most popular entertainment was chariot racing at the circus. Spectators betted on the outcome and shouted wildly as their favorite team careered round the course.

The huge racing track was long and narrow. Along its center, the *spina*, or spine, were statues of gods, seating for the marshals or judges, and the lap counters. At the ends were huge columns which marked the turning point. Remains of the circus at Lepcis Magna in Libya show that three cones over 16 feet tall (far right) stood at each end of the *spina*.

This Roman lamp, made in Italy about AD 30-70, shows a victorious charioteer holding a palm leaf.

Our evidence of chariots includes this chariot pole end (far right) and a bronze model of a two-horse chariot (above).

The relief above shows chariots racing at the Circus Maximus in Rome. Nothing survives of the Circus Maximus, but there are many remains at Lepcis Magna in Libya. The scene at the top of the page shows the starting gates at Lepcis Magna, which were about 10 feet wide. There is some evidence that the gates could be opened simultaneously by pulling a lever.

Chariots burst from the starting gates at Lepcis Magna. One team has broken through the wooden gates and is sweeping its rider out of control. The teams had to race around seven laps of the circus (below), over 4 miles, before finishing in the straight.

The statesman and writer Cassiodorus told how 'the gates are suddenly opened all at the same time' with a loud snap that could be heard by all the spectators.

The chariots had two wheels and were very light. They were usually pulled by four horses, but sometimes by just two. Like many modern sports teams, some chariot-racing teams had their own 'supporters' clubs' and a team color.

The poet Ovid wrote how each charioteer, a slave, stood alone behind the starting gate, 'the horses panting impatiently against the bars of the gate'. The charioteer needed a great deal of skill to control his horses and to balance the small chariot while moving at high speed. There were often terrible crashes, especially at the turns: many charioteers were killed. However, a few successful racers became famous and earned enough money to buy their freedom.

The picture (left) is based on a mosaic (above left) and an ivory figure (above). A palm branch was given to the winning charioteer as a symbol of his victory.

A Roman Milecastle

By the 1st century AD, the Roman Empire had expanded and many new provinces had been conquered. This success came because the army was well organized, trained, equipped and disciplined. At the height of its power, the army had 29 legions, each of about 6,000 soldiers and craftsmen such as smiths and tent-makers. These were supported by units of foreign soldiers called auxiliaries.

The cavalry was a small but important part of the Roman army. Above and left are a bit (a mouth-piece for controlling a horse), a horseshoe, and a cavalryman's parade helmet.

Left is the tombstone of an auxiliary cavalryman from Thrace, Greece. It shows a cavalryman riding down a barbarian enemy.

Foot soldiers (right) were armed with a javelin (called a pilum*) for throwing at the enemy. They then fought with a stabbing sword like the one on the left. Until the second century AD, soldiers were also issued with daggers (above, shown with its scabbard). Cavalry (above right) carried large spears and a longer sword.*

Boots (above) were very important to a soldier. The boots had heavy hobnails to make the soles hard-wearing.

The legionaries were citizens of Rome. They enlisted in the army for 25 years, after which they retired and received a pension or perhaps some land and a small villa. Some auxiliaries had been captured and forced to be in the army; others were volunteers. They looked forward to the day when they could leave the army, becoming a Roman citizen.

When they were away from their main bases, the legions built camps for the night. First they would dig a large ditch around the site, piling the earth to form a bank. Next the soldiers put up wooden stakes to make a protective fence, called a palisade, before pitching their tents for the night.

To mark the boundaries of the empire, the Romans built defensive walls, such as Hadrian's Wall in northern England. In the scene on the left we can see soldiers at a guard post on the wall. Mostly the soldiers did boring guard duties, such as watching traders cross the frontier. Occasionally there might be a raid across the border, perhaps to seize cattle. On the left we can see the milecastle under attack from Celtic warriors charging across the snow.

SIEGE!

ROMANIA AD 101

Every aspect of warfare was planned carefully by the army. In open battle, the Roman soldiers nearly always defeated their opponents. Armies of Celts, like the Germans and Dacians, fought as unruly mobs, yelling war cries as they charged the tight Roman formations. The Romans waited calmly before throwing their deadly javelins, then fought shoulder-to-shoulder, shield-to-shield, cutting down the enemy with their short stabbing swords.

Sometimes the Romans would have to lay siege to an enemy stronghold. The legionaries dug ditches and erected walls to keep the enemy contained, and to protect their own camps. Skilled craftsmen and ingenious engineers then built siege machines. There were battering rams for breaking down gates, and tall towers with a drawbridge on top so the soldiers could run on to the top of the walls. Catapults and ballistas threw stones or arrows with terrifying accuracy and deadly results.

The army looked after its wounded soldiers. Medical instruments included a surgical knife, a blunt probe, glass dropper, tweezers, hook, spatula, scalpel and forceps.

This gold medallion dating from AD 296 was found in France. It shows a galley ship and the freeing of London from rebels.

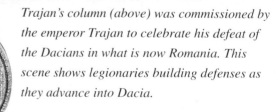

Trajan's column (above) was commissioned by the emperor Trajan to celebrate his defeat of the Dacians in what is now Romania. This scene shows legionaries building defenses as they advance into Dacia.

28

Four-pointed caltrops and iron spikes (below) were spread on the ground to stop cavalry attacks. They could inflict terrible injuries on horses. The metal spike was buried with only the barb sticking up.

These pieces from a ballista (left) were found in Romania. Many of Rome's enemies did not have weapons like the ballista and catapult shown in the above scene.

Siege engines fired a variety of missiles, such as these catapult stones (below), rocks and large spears.

Each unit of the army had a standard (below right), which was taken out of the main camp only when the army marched into battle. The loss in battle of its eagle standard, called an aquila, was a great humiliation for a Roman legion.

The signal to attack, or retreat, would be given on war-horns like the one below.

Even though a siege could last for months, the Romans nearly always broke through. As the defenses crumbled, the foot soldiers attacked. To protect themselves from volleys of stones, spears and arrows, most of the soldiers held their shields above their heads, overlapped to form a protective roof. The men at the sides held their shields to form walls. Advancing slowly, this 'tortoise' formation was hard to break up. The writer Sallust said of Marius, an army commander:

'He went outside the mantlets [protective shields], formed the tortoise-shield, and advanced to the wall.'

Once inside the defenses, the Romans ruthlessly hunted down their foes.

GLOSSARY

This list explains the meaning of some of the words
and terms used in the book.

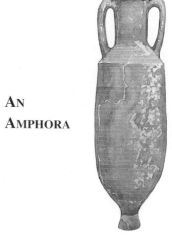

AN
AMPHORA

ATRIUM	The main reception area of a Roman home.
AMPHITHEATRE	A circular or oval building which was used for spectator sports such as animal and gladiator fights.
AMPHORA	A large two-handled pottery vessel used for storing goods.
ARCHAEOLOGIST	A person who finds and studies the remains of past cultures.
ARTIFACT	An object from the past that has been made by people.
AUXILIARIES	Soldiers of the Roman army who were not Roman citizens.
BALLISTA	A weapon similar to a crossbow, but much larger. It fired large stones or spears.
CAVALRY	Soldiers of the Roman army who fought on horseback. The cavalry were auxiliaries.
CELTS	The name given to native peoples of western and central Europe, including Britain and Gaul.
CHRISTIANITY	The belief in the life, death and teachings of Jesus Christ.
CITIZEN	A free person who had rights in their own city or region. At first, Roman citizens had to be born in Rome, but later other people in the empire could become Roman citizens.
DACIA	A country in the area that is now Romania. Dacia was conquered by the emperor Trajan in AD 101-106.
DENDRO-CHRONOLOGY	System of dating wood from the patterns of its annual growth rings. The system can date wood accurately to more than 7,000 years ago.
EMPEROR	Ruler of the Roman empire. The first emperor was Augustus, who reigned from 27 BC to AD 14.
FRESCO	A wall-painting that is painted directly onto plaster.
GLADIATORS	Slaves who were trained to fight in the amphitheatre. A gladiator who won many combats could earn his freedom.
GROMA	A surveying instrument used by the engineers of the army.
JAVELIN	A long, thin spear that was thrown at the enemy.
LEGION	A large unit of the Roman army, which contained approximately 5,000 soldiers plus craftsmen.
MILECASTLE	A small military stronghold, which usually protected a gateway through a defensive wall.
MOSAIC	A pattern or picture made out of tiny colored pieces of stone or pottery called *tessarae*.
PATRICIANS	The Roman nobles, who were wealthy landowners. The ordinary free citizens were Plebeians.
POMPEII	An Italian town buried by ash from the volcano Vesuvius in AD 79.
PROVINCE	An area outside Italy that was ruled by the Romans.
RELIEF	A scene carved out of stone.

AUXILIARY
CAVALRY
PARADE
HELMET

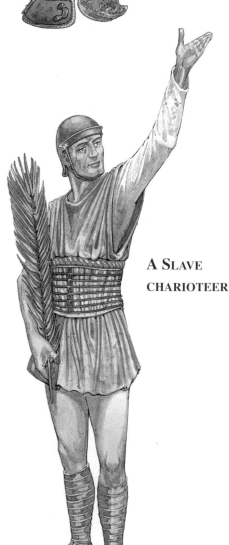

A SLAVE
CHARIOTEER

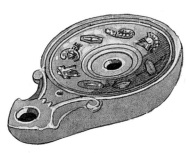

A TERRACOTTA
LAMP

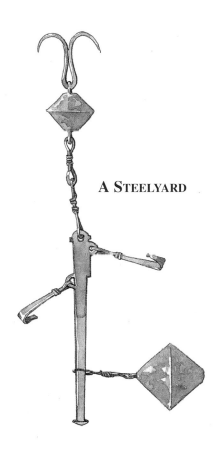

A STEELYARD

SAMIAN
POTTERY

SAMIAN	The common name for a form of red pottery made in Gaul.
SLAVE	People who were owned by other people. Slaves had very few rights, but could be treated almost as part of the family: many educated Greek slaves were bought as tutors for Roman children, and some slaves eventually earned their freedom.
STEELYARD	A device for weighing heavy objects.
TERRACOTTA	Baked clay, used for statuettes and pottery
TESSARAE	Small colored stones used to make a mosaic.

KEY DATES

BC

753	The city of Rome is founded, according to legend. It is ruled by kings.
509	Rome becomes a Republic, ruled by people elected by Roman citizens.
340-280	Rome expands her lands in Italy, at the expense of the Etruscans, Samnites and the Celts of northern Italy.
264-146	Rome defeats the Carthaginians in three long wars, and sets up provinces in Spain, Greece, southern France and North Africa.
58-44	Julius Caesar conquers France (Gaul), then becomes ruler of Rome.
30	Egypt becomes the Roman province of Aegyptus.
27	The end of the Republic; Augustus becomes the first emperor.

AD

43-84	Conquest of Britain (Britannia).
70	Roman armies put down a revolt in Jerusalem in Judaea.
79	Mount Vesuvius erupts, burying the towns of Pompeii and Herculaneum in ash.
98-117	The reign of the emperor Trajan, who conquers Dacia. The empire is at its most powerful.
306-337	The reign of the emperor Constantine, who makes Constantinople the Christian capital and declares Christianity to be the official religion.
395	The empire is divided into two parts: the West, which is ruled from Rome, and the East, ruled from Constantinople.
406-476	The Western empire is overrun and collapses.
1453	The Eastern empire (Byzantium) collapses when Constantinople is captured by the Ottoman Turks.

QUOTATIONS

The quotations are taken from some of the many books, diaries and letters written by Romans. The lawyer and poet Juvenal (Decimus Junius Juvenalis) was very critical of the Roman upper classes. His *Satires*, written in the early second century AD, provide many realistic details of everyday life. Lucius Seneca was a philosopher and an adviser to the emperor Nero, who reigned from AD 54 to 68. Diodorus Siculus was a Greek historian who wrote a history of the world called the *Bibliotheca Historica*. The poet Ovid (Publius Ovidius Naso) trained as a lawyer but devoted his time to writing. The politician and historian Sallust (Gaius Sallustius Crispus) wrote about the general Marius, who he had never met, many years after Marius' death. However, Sallust would have known about the 'tortoise' formation (page 29) because he had served in the army in Africa.

INDEX